Generalized Functions and Partial Differential Equations

AVNER FRIEDMAN
The Ohio State University

DOVER PUBLICATIONS, INC.
Mineola, New York

Bibliographical Note

This Dover edition, first published in 2005, is an unabridged republication of the work originally published by Prentice-Hall, Inc., Englewood Cliffs, New Jersey, in 1963.

Library of Congress Cataloging-in-Publication Data

Friedman, Avner.
Generalized functions and partial differential equations / Avner Friedman.
p. cm.
Originally published: Englewood Cliffs, N.J. : Prentice-Hall, 1963.
Includes bibliographical references and indexes.
ISBN 0-486-44610-7 (pbk.)
1. Differential equations, Partial. 2. Theory of distributions (Functional analysis) I. Title.

QA377.F74 2005
515'.353—dc22

2005051302

Manufactured in the United States of America
Dover Publications, Inc., 31 East 2nd Street, Mineola, N.Y. 11501

To My Parents

PREFACE

In recent years important progress has been made in the theory of partial differential equations due, to a large extent, to the development of new methods. The purpose of this book is to give a self-contained account of one of the main directions of this recent progress. The theory of generalized functions and, in particular, the theory of distributions, will be developed in detail. This theory will then be applied to solve a variety of problems in partial differential equations. The use of generalized functions (and of distributions) in solving these problems is not merely a matter of convenience. In fact, the need to use generalized functions arises in a natural way, and an attempt to avoid them will result in far more complicated proofs and in less decisive theorems.

The theory of generalized functions has been successfully applied also in other fields of mathematics, e.g., representation of locally compact groups, exterior differential forms, operational calculus, probability, etc. Beside the topics in differential equations covered in this book, generalized functions occur also in questions of eigenvalue expansions, approximation theorems, division problems, etc.

The theory of distributions was developed by L. Schwartz in 1945–48. A detailed exposition appears in his books [54]. The introduction of generalized functions of any class and their use in solving the Cauchy problem were given by Gelfand and Shilov in 1953. In 1956–58 they published three volumes (in Russian) on the subject [24–26]. A large part of the present book is based upon material from the books of Schwartz and of Gelfand and Shilov. In addition, the book contains more recent applications to differential equations as well as a more complete treatment (than that of Gelfand and Shilov) of the Cauchy problem.

The reader is assumed to have a sound knowledge of both real variables and complex variables. Familiarity with the basic theory of functional analysis (in particular, normed spaces) is desirable but not necessary.

We briefly describe the contents of the book. Chapter 1 gives the functional analysis background. In Chap. 3 the theory of distributions is presented in detail. Generalized functions are studied in Chaps. 2 and 4. In Chaps. 5 and 9 we introduce special classes of fundamental functions. The fundamental functions introduced in Chap. 5 are used in Chap. 6 in finding the structure of the Fourier transforms of entire functions. In

Chapter 7 the Cauchy problem is solved for any differential system with time-dependent coefficients of the form $\partial u/\partial t = P(t, \partial/\partial x)u + f(t, x)$. The results of Chaps. 2, 4, 5 and, in particular, 6 are then systematically applied. Other applications to differential equations are given in Chaps. 8, 10 and 11. In Chap. 8 the results of Chap. 7 are extended to more general systems and, furthermore, the Goursat problem is solved. Chapter 10 contains several independent topics: a Phragmén-Lindelöf type theorem and a Liouville type theorem for differential systems, the existence of a fundamental solution for any differential equation with constant coefficients, and an explicit construction of a fundamental solution for hyperbolic equations. In Chap. 11 we characterize the polynomials $P(x)$ such that all the solutions of $P(\partial/\partial x)u = 0$ are infinitely differentiable functions, and also solve, in the whole space, the equation $P(\partial/\partial x)u = f(x)$.

At the end of each chapter some easy problems are given with the purpose of familiarizing the reader with the material. Bibliographical remarks and bibliography will be found at the end of the book.

It is a pleasure to express my gratitude to my wife, for doing an excellent job of typing the manuscript.

While working on the book I was partially supported by ONR contracts and by the Alfred P. Sloan Foundation.

Avner Friedman

CONTENTS

CHAPTER 1

LINEAR TOPOLOGICAL SPACES

In this chapter we give the functional analysis background that will be needed in this book. In addition to a brief summary of some fundamental concepts and theorems from the theory of linear topological spaces and of normed spaces, we introduce the concepts of *countably normed* spaces and of *perfect* spaces and prove in detail some basic results concerning these spaces and their conjugate spaces. The concept of *inductive limit* of spaces and some related theorems are also given in detail.

1. Topological Spaces and Metric Spaces

Let X be a set of points and let T be a system of subsets of X having the following properties:

1. the empty set $\emptyset$ and the set X belong to T,
2. the union of any number of sets of T belongs to T, and
3. the intersection of any finite number of sets of T belongs to T.

We then say that T defines a *topology* on X and we call the pair (X, T) a *topological space*. When there is no confusion we designate (X, T) by X. The sets of T are called *open* sets. If T_1 and T_2 define two topologies on X, we say that the topology defined by T_1 is *weaker* (or *coarser*) than the topology defined by T_2 or, equivalently, that the topology defined by T_2 is *stronger* (or *finer*) than the topology defined by T_1, if every set of T_1 contains a set of T_2. If the T_1 topology is both weaker and stronger than the T_2 topology, then we say that the topologies are *equivalent*.

A system of open sets is said to constitute a *basis* of a topology if every open set is the union of sets of the system. A *neighborhood* of a point $x \in X$ (of a set $A \subset X$) is any set of X which contains an open set containing x (including A). A system W of neighborhoods of a point x is called a *basis of neighborhoods* (or a *neighborhood basis*) at x if every neighborhood of x contains some set of W. Two neighborhood bases at the same point x are said to be *equivalent* if each neighborhood of either one of them contains a neighborhood of the other. A similar definition holds for bases of a topology. Taking a neighborhood basis at each point of X, the set whose elements are all the neighborhoods thus obtained is clearly a basis of the topology. X is said to satisfy the *first countability axiom* if there exists a countable neighborhood basis at each point of X. X is said to satisfy the *second countability axiom* if there exists a countable basis for X.

The complement in X of an open set is called a *closed set.* x is an *interior point* of a set M if there exists a neighborhood of x contained in M. The set of all interior points of M is called the *interior of* M and is denoted *int* M. *Int* M is an open set, and M is open if and only if $M = int\ M$. x is a *limit point* of M if for every neighborhood V of x, $V - x$ intersects M, i.e., $(V - x) \cap M \neq \emptyset$. The *closure* of M (denoted by $\bar{M}$) is the union of M and the set of all limit points of M. $\bar{M}$ is a closed set and M is closed if and only if $\bar{M} = M$. $\bar{M} - int\ M$ is called the *boundary* of M.

A sequence $\{x_n\}$ is said to be *convergent* to x (we write $x_n \to x$ or $\lim_{n\to\infty} x_n = x$) if every neighborhood of x contains all the x_n with $n \geqslant n_0$, where n_0 depends on the neighborhood. A set M is said to be *dense* in X if $\bar{M} = X$, and *nowhere dense* if $\bar{M}$ does not contain interior points. X is called a *separable space* if there exists in X a dense countable set.

X is a *Hausdorff space* if each two disjoint points of X have disjoint neighborhoods. In a Hausdorff space X, a sequence cannot converge to two different points. Also, if $\{V_\alpha\}$ is a neighborhood basis at x, then $\bigcap_\alpha V_\alpha = x$. X is a *regular* space if for every point x and a closed set M, $x \notin M$, x and M have disjoint neighborhoods. X is *normal* if each two disjoint closed sets have disjoint neighborhoods. X is called a *compact* space if every family of open sets which covers X contains a finite subfamily which covers X. X is called *sequentially compact* if every (infinite) sequence of X contains a convergent subsequence. Thus, every space consisting of a finite number of points is both compact and sequentially compact.

A subset Y of X is called a *topological subspace* of X if a topology is defined on Y in the following way: a set $M \subset Y$ is open if and only if M is the intersection of Y with some open set of X. This rule indeed defines a topology on Y which is called the *induced topology.* A subset $Y \subset X$ is said to be (*sequentially*) *compact* if Y as a topological subspace is (sequentially) compact. Y is *relatively* (*sequentially*) *compact* if $\bar{Y}$ is (sequentially) compact.

A set of points X is called a *metric space* if to each pair of points x, y there corresponds a non-negative number $\rho(x, y)$ (called the *distance* from x to y) which satisfies

1. $\rho(x, y) = 0$ if and only if $x = y$,
2. $\rho(x, y) = \rho(y, x)$, and
3. $\rho(x, z) \leqslant \rho(x, y) + \rho(y, z)$.

The set of points x satisfying $\rho(x, x_0) < r$ ($\rho(x, x_0) \leqslant r$) is called the *open* (*closed*) *ball* of center x_0 and radius r. In each metric space we define a topology by taking the set of all open balls to form a basis for the open sets of X. X is then a Hausdorff space satisfying the first countability axiom. X is separable if and only if it satisfies the second countability axiom.

Clearly, $x_n \to x$ if and only if $\rho(x_n, x) \to 0$. A *Cauchy sequence* $\{x_n\}$ is a sequence for which $\rho(x_n, x_m) \to 0$ as $n \to \infty$, $m \to \infty$ independently. X is called a *complete* space if every Cauchy sequence is convergent. If X is not complete, then there exists a complete metric space $\hat{X}$ which contains X as a dense subset, and the metric functions ρ of X and $\hat{\rho}$ of $\hat{X}$ coincide on X. $\hat{X}$ is called the *completion* of X. It is uniquely determined up to an isometry; more precisely, if X^* is another completion, then there exists a one-to-one correspondence $\hat{x} \leftrightarrow x^*$, where $\hat{x}$ and x^* vary on the entire spaces $\hat{X}$ and X^* respectively, such that (1) $\hat{x} = x^*$ on X, and (2) $\hat{\rho}(\hat{x}, \hat{x}_0) = \rho^*(x^*, x_0^*)$ if $\hat{x}_0 \leftrightarrow x_0^*$, ρ^* being the distance function of X^*. A complete metric space cannot be written as a countable union of nowhere dense sets. Any topological space possessing this property is said to be of the *second category*.

A metric space is compact if and only if it is sequentially compact. A compact metric space is separable and complete.

Let f be a mapping from a topological space X into a topological space Y. Thus to every $x \in X$ there corresponds $y = f(x)$ in Y. y is called the *image* of x and x is called the *inverse image* (or *source*) of y. We denote by $f(A)$ the set $\{f(x); x \in A\}$ and by $f^{-1}(B)$ the set $\{x; f(x) \in B\}$, i.e., the set of all inverse images of the points of B. f is said to be one-to-one if $f(x) = f(x_0)$ implies $x = x_0$, and f is said to map X onto Y if $f(X) = Y$. f is *continuous* at x if the inverse image of any neighborhood of $f(x)$ is a neighborhood of x. f is said to be *continuous on* X if f is continuous at each point of X. f is continuous on X if and only if the inverse image of any open set is an open set. If f is a one-to-one mapping of X onto Y and if f and f^{-1} are continuous, then f is called a *topological mapping* (or *homeomorphism*) of X onto Y.

2. Linear Topological Spaces

A *linear topological space* is a vector space X (either over the real numbers or over the complex numbers) which is at the same time a Hausdorff topological space such that the operations of addition and scalar multiplication are continuous operations. We shall denote the elements of X by Latin letters x, y, z, etc., and the scalars by Greek letters λ, μ, ν, etc. Then, for any x_0, y_0 in X and for any neighborhood W_0 of $x_0 + y_0$ there correspond neighborhoods $V(x_0)$, $V(y_0)$ of x_0, y_0 such that the set

$$V(x_0) + V(y_0) = \{x + y; x \in V(x_0), y \in V(y_0)\}$$

is contained in W_0. Also, for any number λ_0 and for any neighborhood W_1 of $\lambda_0 x_0$ there corresponds a neighborhood $V_1(x_0)$ of x_0 and a neighborhood $\Lambda_0 = \{\lambda; |\lambda - \lambda_0| < \epsilon\}$ of λ_0 such that the set

$$\Lambda_0 V_1(x_0) = \{\lambda x; \lambda \in \Lambda_0, x \in V_1(x_0)\}$$

is contained in W_1. Taking, in particular, $\lambda_0 = 0$, $x_0 = 0$ we conclude that to every neighborhood U of 0 there corresponds an $\epsilon > 0$ and a neighborhood V of 0 such that the neighborhood of 0

$$U' = \bigcup_{|\lambda|<\epsilon} \lambda V$$

is contained in U. We say that W is a *normal* neighborhood of 0 if $\alpha W \subset W$ for all $|\alpha| \leqslant 1$. U' is therefore a normal neighborhood of 0. If we replace each U by U', we obtain from any given neighborhood basis at 0 an equivalent neighborhood basis at 0. Therefore, from now on we shall assume, without loss of generality, that all the neighborhoods of 0 are normal.

A mapping $x \to x + x_0$ is called a *translation*. It is a topological mapping from X onto X. A translation $x \to x + x_0$ maps a neighborhood basis at a point y into a neighborhood basis at the point $y + x_0$. Hence, the topology of the space is determined by giving a neighborhood basis B at the origin 0. It is easily verified that the sets of B satisfy the following properties:

1. $0 \in V$ for any $V \in B$.
2. If V_1, V_2 belong to B then there exists in B a set $V_3 \subset V_1 \cap V_2$.
3. If $x \neq 0$ then there exists a set $V \in B$ such that $x \notin V$.
4. For any $V \in B$ there exists a set $W \in B$ such that $W + W \in V$.
5. If $x \in V$ for some $V \in B$ then there exists a set $U \in B$ such that $x + U \in V$.
6. For any $V \in B$ and $\lambda \neq 0$ there exists a set $U \in B$ such that $\lambda U \subset V$.
7. For any $V \in B$ and $x \in X$ there exists an $\epsilon > 0$ such that $\lambda x \in V$ if $|\lambda| < \epsilon$.
8. For any $V \in B$ there exists an $\epsilon > 0$ such that $\lambda V \subset V$ if $|\lambda| < \epsilon$.

It can be shown (the proof is left to the reader) that if X is a linear space and B a system of subsets of X for which the properties 1 through 8 are satisfied, then X can be topologized, uniquely up to equivalence, in such a way that in the linear topological space obtained B is a neighborhood basis at 0.

A set $K \subset X$ is said to be *convex* if whenever x_1, x_2 belong to K then $\theta x_1 + (1 - \theta)x_2$ also belongs to K for all $0 < \theta < 1$. K is said to be *symmetric* if $K = -K$, i.e., if $x \in K$ implies $-x \in K$. A linear topological space X is said to be *locally convex* (and it is then called a *locally convex space*) if there exists a neighborhood basis at 0 in X all of whose elements are symmetric convex sets.

A sequence $\{x_n\}$ is called a *Cauchy sequence* if for every neighborhood V of 0 in X, $x_n - x_m \in V$ if m and n are sufficiently large. X is called a *sequentially complete* space if all its Cauchy sequences are convergent. In the future we shall consider only this concept of completeness and therefore omit the word "sequentially." A set $M \subset X$ is said to be *bounded* if, for any neighborhood V of 0, $M \subset \lambda V$ for all $|\lambda|$ sufficiently large (or, since V is normal, for some λ). The sum and the union of a finite number of bounded sets are bounded sets. The closure of a bounded set is a bounded set. If F is an open (closed) set then λF is an open (closed) set if $\lambda \neq 0$. For any $A \subset X$, $\overline{\lambda A} = \lambda \bar{A}$. A convergent sequence is bounded. Proofs are left to the reader.

A linear space X is called a *normed space* if there is defined on it a non-negative function $\|x\|$ (called the *norm* of x) which satisfies

1. $\|x\| = 0$ if and only if $x = 0$,
2. $\|\lambda x\| = |\lambda| \, \|x\|$, and
3. $\|x + y\| \leqslant \|x\| + \|y\|$ (the triangle inequality).

We define a distance function $\rho(x, y) = \|x - y\|$, and X then becomes a metric space. In the topological space thus determined, $x_n \to x$ if and only if $\|x_n - x\| \to 0$.

Let $\|x\|_1$ and $\|x\|_2$ be two norms defined on X. The second norm is said to be *stronger* than the first norm (and the first norm is said to be *weaker* than the second norm) if

$$(2.1) \qquad \|x\|_1 \leqslant C\|x\|_2 \qquad (C \text{ a constant}).$$

If one of the norms is both stronger and weaker than the other, then the two norms are said to be *equivalent*.

A complete normed space is called a *Banach space*. Every normed space can be extended to a Banach space by completing it as a metric space. If X becomes a Banach space by a norm $\|x\|_1$ and by a norm $\|x\|_2$, and if (2.1) holds, then the norms are necessarily equivalent.

Two norms $\|\cdot\|_1$ and $\|\cdot\|_2$ on a linear space X are said to be *in concordance* if, whenever $\{x_n\}$ is a Cauchy sequence in X in both norms and $\{x_n\}$ is convergent to 0 in one of the norms, it is then also convergent to 0 in the other norm. Let $\|\cdot\|_1$, $\|\cdot\|_2$ be in concordance and let (2.1) be satisfied. Denote by X_i $(i = 1, 2)$ the completion of X with respect to the norm $\|\cdot\|_i$. The mapping $x \to x$ from X onto X, where the image space is provided with the norm $\|\cdot\|_1$ whereas the source space is provided with the norm $\|\cdot\|_2$, can be extended as a one-to-one mapping from X_2

into X_1. We can therefore identify X_2 with a linear subspace of X_1. We then have $X \subset X_2 \subset X_1$.

If $\|\cdot\|_1$ and $\|\cdot\|_2$ are in concordance (and (2.1) does not necessarily hold), then the norm $\|\cdot\|_3 = \max(\|\cdot\|_1, \|\cdot\|_2)$ is also in concordance with $\|\cdot\|_1$ and with $\|\cdot\|_2$.

3. Countably Normed Spaces

Let X be a linear space and let $\{\|x\|_n\}$ be a sequence of norms defined on X. We define a topology on X by giving a neighborhood basis at 0. The sets of the basis are

$$U_{p,\epsilon} = \{x; \|x\|_1 < \epsilon, \|x\|_2 < \epsilon, \ldots, \|x\|_p < \epsilon\}$$

for any integer $p \geqslant 1$ and $\epsilon > 0$. One easily verifies that the requirements 1 through 8 of Sec. 2 are satisfied. Hence, X is a linear topological space and, in fact, a locally convex space satisfying the first countability axiom. If in each pair of norms of the sequence the norms are in concordance, we call X a *countably normed* space.

It is clear that $x_n \to 0$ if and only if $\|x_n\|_p \to 0$ for any p. Also, $\{x_n\}$ is a Cauchy sequence if and only if, for any p, $\|x_n - x_m\|_p \to 0$ as $m \to \infty$, $n \to \infty$ independently. Without loss of generality, the norms in a countably normed space can be taken to be monotone increasing, i.e.,

$$\|x\|_1 \leqslant \|x\|_2 \leqslant \cdots \leqslant \|x\|_p \leqslant \cdots, \tag{3.1}$$

since otherwise we can use a different sequence, namely

$$\|x\|_p^* = \max(\|x\|_1, \ldots, \|x\|_p),$$

which yields the same topology; in each pair of the new sequence the norms are also in concordance. From now on we always assume that (3.1) is satisfied.

Let X_n be the completion of X by $\|\cdot\|_n$. Then

$$X_1 \supset X_2 \supset \cdots \supset X_p \supset \cdots \supset X.$$

Theorem 1. *X is complete if and only if* $X = \bigcap_{p=1}^{\infty} X_p$.

Proof. Suppose $X = \bigcap_{p=1}^{\infty} X_p$, and let $\{x_n\}$ be a Cauchy sequence in X. Then $\{x_n\}$ is also a Cauchy sequence in each X_p and hence $x_n \to x_n^{(p)}$ for some $x_n^{(p)}$ in X_p. Since we identify the limits of the same Cauchy sequence in X_p and X_{p+1}, $x_n^{(p)} = x^*$ for all p. Hence, $x^* \in \bigcap_p X_p = X$. Since $\|x_n - x^*\|_p \to 0$ for every p, $x_n \to x^*$ in X and X is complete. Conversely, let X be complete and let $x \in \bigcap_p X_p$. Then, for any p there exists an $x_p \in X$ such that $\|x_p - x\|_p < \frac{1}{p}$. For any $k \leqslant p$,

$$\|x - x_p\|_k \leqslant \|x - x_p\|_p < \frac{1}{p};$$

hence $x_p \to x$ in X_k and it follows that $\{x_p\}$ is a Cauchy sequence in each X_k, and therefore in X. Let $\bar{x} = \lim x_p$ in X. Then $\|x_p - \bar{x}\|_k \to 0$ for any k and we get $\|x - \bar{x}\|_k = 0$, thus proving that $x = \bar{x} \in X$.

Example. Let $x = (x_1, \ldots, x_n)$ be a variable point in the real n-dimensional Euclidean space R^n, and let N be a compact set in R^n. Denote by (D_N) the set of all infinitely differentiable complex-valued functions on R^n which vanish outside N, and by (D_N^m) the set of all m-times continuously differentiable functions on R^n which vanish outside N. (D_N^p) is a Banach space if the norm is defined by

$$\|f\|_p = \max_{|\alpha| \leqslant p} \max_{x \in N} \left| \frac{\partial^{|\alpha|} f(x)}{\partial x_1^{\alpha_1} \cdots \partial x_n^{\alpha_n}} \right|. \tag{3.2}$$

(D_N) is a countably normed space with the norms (3.2), and it is complete since $(D_N) = \bigcap_{p=1}^{\infty} (D_N^p)$.

In a countably normed space X we can introduce a metric

$$\rho(x, y) = \sum_{p=1}^{\infty} \frac{1}{2^p} \frac{\|x - y\|_p}{1 + \|x - y\|_p}.$$

One easily verifies that the topology of the metric space is equivalent to the topology of X. In addition to the metric axioms, ρ also satisfies

4. $\rho(x, y) = \rho(x - y, 0)$, and
5. $\rho(\lambda_n x, 0) \to 0$ if $x \in X$, $\lambda_n \to 0$, and $\rho(\lambda x_n, 0) \to 0$ if λ is any number and $x_n \to 0$.

A linear topological space X which can be metrized so that its metric satisfies requirements 4 and 5 is called a *linear metric space.* If X is also locally convex and complete, it is called a *Fréchet space.* Thus, complete countably normed spaces are Fréchet spaces.

A set F in a linear topological space X is said to be *absorbing* if for any $x \in X$ there exists an $\epsilon > 0$ such that $\lambda x \in F$ if $|\lambda| < \epsilon$.

Theorem 2. *If F is a closed, convex, symmetric and absorbing set in a Fréchet space X, then F contains a neighborhood of 0 in X.*

Proof. Since F is an absorbing set, $\bigcup_{m=1}^{\infty} (mF)$ covers X. Recalling that X is a space of the second category, it follows that F cannot be nowhere dense. Hence, $F = \bar{F}$ contains interior points, i.e., there exists an $x_0 \in F$ and a normal neighborhood U of 0 such that $x_0 + U \subset F$. Since F and U are symmetric, $-x_0 + U = -x_0 - U \subset -F = F$. Using the convexity of F it follows that F contains

$$\frac{(x_0 + x) + (-x_0 + x)}{2}$$

for any $x \in U$, i.e., $F \supset U$.

We mention the following theorem (for proof see Problem 1).

Theorem 3. *A complete countably normed space X cannot be made into a normed space with the same topology if and only if there exists a subsequence of non-equivalent norms in the sequence of norms of X.*

Example. (D_N) cannot be normed.

Let X be a countably normed space under two sequences of norms

$$(3.3) \qquad \|x\|_1 \leqslant \|x\|_2 \leqslant \cdots \leqslant \|x\|_p \leqslant \cdots,$$

$$(3.4) \qquad \|x\|_1' \leqslant \|x\|_2' \leqslant \cdots \leqslant \|x\|_p' \leqslant \cdots.$$

The second sequence is said to be *stronger* than the first one (and the first sequence is said to be *weaker* than the second one) if for every p there exists a $q = q(p)$ such that $\|x\|_p \leqslant C\|x\|_q'$. In that case, if $x_n \to x$ in the topology of (3.4), then $x_n \to x$ in the topology of (3.3). Conversely:

Theorem 4. *If $x_n \to x$ in the topology of* (3.4) *implies $x_n \to x$ in the topology of* (3.3), *then the sequence* (3.4) *is stronger than the sequence* (3.3).

Proof. If the assertion is not true, then for some p there exists a sequence $\{x_n\} \subset X$ such that $\|x_n\|_p = 1$ and $\|x_n\|_n' \to 0$. Hence, $\|x_n\|_k' \to 0$ for any k, i.e., $x_n \to 0$ in the topology of (3.4). Since $\{x_n\}$ does not converge to 0 in the topology of (3.3), we have derived a contradiction.

If (3.3) is both stronger and weaker than (3.4), we say that the two sequences are *equivalent.*

Obviously, a set $B \subset X$ is bounded if and only if

$$\|x\|_p \leqslant C_p \qquad (C_p \text{ a constant})$$

for all p and $x \in B$. If X cannot be normed then, by Theorem 3, any set $\{x; \|x\|_p < C\}$ is unbounded. Hence any bounded set cannot contain interior points (since otherwise some translation of such a set would contain a neighborhood of 0). Since the closure of a bounded set is also bounded, we conclude that *if X cannot be normed then bounded sets are nowhere dense.*

4. Continuous Linear Functionals

Let X be a real (complex) linear space. A real (complex) linear functional f is a mapping $f(x) = (f, x)$ from X into the real (complex) numbers such that $f(\lambda x + \mu y) = \lambda f(x) + \mu f(y)$. Clearly, f is continuous if and only if to every $\epsilon > 0$ there corresponds a neighborhood U of 0 in X such that $|f(x)| < \epsilon$ if $x \in U$. Note that f is continuous if and only if f is bounded on some neighborhood V of 0 (i.e., $|f(x)| < C$ for all $x \in V$). f is said to be *bounded* if it is bounded on bounded sets of X. If X is a normed space, f is continuous if and only if f is bounded. For general linear topological spaces the following holds.

Theorem 5. *If a linear functional f is continuous then f is bounded. If X satisfies the first countability axiom and f is a bounded linear functional then f is continuous.*

Proof. The first part of the theorem is rather obvious, using the definition of a bounded set. Suppose now that f is bounded and let $U_1 \supset U_2 \supset \cdots$ be a neighborhood basis at 0. If f is not continuous, then for each n there exists an $x_n \in U_n$ such that $f(x_n) \to \infty$ as $n \to \infty$. Since $x_n \to 0$, $\{x_n\}$ is a bounded set, and our assumption is contradicted.

The above proof with trivial modifications establishes also the second part of the following theorem.

Theorem 6. *If f is a continuous linear functional then $f(x_n) \to 0$ whenever $x_n \to 0$. If X satisfies the first countability axiom and if f is a linear functional such that $f(x_n) \to 0$ whenever $x_n \to 0$, then f is a continuous functional.*

The first part is obvious.

Theorem 7. *Let f be a continuous linear functional on a dense linear subspace X_0 of a linear topological space X satisfying the first countability axiom. Then f can be extended, by continuity, into a continuous linear functional on X. This extension is unique.*

Proof. We define $f(x)$ for any $x \in X$, $x \notin X_0$ by $f(x) = \lim f(x_n)$, where $\{x_n\}$ is any sequence in X_0 converging to x. This definition uniquely determines $f(x)$ since if $x_n \to 0$ then $f(x_n) \to 0$. To prove that f is continuous it suffices to show (in view of Theorem 6) that if $\{y_m\} \subset X$, $y_m \to 0$, then $f(y_m) \to 0$. For any n take $x_n \in X_0$ such that $x_n - y_n \in U_n$ and $|f(x_n) - f(y_n)| < 1/n$. Then $x_n \to 0$ and therefore

$$|f(y_n)| \leqslant |f(y_n) - f(x_n)| + |f(x_n)| \to 0.$$

The uniqueness assertion is trivial.

A function $p(x)$ satisfying the properties

1. $p(x) \geqslant 0$,
2. $p(x + y) \leqslant p(x) + p(y)$, and
3. $p(\lambda x) = \lambda p(x)$ if $\lambda \geqslant 0$,

is called a *convex* functional. If $p(\lambda x) = |\lambda| p(x)$ for any complex λ, p is called a *symmetric convex* functional or a *semi-norm.*

Hahn-Banach Lemma. *Let $p(x)$ be a convex (symmetric convex) functional on a real (complex) linear space X and let $f(x)$ be a real (complex) linear functional on a linear subspace M of X, such that $f(x) \leqslant p(x)$ ($|f(x)| \leqslant p(x)$). Then f can be extended into a real (complex) linear functional F on X such that $F(x) \leqslant p(x)$ ($|F(x)| \leqslant p(x)$) on X.*

From this important lemma one deduces without difficulty the following three theorems.

From this important lemma one deduces the following three theorems.

Theorem 8a. (*Hahn-Banach Theorem*). *Let H be a linear subspace of a normed space X and let f be a continuous linear functional on H. Then f can be extended into a continuous linear functional F on X and*

$$\sup_{x \in H, \|x\| = 1} |f(x)| = \sup_{x \in X, \|x\| = 1} |F(x)|. \tag{4.1}$$

Theorem 8b. *Let X be a normed space and let $x_0 \in X$. Then there exists a continuous linear functional f on X satisfying*

$$f(x_0) = \|x_0\|, \qquad \sup_{\|x\| = 1} |f(x)| = 1. \tag{4.2}$$

Theorem 8c. *Let X_0 be a closed linear subspace of a normed space X and suppose that $X_0 \neq X$. Then there exists a continuous linear functional f such that $f \neq 0$ and $f(x_0) = 0$ for every $x_0 \in X_0$.*

We now turn to a countably normed space X. If f is a continuous linear functional, then it is bounded on some neighborhood of 0 in X; hence, for some $p \geqslant 1$,

$$|f(x)| \leqslant C\|x\|_p \qquad (C \text{ a constant}). \tag{4.3}$$

Conversely, if (4.3) is satisfied, then f is bounded on some neighborhood of 0 in X and hence it is continuous. The smallest p for which (4.3) holds is called the *order* of f.

Next, if f is defined only on a linear subspace H of X and if (4.3) holds, then, by Theorem 8a, f can be extended and the extension F satisfies (4.3) on the entire space X.

In the space of all continuous linear functionals over a linear topological space X we introduce the definition $\lambda f + \mu g$ in an obvious manner, namely, $(\lambda f + \mu g)(x) = \lambda f(x) + \mu g(x)$. The vector space thus obtained is called the *conjugate* (or *dual*) space of X and is denoted by X'. If X is a normed space, X' becomes a normed space when we define

$$\|f\| = \sup_{\|x\|=1} |f(x)|.$$

Note that (4.1) can be written as $\|f\|_H = \|F\|_X$ when $\|\cdot\|_H$ and $\|\cdot\|_X$ refer to the norms in the spaces H' and X' respectively.

Let X be a countably normed space. Denote by X'_p the conjugate of X_p. The elements of X'_p can be identified with the continuous linear functionals, on X, of order $\leqslant p$. Clearly,

$$X'_1 \subset X'_2 \subset \cdots \subset X'_p \subset \cdots \subset X', \qquad X' = \bigcap_{p=1}^{\infty} X'_p. \tag{4.4}$$

If f is of order p, then the norms

$$\|f\|_q = \sup_{\|x\|_q=1} |f(x)|$$

exist for all $q \geqslant p$, and

$$\|f\|_p \geqslant \|f\|_{p+1} \geqslant \cdots.$$

In a normed space X, a set B is bounded if and only if

$$|f(B)| \equiv \sup_{x \in B} |f(x)| < \infty$$

for any $f \in X'$. Analogously we have the following:

Theorem 9. *A set B in a countably normed space X is bounded if and only if $|f(B)| < \infty$ for any $f \in X'$.*

Proof. Let B be bounded. If $f \in X'$ then $f \in X'_p$ for some p. Since B is a bounded set in each normed space X_p, $|f(B)| < \infty$. Conversely, if for any f in any X'_p, $|f(B)| < \infty$, then B is a bounded set in each X_p and hence it is bounded.

5. Weak and Strong Topologies

Let X be a linear topological space. We define *strong neighborhoods* of 0 in X' as sets

$$V(A, \epsilon) = \{f; |f(A)| < \epsilon\},$$

where A is an arbitrary bounded set in X and ϵ is an arbitrary positive number. The properties 1 through 8 of Sec. 2 are easily verified so that a topology is defined on X' with the strong neighborhoods as a neighborhood basis at 0. We call this topology the *strong topology*. For normed spaces it coincides with the topology of the norms $\|f\|$. A sequence which is convergent in the strong topology is called a *strongly convergent* sequence. Concepts such as strong boundedness, strong closure, and so on are similarly defined.

Theorem 10. *If X satisfies the first countability axiom then X' is complete in the strong topology.*

Proof. If $\{f_n\}$ is a Cauchy sequence in the strong topology then, for any x, $\{f_n(x)\}$ is a Cauchy sequence. Let $f(x) = \lim f_n(x)$. Then f is linear and for every bounded set $A \subset X$, $|f(A)| < \infty$ since $\{|f_n(A)|\}$ is a bounded sequence. By Theorem 5 f is continuous. It is finally clear that $f_n \to f$ in the strong topology.

Theorem 11. *A set $B' \subset X'$ is strongly bounded if and only if B' is bounded on every bounded set $A \subset X$, i.e., if and only if*

$$|B'(A)| \equiv \sup_{f \in B'} |f(A)| < \infty \tag{5.1}$$

for any bounded set $A \subset X$.

Proof. Let B' be strongly bounded. Then for any neighborhood $V(A, 1)$, $B' \subset \lambda V(A, 1)$ for some $\lambda > 0$, thus implying that $|B'(A)| < \lambda$. Conversely, if for any bounded set $A \subset X$, $|B'(A)| < \infty$, then for any neighborhood $V = V(A, \epsilon)$ of 0 in X', B' is contained in μV for any $\mu > |B'(A)|/\epsilon$ and it is therefore bounded.

Theorem 12. *If X satisfies the first countability axiom then every strongly bounded set $B' \subset X'$ is bounded on some neighborhood of 0 in X.*

Proof. If the assertion is not true, there exists a sequence $\{x_n\}$, $x_n \in U_n$, where $U_1 \supset U_2 \supset \cdots$ is a neighborhood basis at 0, and a sequence $\{f_n\} \subset B'$ such that $|f_n(x_n)| \to \infty$. Since $\{x_n\}$ is a bounded set, we obtain a contradiction to Theorem 11.

We now turn to countably normed spaces.

Theorem 13. *Let X be a countably normed space. A set $B' \subset X'$ is strongly bounded if and only if it is contained in some X'_p and it is bounded in the norm $\|\cdot\|_p$ of X'_p.*

Proof. If $B' \subset X'_p$ and is bounded in the norm of X'_p then B' is bounded on the set $\{x; x \in X, \|x\|_p \leqslant 1\}$ which is a neighborhood of 0 in X. Hence it is bounded on any bounded set of X, and use of Theorem 11 completes the proof. Conversely, if B' is strongly bounded, then, by Theorem 12, B' is bounded on some neighborhood U of 0 in X defined by, say, $\|x\|_p < \delta$. Let $|B'(U)| < M$. Then $B' \subset X'_p$ and $\|f\|_p \leqslant M/\delta$ for any $f \in B'$.

The *weak topology* in X' is defined in terms of a neighborhood basis at 0 given by $V = V(x_1, \ldots, x_m; \epsilon): f \in V$ if and only if

$$|f(x_1)| < \epsilon, \ldots, |f(x_m)| < \epsilon.$$

The concepts of weakly convergent sequences, weak closure, weak boundedness, and so on are defined in an obvious way. The weak topology is obviously weaker than the strong topology. Weak convergence of $\{f_n\}$ to f clearly means that $f_n(x) \to f(x)$ for any $x \in X$.

Theorem 14. *Let X be a countably normed space and let $\{f_n\}$ be a strongly bounded sequence such that $f_n(x) \to 0$ for any x in a dense subset A of X. Then $f_n(x) \to 0$ for any $x \in X$.*

Proof. By Theorem 13 the f_n form a bounded set in some X'_p and $f_n(x) \to 0$ for any x in A where A is dense in X and hence in X_p. The proof now follows as for normed spaces.

Note that a set B' is weakly bounded if and only if

$$\sup_{f \in B'} |f(x)| < \infty$$

for any $x \in X$. A strongly bounded set is obviously also weakly bounded. For complete countably normed spaces the converse is also true.

Theorem 15. *For a complete countably normed space X, every weakly bounded set of X' is strongly bounded.*

Proof. Let B' be a weakly bounded set and let F be the set of all points $x \in X$ such that $|f(x)| \leqslant 1$ for all $f \in B'$. F is closed, convex, symmetric, and absorbing (since B' is weakly bounded). By Theorem 2, F contains a neighborhood U of 0 in X. Since B' is bounded on U, it is bounded on any bounded set of X, and the strong boundedness of B' follows by using Theorem 11.

Corollary. *For a complete countably normed space X, a weakly convergent sequence $\{f_n\}$ in X' is strongly bounded.*

Theorem 16. *Let X be a complete countably normed space. Then X' is complete with respect to the weak topology.*

Proof. Let $\{f_n\}$ be a weak Cauchy sequence. Then $f(x) = \lim f_n(x)$ exists for any $x \in X$. By Theorem 15, $|f_n(A)| < C$ for any bounded set $A \subset X$. Hence $|f(A)| \leqslant C$ and f is continuous, by Theorem 5. Finally, $f_n \to f$ weakly.

Theorem 17. *Let X be a complete countably normed space. Then any sequence $\{f_n\}$ in X' is weakly convergent to f if and only if the f_n and f belong to some X'_p and $f_n \to f$ weakly in X'_p.*

Proof. Let $f_n \to f$ weakly in X'. Then the f_n form a strongly bounded set and, by Theorem 13, they belong to some X'_p and form in it a bounded set. Without loss of generality we may assume that f also belongs to the same X'_p. Since $f_n(x) \to f(x)$ on X which is dense in X_p, the convergence for any $x \in X_p$ follows. The converse is trivial.

By means of the strong and weak topologies of X', we can define two topologies on X: (a) the *strong topology*, which is given in terms of *strong neighborhoods* $V = V(B', \epsilon)$ of 0 in X; $x \in V$ if and only if

$$|f(x)| < \epsilon \qquad \text{for any } f \in B',$$

where B' is an arbitrary strongly bounded set in X' and ϵ is an arbitrary positive number; (b) the *weak topology*, which is given in terms of *weak neighborhoods* $W = W(f_1, \ldots, f_m; \epsilon)$ of 0 in X; $x \in W$ if and only if

$$|f_1(x)| < \epsilon, \ \ldots, \ |f_m(x)| < \epsilon,$$

where $f_1, \ldots, f_m$ vary in X', m is an arbitrary positive integer, and ϵ is an arbitrary positive number. Clearly, the weak topology is weaker than the strong topology.

Theorem 18. *If X is a complete countably normed space, then the strong topology of X is equivalent to the topology of X.*

Proof. Each neighborhood $V = \{x; \|x\|_p < \epsilon\}$ of 0 in X is also a strong neighborhood of 0. Indeed, if $x \in V$ then $|f(x)| < \epsilon$ for all f in the set

$$B' = \{f; f \in X'_p, \|f\|_p = 1\}.$$

Conversely, if $|f(x)| < \epsilon$ for all $f \in B'$, then, taking f such that $f(x) = \|x\|_p$ (see Theorem 8b), it follows that $x \in V$. Since B' is strongly bounded, V is a strong neighborhood of 0. Conversely, let U be a strong neighborhood of 0 in X, i.e.,

$$U = \{x; |f(x)| < \epsilon \qquad \text{for all } f \in B'\},$$

where B' is a strongly bounded set in X'. By Theorem 13, $B' \subset X'_p$ and $\|f\|_p < C$ for all $f \in B'$. Hence U contains the set

$$U_0 = \left\{x; |f(x)| < \frac{\epsilon}{C} \qquad \text{for all } f \in X'_p, \|f\|_p = 1\right\},$$

which is identical with the set $\{x; \|x\|_p < \epsilon/C\}$, and this set is a neighborhood of 0 in X.

Theorem 19. *If X is a complete countably normed space, then a set in X is (strongly) bounded if and only if it is weakly bounded.*

Proof. That (strong) boundedness implies weak boundedness is obvious. Conversely, let A be a weakly bounded set. Then λA is contained in any given weak neighborhood of 0 provided λ is sufficiently small. Hence, for any $f \in X'$, $|f(x)| < C$ for all $x \in A$, where C is a constant depending on f. By Theorem 9, A is bounded.

Corollary. *In a complete countably normed space, weakly convergent sequences are (strongly) bounded.*

6. Perfect Spaces

In a linear topological space X, *if* $A \subset X$ *is sequentially compact, then* A *is bounded.* Indeed, if A is not bounded then there exists a neighborhood U of 0 and a sequence $\{x_n\} \subset A$ such that $(1/n)x_n \notin U$. But then there is no subsequence $\{x_{n'}\}$ of $\{x_n\}$ which is convergent. Indeed, if $x_{n'} \to x$, let V be a normal neighborhood of 0 such that $V + V \subset U$. Since $(1/n')x \in V$ and $(1/n')(x - x_{n'}) \in V$ for large n', it follows that $(1/n')x_{n'} \in U$, which is a contradiction.

If $A \subset X$ *is compact, then* A *is bounded.* Indeed, for any normal neighborhood U of 0, $x \in \lambda U$ for any given x provided $\lambda = \lambda_x$ is sufficiently large. From the sets $\lambda_x U$ which cover A pick a finite family $\lambda_1 U, \ldots, \lambda_m U$ which still covers A. Then $A \subset \lambda U$ if $\lambda = \max |\lambda_i|$.

We shall be interested in spaces where the converse of each of the above two propositions is true.

X is called a *perfect space* if X is a complete countably normed space having the property that the bounded sets of X are relatively sequentially compact. Since X is a complete metric space, if we replace "sequential compactness" by "compactness" we get an equivalent definition.

A *Montel space* is a locally convex space which has the property asserted in Theorem 2, Sec. 3, and in which bounded sets are relatively compact. Hence, a perfect space is a Montel space.

Theorem 20. *Let X be a complete countably normed space and let $\{p_n\}$ be an increasing sequence of positive integers. If from any (infinite) sequence $A \subset X$ which is bounded in $\|\cdot\|_{p_{n+1}}$ one can choose a Cauchy sequence in $\|\cdot\|_{p_n}$, then X is a perfect space.*

The proof follows by the standard diagonalization procedure.

Example. The space (D_N) is a perfect space. (In proving the property assumed in Theorem 20 we use the lemma of Ascoli-Arzela.) The spaces (D_N^m) are not perfect spaces.

Theorem 21. *In a perfect space X, weak convergence implies strong convergence.*

Proof. Let $x_n \to 0$ weakly. By the corollary to Theorem 19, it is strongly bounded, and therefore relatively sequentially compact. Hence if $\{x_n\}$ does not converge strongly to 0 there must exist a subsequence $\{x_{n'}\}$ which converges strongly, hence weakly, to some $x \neq 0$, which is impossible.

Theorem 22. *A perfect space X is complete with respect to the weak topology, i.e., weak Cauchy sequences are (weakly) convergent.*

The proof is similar to the proof of Theorem 21 and is left to the reader.

Theorem 23. *If X is a perfect space then weak convergence in X' implies strong convergence in X'.*

Proof. Let $f_n(x) \to 0$ for any $x \in X$. We have to show that $f_n(x) \to 0$ uniformly in x in bounded sets of X. If this is not the case, then there exists a sequence $\{x_n\}$ in some bounded set A such that $|f_n(x_n)| > \epsilon$ for some $\epsilon > 0$. Since the sequence is bounded, one of its subsequences $\{x_{n'}\}$ is (strongly) convergent, say, to x. Thus, $f(x_{n'} - x) \to 0$ uniformly with respect to f in bounded sets B' of X'. Taking $B' = \{f_{n'}\}$ we get

$$f_{n'}(x_{n'}) = f_{n'}(x_{n'} - x) + f_{n'}(x) \to 0$$

as $n' \to \infty$, which is a contradiction.

Theorem 24. *Let X satisfy the assumptions of Theorem 20. If $f_n \to 0$ weakly (or strongly), then, for some p, $f_n \to 0$ in the norm of X'_p.*

The converse is trivial.

Proof. $\{f_n\}$ is a bounded set and hence $\|f_n\|_r < C$ for some r and, since X is dense in X_r, $f_n \to 0$ weakly in X'_r. Let p be such that every bounded set in X_p is relatively sequentially compact in X_r. Clearly $\|f_n\|_p < C$ and $f_n \to 0$ weakly also in X'_p. We claim that $\|f_n\|_p \to 0$. Indeed, if this is not the case, then there exists a sequence $\{x_n\}$ in some bounded set of X_p, such that $|f_n(x_n)| > \epsilon$ for some $\epsilon > 0$. Taking $\{x_{n'}\}$ to be a convergent subsequence in X_r and denoting its limit by x we obtain

$$|f_{n'}(x_{n'})| \leqslant |f_{n'}(x_{n'} - x)| + |f_{n'}(x)| \leqslant \|f_{n'}\|_r \|x_{n'} - x\|_r + |f_{n'}(x)| \to 0,$$

which is a contradiction.

We want to show that if X is a perfect space then the bounded sets of X' are relatively sequentially compact. This follows from the next two theorems.

Theorem 25. *If X is a separable complete countably normed space then every bounded set in X' is relatively sequentially compact in the weak topology.*

Proof. Let $\{x_m\}$ be a dense sequence in X. Given any infinite bounded set $B' \subset X'$ pick an infinite sequence $\{f_n\}$ which converges on each x_m (this can be done by the standard procedure of diagonalization). Since $\|f_n\|_p < C$ for some p, $\{f_n\}$ is weakly convergent on X and its weak limit f belongs to X'.

Theorem 26. *Perfect spaces are separable.*

Proof. If all the X_p are separable then we take in each X_p a countable dense set S_p belonging to X. $\bigcup_{p=1}^{\infty} S_p$ is dense in X since if $x \in X$ there exists an $x_p \in S_p$ such that $\|x_p - x\|_p < 1/p$, for any p, and it is clear that $x_p \to x$ in X, since $\|x_p - x\|_k \leqslant \|x_p - x\|_p$ if $p \geqslant k$.

To complete the proof we shall show that all the X_p must be separable spaces. If this is not true, then there exists a nonseparable space X_{p_0}, and for simplicity we may take $p_0 = 1$. If for any n $(n = 1, 2, \ldots)$ there exists a sequence $\{x_m\}$ in X such that $\|x - x_i\|_1 < 1/n$ for any $x \in X_1$ and for some i, then X_1 would be separable. Hence, there exists an $\epsilon = 1/n_0$ such that no sequence $\{x_m\}$ has the above property. Consider now all the sets $A \subset X$ having the following property: for any pair x, y in A, $\|x - y\|_1 \geqslant \epsilon$. Define a partial ordering $A \prec B$ by $A \subset B$. Then, every ordered set of sets A has an upper bound, namely, their union. By Zorn's lemma there exists a maximal set Z, i.e., $Z \prec A$ implies $Z = A$. Clearly $\|x - y\|_1 \geqslant \epsilon$ for any $x, y \in Z$, and for any $t \in X$ $\|t - x\|_1 < \epsilon$ for some $x \in Z$. Z is therefore nondenumerable.

Since X is contained in the union of the balls $\|x\|_1 < m$ $(m = 1, 2, \ldots)$, $Z_1 = Z \cap \{x; \|x\|_1 < m_1\}$ is nondenumerable for some m_1. Similarly, $Z_2 = Z_1 \cap \{x; \|x\|_2 < m_2\}$ is nondenumerable for some m_2, etc. In this way we obtain, for any integer $p > 0$, a nondenumerable set Z_p which is bounded in X_p. Taking, for each p, x_p in Z_p such that $x_p \neq x_q$ if $p \neq q$, we obtain a sequence which is bounded in X but has no convergent subsequence. This is in contradiction with the assumption that X is a perfect space.

Combining Theorems 26, 25 and 23, we obtain:

Theorem 27. *If X is a perfect space then bounded sets in X' are relatively sequentially compact in both the weak and the strong topologies.*

Let $x(t)$ be a continuous function from a real interval $a \leqslant t \leqslant b$ into a perfect space X. For any $f \in X'$, the complex-valued function $f[t] \equiv f(x(t))$ is continuous, and the integral

$$\int_a^b f[t]\,dt = \lim_{\max \Delta t_\nu \to 0} \Sigma f(x(t_\nu))\,\Delta t_\nu$$

therefore exists. Since the right side is equal to $\lim f(\Sigma\, x(t_\nu)\, \Delta t_\nu)$, $\Sigma\, x(t_\nu)\, \Delta t_\nu$ is weakly convergent and, by Theorem 21, $\Sigma\, x(t_\nu)\, \Delta t_\nu$ is convergent in X. We denote the limit by $\int_a^b x(t)\, dt$ and call it the *integral* of $x(t)$ from a to b. Thus

$$\int_a^b x(t)\, dt = \lim_{\max \Delta t_\nu \to 0} \Sigma\, x(t_\nu)\, \Delta t_\nu. \tag{6.1}$$

The left side satisfies all the basic properties which the standard integrals satisfy. In particular,

$$\lim_{b \to a} \frac{1}{b - a} \int_a^b x(t)\, dt = x(a), \tag{6.2}$$

and

$$\left\| \int_a^b x(t)\, dt \right\|_m \leqslant \int_a^b \|x(t)\|_m\, dt. \tag{6.3}$$

The first formula follows by applying any $f \in X'$ to both sides, and the second formula follows directly from (6.1).

The definition of the integral can immediately be extended to the case where $t = (t_1, \ldots, t_n)$. The basic properties of the standard multiple integrals, and in particular (6.3), are valid for the present integral.

7. Linear Operators

An operator (or transformation, mapping) A from one linear topological space X into another linear topological space Y is called a *linear operator* if $A(\lambda x + \mu y) = \lambda Ax + \mu Ay$. Clearly, A is continuous if and only if the inverse image of any neighborhood W of 0 in Y is a neighborhood of 0 in X. A is called a *bounded operator* if it maps bounded sets into bounded sets. A continuous linear operator is easily seen to be bounded. Conversely:

Theorem 28. *If X and Y are linear topological spaces and X satisfies the first countability axiom, then every bounded linear operator A from X into Y is continuous.*

Proof. If A is not continuous, then there exists a neighborhood W of 0 in Y such that for any λ $A(\lambda U_n) \not\subset W$, where the U_n form a neighborhood basis of X at 0 such that $U_n \supset U_{n+1}$ for any n. Thus, for some sequence $\{x_n\}$, $x_n \in U_n$, we have $A(nx_n) \notin W$. It follows that $\{Ax_n\}$ is an unbounded set in Y. Since $\{x_n\}$ is a bounded set, we have derived a contradiction to the assumption that A is bounded.

Theorem 29. *For a linear operator A, which maps a linear topological space X into a linear topological space Y, to be continuous it is necessary* *and, if X satisfies the first countability axiom, also sufficient that $Ax_n \to 0$ whenever $x_n \to 0$.*

Proof. The "necessary" part is clear. As for the sufficiency, suppose A is not continuous. Then, by Theorem 28, for some bounded set $B \subset X$, $A(B)$ is not bounded and, consequently, there exists a sequence $\{y_n\}$, $y_n = Ax_n$, such that $x_n \in B$, $(1/n)y_n \notin W$, where W is some neighborhood of 0 in Y. Since $(1/n)x_n \to 0$ whereas $A((1/n)x_n) = (1/n)y_n$ does not converge to 0, we have derived a contradiction.

Theorem 30. *Let X, Y be complete countably normed spaces and let A be a one-to-one continuous linear operator from X onto Y. Then A^{-1} is also continuous.*

In fact, this theorem was established by Banach in the more general case where X and Y are Fréchet spaces.

Taking $Ax = x$ we conclude that if X is a complete countably normed space with respect to each of two different sequences of norms, and if one sequence is stronger than the other, then the sequences are equivalent. The corollary to the following theorem gives a sharper result.

Theorem 31. *Let X, Y be two complete countably normed spaces and let $Y \subset X$. Suppose that the topologies of X and Y are in concordance, i.e., if $y_n \to y^*$ in Y and $y_n \to x^*$ in X then $x^* = y^*$. Then whenever $x_n \to x$ in Y, also $x_n \to x$ in X.*

Proof. Taking max $(\|\cdot\|_p, \|\cdot\|'_p)$ to be new norms in Y (where $\|\cdot\|'_p$ are the original norms in Y), Y becomes a complete space, and by the previous remark the new sequence of norms of Y is equivalent to each of the previous two sequences.

Corollary. *If X is a complete countably normed space by either one of two sequences of norms and if the two corresponding topologies are in concordance, then the sequences of norms are equivalent.*

The space of continuous linear operators from X into Y is a linear space when addition and scalar multiplication are defined in a natural way. If $X = Y$, the product AB of operators A, B is defined by

$$(AB)x = A(Bx).$$

The adjoint A^* of a continuous linear operator A mapping X into Y is defined by

$$(A^*g, x) = (g, Ax)$$

for all $x \in X$, $g \in Y'$. A^* is a linear operator from Y' into X'. One easily sees that A^* is a continuous operator from Y' into X', when both X', Y' are provided with either the strong topologies or the weak topologies.

If A is a topological mapping from X onto Y then A^* is a topological mapping from Y' onto X'.

We say that $A_n \to A$ in the strong (weak) sense if, for any $x \in X$, $A_n x \to Ax$ in the strong (weak) topology.

Theorem 32. *Let X, Y be complete countably normed spaces and let A_n be continuous linear operators from X into Y. If $A_n \to A$ weakly, then A is also a continuous linear operator.*

Proof. A is clearly linear. For any $g \in Y'$ define

$$f_n(x) = g(A_n x), \qquad f(x) = g(Ax).$$

Then $f_n \in X'$ and $f_n \to f$ weakly. By the corollary to Theorem 15, f is bounded on bounded sets $B \subset X$, i.e., the set $\{g(Ax); x \in B\}$ is bounded for any $g \in Y'$. By Theorem 9 the set $\{Ax; x \in B\}$ is then bounded. Since this holds for any bounded set $B \subset X$, A is a bounded operator and, by Theorem 28, it is continuous.

8. Inductive Limits and Unions of Topological Spaces

A partially ordered (by "$<$") set I is called a *directed set* (more precisely, an *upward-directed set*) if to any two elements α, $\beta \in I$ there corresponds some $\gamma \in I$ such that $\alpha < \gamma$, $\beta < \gamma$. Let $X^{(\alpha)}$ be locally convex spaces, where α varies in a directed set I, and let X be a locally convex space such that $X = \bigcup_{\alpha \in I} X^{(\alpha)}$. Suppose that $\alpha < \beta$ if and only if $X^{(\alpha)} \subset X^{(\beta)}$ and that the topology of $X^{(\alpha)}$ is then stronger than the topology of $X^{(\beta)}$ (i.e., the mapping $x \to x$ from $X^{(\alpha)}$ into $X^{(\beta)}$ is continuous). Suppose further that the following property holds: for any convex set $V \subset X$, V is a neighborhood of 0 in X if and only if, for each $\alpha \in I$, $V \cap X^{(\alpha)}$ is a neighborhood of 0 in $X^{(\alpha)}$ (i.e., the mapping $x \to x$ from any $X^{(\alpha)}$ into X is continuous, and of all the topologies on X which have this property the present topology of X is the strongest). Then we say that X is the *inductive limit* of the $X^{(\alpha)}$.

Theorem 33. *A linear operator A from X into a locally convex space Y is continuous if and only if its restriction to each $X^{(\alpha)}$ is continuous.*

Proof. Let A be continuous on X. Then, for any convex neighborhood N of 0 in Y, the convex set $A^{-1}(N)$ is a neighborhood of 0 in X. Hence $A^{-1}(N) \cap X^{(\alpha)}$ is a neighborhood of 0 in $X^{(\alpha)}$, thus proving that A, restricted to $X^{(\alpha)}$, is continuous. Suppose conversely that A restricted to each $X^{(\alpha)}$ is continuous. Then, with N as above, $A^{-1}(N) \cap X^{(\alpha)}$ is a neighborhood of 0 in $X^{(\alpha)}$, for any α, and therefore $A^{-1}(N)$ is a neighborhood of 0 in X; hence A is continuous on X.

Consider now the special case where I is a sequence and

$$X^{(1)} \subset X^{(2)} \subset \cdots \subset X^{(m)} \subset \cdots,$$

and suppose further that the topology of each $X^{(n)}$ is the topology induced by $X^{(n+1)}$. In that case we call X the *strict inductive limit* of the $X^{(m)}$. The following theorems are then valid.

Theorem 34. *If $x_n \to x$ in X then x and the x_n belong to some $X^{(m)}$ and $x_n \to x$ in $X^{(m)}$.*

Theorem 35. *If B is a bounded set in X then B is contained in some $X^{(m)}$ and is a bounded set in $X^{(m)}$.*

Theorem 36. *If each $X^{(m)}$ is complete, then X is complete.*

The converse of Theorems 34, 35 (and of Theorem 36, provided each $X^{(m)}$ is a closed subspace of X) is obvious.

Proof of Theorem 34. Without loss of generality we may take $x = 0$ If the x_n do not belong to some $X^{(m)}$, then there exist two sequences $\{x_{n'}\}$ and $\{X^{(m')}\}$ ($m' = m'(n')$, and m', n' increase monotonically with m and n respectively) such that $x_{k'} \notin X^{(m')}$ if $k' > n'$ and $x_{k'} \in X^{(m')}$ if $k' \leqslant n'$. For each m', let $V^{m'}$ be a neighborhood of 0 in $X^{(m')}$ which does not contain the $x_{k'}$ for $k' \leqslant n'$. Then $V^{m'} \cap \{x_{k'}\} = \emptyset$. Since $V = \bigcup_{m'} V^{m'}$ is a neighborhood of 0 in X which does not intersect the set $\{x_{n'}\}$, we obtain a contradiction to the assumption that $x_n \to 0$ in X.

Let then $\{x_n\} \subset X^{(m)}$ for all $m \geqslant m_0$. We claim that $x_n \to 0$ in $X^{(m_0)}$. Indeed, if this is not the case, then there exists a neighborhood W^{m_0} of 0 in $X^{(m_0)}$ which does not intersect some subsequence of $\{x_n\}$; denote this sequence, for simplicity, again by $\{x_n\}$. For $m > m_0$, let W^m be neighborhoods of 0 in $X^{(m)}$ such that $W^m = X^{(m)} \cap W^{m+1}$ if $m \geqslant m_0$. Then $W^m \cap \{x_n\} = \emptyset$ since $\{x_n\} \subset X^{(m_0)}$. We conclude that $W = \bigcup_{m \geqslant m_0} W^m$ is a neighborhood of 0 in X which does not intersect $\{x_n\}$, and this is in contradiction with the assumption that $x_n \to 0$ in X.

Proof of Theorem 35. If B is not contained in some $X^{(m)}$, then there exist two sequences $\{x_{n'}\}$ and $\{X^{(m')}\}$ ($m' = m'(n')$, and m', n' increasing monotonically with m and n respectively) such that $\{x_{n'}\} \subset B$, $x_{k'} \notin X^{(m')}$ if $k' > n'$, and $x_{k'} \in X^{(m')}$ if $k' \leqslant n'$. Let $V^{m'}$ be a neighborhood of 0 in $X^{(m')}$ such that $m'V^{m'}$ does not contain the $x_{k'}$ if $k' \leqslant n'$. Then $m'V^{m'} \cap \{x_{k'}\} = \emptyset$. The set $V = \bigcup_{m'} V^{m'}$ is a neighborhood of 0 in X and, since

$$k'V \subset [k' \bigcup_{m' \leqslant k'} V^{m'}] \cup [\bigcup_{m' > k'} m'V^{m'}],$$

$\{x_{n'}\}$ is not contained in $k'V$, for any integer k'. Since V is a convex neighborhood, the bounded set $\{x_{n'}\}$ is not contained in λV for any λ, which is impossible.

Let then $B \subset X^{(m)}$ for all $m \geqslant m_0$. If B is not bounded in $X^{(m_0)}$, then for some neighborhood W^{m_0} of 0 in $X^{(m_0)}$, B is not contained in any λW^{m_0}. Construct W^m from W^{m_0} as in the proof of Theorem 34. Then B is not contained in any λW^m and hence B is not contained in any λW where $W = \bigcup_{m \geqslant m_0} W^m$, which is impossible since W is a neighborhood of 0 in X.

Theorem 36 can be proved by the method of the previous proofs. Details are left to the reader.

If each $X^{(m)}$ satisfies the first countability axiom then a linear functional f on X is continuous if and only if $f(x_n) \to 0$ whenever $x_n \to 0$ in X; also, if and only if f is bounded on bounded sets of X. Theorem 7 also remains valid for the space X. The proofs follow by combining Theorems 33–35, and 5–7.

Theorem 37. *If each $X^{(m)}$ is a complete countably normed space then the conjugate X' of X is a complete space in the weak topology.*

Proof. Indeed, if $\{f_n\}$ is a weak Cauchy sequence, then its limit f is a linear continuous functional on each $X^{(m)}$. Now apply Theorem 33.

If each of the $X^{(m)}$ is a complete countably normed space and if a set $B' \subset X'$ is weakly bounded (i.e., for any $x \in X$, $\sup |f(x)| < \infty$ when $f \in B'$), then B' is strongly bounded, i.e., for any bounded set A in X,

$$\sup |f(x)| < \infty \qquad \text{when } x \in A, f \in B'.$$

Indeed, by Theorem 35, A is contained in some $X^{(m)}$, whereas B' is weakly bounded in $X^{(m)\prime}$. Now apply Theorems 15 and 11.

If A is a linear operator from X into a locally convex space Y, and if each $X^{(m)}$ satisfies the first countability axiom, then A is continuous if and only if $Ax_n \to 0$ whenever $x_n \to 0$; also, if and only if A maps bounded sets into bounded sets. The proofs follow by combining Theorems 33–35, 28 and 29.

Definition. Let

$$X^{(1)} \subset X^{(2)} \subset \cdots \subset X^{(m)} \subset \cdots, \qquad X = \bigcup_{m=1}^{\infty} X^{(m)}$$

where the $X^{(m)}$ are linear topological spaces and suppose that the topology of each $X^{(n)}$ is stronger than the topology induced by $X^{(n+1)}$. We introduce in X the concepts of convergence and boundedness:

$x_n \to x$ if and only if the x_n and x belong to some $X^{(m)}$ and $x_n \to x$ in $X^{(m)}$;

B is a bounded set in X if and only if B is contained in some $X^{(m)}$ and is bounded in $X^{(m)}$.

We then call X the *countable union* space of the spaces $X^{(m)}$.

Note that we have not defined a topology on X. Nevertheless, we can proceed to define the concepts of continuity and boundedness for both linear functionals and linear operators in a natural way, as follows: a linear functional f is said to be continuous (bounded) if and only if f is continuous (bounded) on each space $X^{(m)}$. A linear operator A from X into a topological space Y is said to be continuous (bounded) if and only if A is continuous (bounded) on each space $X^{(m)}$. The set of all continuous linear functionals is denoted by X'. In X' we introduce two concepts of convergence: $f_n \to 0$ weakly if $f_n(x) \to 0$ for any $x \in X$, and $f_n \to 0$ strongly if $f_n(x) \to 0$ uniformly with respect to x in bounded sets of X.

A set $B' \subset X'$ is said to be weakly (strongly) bounded if for any $x \in X$ (bounded set $B \subset X$) the set of complex numbers $\{f(x); f \in B'\}$ ($\{f(x); f \in B', x \in B\}$) is bounded. If the $X^{(m)}$ are complete countably normed spaces then weak boundedness in X' implies strong boundedness.

The assertions of Theorems 33, 34, 35 are now true by definition. Theorems 36, 37 and the statements preceding and following Theorem 37 also remain true for countable union spaces. (In this connection, $\{x_n\}$ is called a Cauchy sequence if it is a Cauchy sequence in some space $X^{(m)}$.) Thus, if each $X^{(m)}$ satisfies the first countability axiom then a linear functional f is continuous if and only if f remains bounded on bounded sets of X, or if and only if $f(x_n) \to 0$ whenever $x_n \to 0$ in X. Also, a linear operator A from X into Y is continuous if and only if A maps bounded sets into bounded sets, or if and only if $Ax_n \to 0$ in Y whenever $x_n \to 0$ in X. Theorem 7 also remains valid.

If Y is also a countable union space of spaces $Y^{(m)}$ and if the $X^{(m)}$ and the $Y^{(m)}$ satisfy the first countability axiom then we say that a linear operator A from X into Y is continuous if and only if $Ax_n \to 0$ whenever $x_n \to 0$. Clearly, A is continuous if and only if A is bounded. If A is continuous, we define the adjoint A^* of A in the obvious way. One easily shows that A^* is a continuous operator from Y' into X' when each of these spaces is provided with either the weak or the strong concept of convergence. We say that $A_n \to 0$ uniformly on bounded sets of X, the A_n being continuous linear operators from X into Y, if for any bounded set $B \subset X$ there exists some $Y^{(m)}$ containing the set $\{A_n x;\ x \in B,$

$n = 1, 2, \ldots\}$ and if any neighborhood of 0 in $Y^{(m)}$ contains the sets $\{A_n x; x \in B\}$ for all n sufficiently large.

The concept of a countable union space can be extended to the case of a set $\{X^{(\alpha)}\}$, α varying in a directed set I. We say that $x_n \to x$ if and only if the x_n and x belong to some $X^{(\alpha)}$ and $x_n \to x$ in $X^{(\alpha)}$. The other definitions and statements can be extended almost word by word. We now call X the *union* space of the $X^{(\alpha)}$.

In this book we shall deal only with complete countably normed spaces and with countable unions of complete countably normed spaces. Even though the latter spaces are not topological spaces, for the sake of brevity we shall sometimes use phrases like "convergent in the topology of X," "convergent in the weak topology of X'" and so on to mean convergent in the sense defined above for union spaces and their conjugates. Also, when we speak of a "topological mapping" from one union space X onto another union space Y we simply mean a one-to-one linear mapping from X onto Y which is continuous (in the sense defined above and whose inverse is also continuous. By "the topology of Y is stronger than the topology of X, Y being a subspace of X" we mean that whenever $x_n \to x$ in the sense of the union space Y, also $x_n \to x$ in the sense of the union space X. Let $\{x_h; 0 < h \leqslant 1\}$ be a subset of X. When we say that "$x_h \to 0$ in the topology of X (or simply, in X) as $h \to 0$" we mean that for any sequence $\{h_m\}$ where $h_m \to 0$, $x_{h_m} \to 0$. Other topological statements should be understood in a similar way.

PROBLEMS

1. Prove Theorem 3.

[Hints: (a) If $\|\cdot\|_{p+1}$ is not equivalent to $\|\cdot\|_p$, for all $p \geqslant 1$, then for any sequence $\{M_p\}$ of positive numbers there exists an $\bar{x}$ such that $\|\bar{x}\|_p > M_p$. Indeed, take $\bar{x} = \sum_{p=1}^{\infty} x_p$ where

$$\|x_p\|_{p-1} < \frac{1}{2^{p-1}}, \qquad \|x_p\|_p > M_p + 1 + \|x_1\|_p + \cdots + \|x_{p-1}\|_p.$$

(b) If there exists a norm $\|\cdot\|$ in X then $\|x\|_p \leqslant N_p$ for all p and x, where $\|x\| < 1$. Take $M_p = pN_p$ and apply (a).]

CHAPTER 2

SPACES OF GENERALIZED FUNCTIONS

In this chapter we introduce the concept of generalized functions and of some elementary operations involving them. Some classes of generalized functions are studied and their structure is derived.

1. Fundamental Spaces and Generalized Functions

We shall consider linear spaces Φ whose elements are real- or complex-valued functions $\varphi(x)$ defined on a set R. We assume that the zero function (i.e., $\varphi_0(x) \equiv 0$ on R) is the zero element of Φ and that for any φ, ψ in Φ and for any numbers (scalars) λ, μ, the function $\lambda\varphi(x) + \mu\psi(x)$ is the element $\lambda\varphi + \mu\psi$ of Φ. We call Φ a *fundamental space* (or a *test space*) if the following properties are satisfied:

1. Φ is either a complete countably normed space or it is a countable union space of complete countably normed spaces;
2. if $\varphi_m \to 0$ in Φ, then $\varphi_m(x_0) \to 0$ for any point $x_0 \in R$.

The elements of Φ are called *fundamental functions* or *test functions*. A test function which has a compact support will be called a *finite* test function.

The set R on which the functions of Φ are defined will always be a set of points in either the real n-dimensional Euclidean space R^n or the complex n-dimensional Euclidean space C^n. We shall use the following notation: $x = (x_1, \ldots, x_n)$ and $\xi = (\xi_1, \ldots, \xi_n)$ are variable points in R^n; $z = (z_1, \ldots, z_n)$ and $\zeta = (\zeta_1, \ldots, \zeta_n)$ are variable points in C^n; $z = x + iy$, $\zeta = \xi + i\eta$, where $z_j = x_j + iy_j$, $\zeta_j = \xi_j + i\eta_j$;

$$|x| = (\sum_{j=1}^{n} x_j^2)^{1/2}, \qquad |z| = (\sum_{j=1}^{n} |z_j|^2)^{1/2},$$

$$D^\alpha = D_1^{\alpha_1} \cdots D_n^{\alpha_n}, \qquad |\alpha| = \alpha_1 + \cdots + \alpha_n$$

where $D_j = \partial/\partial x_j$, $\alpha = (\alpha_1, \ldots, \alpha_n)$;

$$x^\alpha = x_1^{\alpha_1} \cdots x_n^{\alpha_n}, \qquad \alpha! = \alpha_1! \cdots \alpha_n!;$$

$$\binom{\alpha}{\beta} = \binom{\alpha_1}{\beta_1} \cdots \binom{\alpha_n}{\beta_n}$$

where $\beta = (\beta_1, \ldots, \beta_n)$;

$$\alpha \geqslant \beta \quad \text{means} \quad \alpha_j \geqslant \beta_j \qquad \text{for } j = 1, \ldots, n;$$

$$\frac{a}{b} = \left(\frac{a_1}{b_1}, \ldots, \frac{a_n}{b_n}\right), \qquad \frac{ak}{b} = \left(\frac{a_1k_1}{b_1}, \ldots, \frac{a_nk_n}{b_n}\right),$$

and

$$\exp\,[a|bz|^k] = \exp\,[a_1|b_1z_1|^{k_1} + \cdots + a_n|b_nz_n|^{k_n}],$$

where $a = (a_1, \ldots, a_n)$, $b = (b_1, \ldots, b_n)$, and $k = (k_1, \ldots, k_n)$. If a, b, or k is a scalar then the above notation is still used with $a_1 = \cdots = a_n = a$, $b_1 = \cdots = b_n = b$ or $k_1 = \cdots = k_n = k$.

We denote by $C^m(G)$ $(0 \leqslant m \leqslant \infty)$ the set of m-times continuously differentiable functions on a set G, if G is open. If G is the closure of an open set, $C^m(G)$ is defined to be the class of those functions for which the first m derivatives exist and are uniformly continuous in the interior of G (and thus can be extended, by continuity, to the boundary of G). We denote by $C_c^m(G)$ (G open) the set of functions of class $C^m(R^n)$ which vanish outside a compact subset of G. $C_c^m(R^n)$ is the set of all functions in $C^m(R^n)$ having a compact support.

We shall now define two general classes of fundamental spaces; the functions of the first class are defined on $R = R^n$, and those of the second class are defined on $R = C^n$.

Consider a sequence of functions

$$(1.1) \qquad 1 \leqslant M_1(x) \leqslant M_2(x) \leqslant \cdots \leqslant M_p(x) \leqslant \cdots,$$

where the $M_p(x)$ are defined on R^n and their values may be finite or ∞. We assume that at each point of R^n either all the M_p are finite or else all of them have the value ∞. We denote by R_M the set of points where the $M_p(x)$ are equal to ∞. We finally assume that, restricted to the complement set of R_M in R^n which we shall denote by R_M^n, all the functions $M_p(x)$ are continuous. To avoid pathological cases we assume that the boundary of R_M^n has a Lebesgue measure 0.

Consider the $C^\infty(R^n)$ complex-valued functions φ which for any $0 \leqslant |\alpha| \leqslant p$, $1 \leqslant p < \infty$ satisfy the following conditions:

$$(1.2) \qquad D^\alpha\varphi(x) = 0 \qquad \text{on } R_M,$$

and

$$(1.3) \qquad M_p(x)\, D^\alpha\varphi(x) \text{ are continuous bounded functions on } R_M^n.$$

This set of functions we denote by $K\{M_p\}$, and it is clearly a linear space.

Topology is defined on $K\{M_p\}$ in terms of the sequence of norms

$$\|\varphi\|_p = \sup_{|\alpha| \leqslant p} \sup_{x \in R^n_M} M_p(x)|D^\alpha \varphi(x)| \qquad (1 \leqslant p < \infty). \tag{1.4}$$

Note that if R^n_M is a closed set, Eq. (1.2) simply means that the support of φ lies in R^n_M.

An example of such a space is the space (D_N) (N compact) introduced in Chap. 1, Sec. 3 (see Eq. (3.2)), when we take

$$M_p(x) = \begin{cases} 1 & \text{if } x \in N, \\ \infty & \text{if } x \notin N. \end{cases} \tag{1.5}$$

We shall use also the notation $K(N) = (D_N)$ and, if N is an n-rectangle defined by $|x_j| \leqslant a_j$ ($j = 1, \ldots, n$), then we denote the space simply by $K(a)$.

Another important example is the space S. The elements of S are the $C^\infty(R^n)$ functions φ for which

$$\lim_{|x| \to \infty} |x|^m |D^\alpha \varphi(x)| = 0 \tag{1.6}$$

for any m and α. The functions of S are called *fast-decreasing* C^∞ functions. The topology of S is given in terms of the norms (1.4) where

$$M_p(x) = (1 + |x|)^p. \tag{1.7}$$

It will be shown in Sec. 2 that the $K\{M_p\}$ spaces are complete countably normed spaces and, under some conditions on the M_p, they are perfect spaces. In particular it would follow that $K(N)$ and S are perfect spaces.

The second class of fundamental spaces is defined in terms of a sequence $\{M_p(z)\}$ where z varies in the complex space C^n. It is assumed that the $M_p(z)$ are continuous functions for all z, and that

$$0 < C(y) \leqslant M_1(z) \leqslant M_2(z) \leqslant \cdots \leqslant M_p(z) \leqslant \cdots \tag{1.8}$$

for some continuous function $C(y)$. The fundamental space consists of all the complex-valued functions $\psi(x)$ which can be extended into entire functions $\psi(z)$ on C^n for which the norms

$$\|\psi\|_p = \sup_{z \in C^n} M_p(z)|\psi(z)| \qquad (1 \leqslant p < \infty) \tag{1.9}$$

are all finite. We denote this space by $Z\{M_p\}$ and define a topology in terms of the norms (1.9).

If we take

$$M_p(z) = e^{-a|y|}(1 + |z|)^p, \tag{1.10}$$

where a is a positive vector (i.e., $a_j > 0$ for $1 \leqslant j \leqslant n$), then we obtain the space of all entire functions of exponential type which satisfy

$$(1.11) \qquad (1 + |z|)^k |\psi(z)| \leqslant C_k e^{a|y|} \qquad (0 \leqslant k < \infty),$$

where the C_k are constants depending on ψ. We denote this space by $Z(a)$.

It will be shown in Sec. 3 that any $Z\{M_p\}$ is a complete countably normed space and that, under some conditions on the M_p, it is a perfect space. It would follow, in particular, that $Z(a)$ is a perfect space.

The following theorem holds for any fundamental space.

Theorem 1. *If a fundamental space Φ is a complete countably normed space which contains all the functions of some space $K(N)$, then whenever a sequence $\{\varphi_m\}$ is convergent to 0 in the topology of $K(N)$, it is also convergent to 0 in the topology of Φ.*

Proof. If $\varphi_m \to 0$ in $K(N)$ and $\varphi_m \to \varphi^*$ in Φ then, by one of the properties of a fundamental space, $\varphi^*(x) = 0$ for any $x \in R$; hence $\varphi^* = 0$ as an element of Φ. Thus the topologies of $K(N)$ and Φ are in concordance and, by Theorem 31 (Chap. 1, Sec. 7), whenever $\varphi_m \to 0$ in $K(N)$ also $\varphi_m \to 0$ in Φ.

The theorem is obviously valid also if Φ is a countable union space of complete countably normed spaces $\Phi^{(m)}$ provided $K(N) \subset \Phi^{(m)}$ for some m.

Definition. A fundamental space Φ is said to be a *derivative space* (or to *admit differentiation*) if for any $\varphi \in \Phi$ the derivatives $D_i\varphi = \partial\varphi/\partial x_i$ $(i = 1, \ldots, n)$ also belong to Φ and the mappings

$$\varphi \to D_i\varphi$$

are continuous mappings from Φ into itself. It is obvious that if Φ is a derivative space then, for any α, the mapping $\varphi \to D^\alpha\varphi$ is a continuous mapping from Φ into itself.

Definition. A function α is called a *multiplier* on (or in) Φ if $\alpha\varphi$ belongs to Φ for any $\varphi \in \Phi$ and the mapping

$$\varphi \to \alpha\varphi$$

from Φ into Φ is continuous.

Definition. A continuous linear functional on a fundamental space Φ is called a *generalized function* over (or on) Φ. Thus, the generalized functions over Φ are precisely the elements of the conjugate space Φ'. When we write $f(x)$ for a generalized function, we merely wish to indicate that f operates on test functions which depend on x; f itself is *not* a function of x. The operation of f on $\varphi \in \Phi$ will be denoted by

$$(f, \varphi), \quad f(\varphi), \quad f\cdot\varphi \quad \text{or} \quad (f(x), \varphi(x)).$$

We make the following important observation: $f \in \Phi'$ *if and only if* f *is a linear functional on* Φ *and* $f\cdot\varphi_m \to 0$ *whenever* $\varphi_m \to 0$ *in* Φ.

If

$$(f, \varphi) = \int_R k(x)\varphi(x)\,dx, \tag{1.12}$$

where $k(x)$ is a locally integrable function, then we say that f is of *function type* $k(x)$.

Clearly, if Φ contains all the $C_c^\infty(R^n)$ functions then k is uniquely determined up to a set of measure 0.

An example of a generalized function which is not of function type is $\delta_{(x^0)}$ defined by

$$\delta_{(x^0)}\cdot\varphi = (\delta_{(x^0)}(x), \varphi(x)) = \varphi(x^0).$$

It is a generalized function over any test space Φ. $\delta = \delta_{(0)}$ is called the *Dirac measure.*

2. The Spaces $K\{M_p\}$

Consider the space $\bar{\Phi}_p$ of all $C^p(R^n)$ functions which vanish on R_M together with their first p derivatives, and for which the continuous functions $M_p(x)\,D^\alpha\varphi(x)$ are bounded on R_M^n for $|\alpha| \leqslant p$. We introduce in this space the norm

$$\|\varphi\|_p = \sup_{|\alpha|\leqslant p}\ \sup_{x\in R^n{}_M} M_p(x)|D^\alpha\varphi(x)|, \tag{2.1}$$

and prove the following:

Lemma 1. $\bar{\Phi}_p$ *is a complete space.*

Proof. Let $\{\varphi_m\}$ be a Cauchy sequence in $\bar{\Phi}_p$. Then, for any $|\alpha| \leqslant p$, $k \geqslant m$, $x \in R_M^n$,

$$\begin{aligned}|D^\alpha\varphi_m(x) - D^\alpha\varphi_k(x)| &\leqslant M_p(x)|D^\alpha\varphi_m(x) - D^\alpha\varphi_k(x)| \\ &\leqslant \|\varphi_m - \varphi_k\|_p \leqslant \epsilon_m,\end{aligned} \tag{2.2}$$

where $\epsilon_m \to 0$ if $m \to \infty$. Since the left side of (2.2) is 0 for x in R_M, we conclude that $D^\alpha\varphi_m \to D^\alpha\varphi_0$ uniformly on R^n. In particular it follows that $\varphi_0 \in C^p(R^n)$ and $D^\alpha\varphi_0(x) = 0$ on R_M for $|\alpha| \leqslant p$. Taking $k \to \infty$ in (2.2) we further obtain, if $x \in R_M^n$,

$$M_p(x)|D^\alpha\varphi_m(x) - D^\alpha\varphi_0(x)| \leqslant \epsilon_m, \tag{2.3}$$

and hence

$$M_p(x)|D^\alpha \varphi_0(x)| \leqslant \epsilon_m + \|\varphi_m\|_p. \tag{2.4}$$

The last inequality shows that $\|\varphi_0\|_p$ exists, i.e., $\varphi_0 \in \bar{\Phi}_p$. The inequality (2.3) shows that $\|\varphi_m - \varphi_0\|_p \leqslant \epsilon_m \to 0$, i.e., $\varphi_m \to \varphi_0$ in $\bar{\Phi}_p$. Thus, $\bar{\Phi}_p$ is a complete space.

If the elements φ_m of the above Cauchy sequence belong to

$$\Phi \equiv K\{M_p\},$$

then the limit is of course also in $\bar{\Phi}_p$. Since the completion of Φ with respect to the norm $\|\cdot\|_p$ is obtained by taking the limits of all the Cauchy sequences with elements in Φ, this completion, which we shall denote by Φ_p, is a linear subspace of $\bar{\Phi}_p$. Since $\Phi = \bigcap_{p=1}^{\infty} \bar{\Phi}_p$ and $\Phi \subset \Phi_p \subset \bar{\Phi}_p$, we thus obtain

$$\Phi \equiv K\{M_p\} = \bigcap_{p=1}^{\infty} \Phi_p. \tag{2.5}$$

We also have, for any $\varphi \in \Phi$,

$$\|\varphi\|_1 \leqslant \|\varphi\|_2 \leqslant \cdots \leqslant \|\varphi\|_p \leqslant \cdots. \tag{2.6}$$

We shall next prove the following:

Lemma 2. *For any integers $p \geqslant 1$, $q \geqslant 1$, the norms $\|\cdot\|_p$, $\|\cdot\|_q$ are in concordance* (see Chap. 1, Sec. 2).

Proof. We have to show that if $\{\varphi_m\} \subset \Phi$, $\|\varphi_m\|_q \to 0$ and $\|\varphi_m - \varphi_k\|_p \leqslant \epsilon_m \to 0$ for $k \geqslant m$, $m \to \infty$, then $\|\varphi_m\|_p \to 0$. Now, since $M_q(x) \geqslant 1$, $\|\varphi_m\|_q \to 0$ implies that $\varphi_m(x) \to 0$ for any $x \in R^n$. On the other hand, if φ_0 is the limit of $\{\varphi_m\}$ in the norm $\|\cdot\|_p$, then $\varphi_m(x) \to \varphi_0(x)$ for any $x \in R^n$, so that $\varphi_0(x) = 0$. Using (2.3) we then obtain

$$M_p(x)|D^\alpha \varphi_m(x)| \leqslant \epsilon_m,$$

i.e., $\|\varphi_m\|_p \leqslant \epsilon_m \to 0$ as $m \to \infty$.

In view of Theorem 1 (Chap. 1, Sec. 3) we have thus proved the following:

Theorem 2. *$K\{M_p\}$ is a complete countably normed space.*

We shall prove below that $K\{M_p\}$ is a perfect space if the following condition is satisfied:

(*P*) *for any integer $p \geqslant 1$ there exists an integer $p' > p$ such that to any $\epsilon > 0$ there corresponds some $N_0 > 0$ such that if $|x| > N_0$ $(x \in R^n_M)$ or if $M_p(x) > N_0$ $(x \in R^n_M)$ then $M_p(x) < \epsilon M_{p'}(x)$.*

If the $M_p(x)$ are finite-valued functions for all x (i.e., $R_M = \emptyset$) then (P) is equivalent to the following statement: *To any integer* $p \geqslant 1$ *there corresponds an integer* $p' > p$ *such that*

$$\lim_{|x| \to \infty} \frac{M_p(x)}{M_{p'}(x)} = 0. \tag{2.7}$$

Note that if (P) holds, then for any $|\alpha| \leqslant p$, $\varphi \in K\{M_p\}$,

$$M_p(x) D^\alpha \varphi(x) \to 0 \quad \text{if} \quad |x| \to \infty \quad \text{or if} \quad M_p(x) \to \infty. \tag{2.8}$$

Indeed, if this is not true, then for some p, α there exists a sequence $\{x_m\}$ with either $|x_m| \to \infty$ or $M_p(x_m) \to \infty$ such that

$$M_p(x_m)|D^\alpha \varphi(x_m)| \geqslant C > 0.$$

But then $M_{p'}(x_m)|D^\alpha \varphi(x_m)| \to \infty$ as $m \to \infty$ which is in contradiction with $\|\varphi\|_{p'} < \infty$.

Definition. A sequence $\{\varphi_m\}$ is said to be *properly convergent* if, for any α, $\{D^\alpha \varphi_m\}$ is a sequence which converges uniformly on bounded sets of R^n.

Lemma 3. *Let* (P) *be satisfied. If* $\{\varphi_m\} \subset K\{M_p\}$, $\varphi_m \to 0$ *properly, and, for any* $p \geqslant 1$, $\|\varphi_m\|_p \leqslant C_p$ *(i.e.,* $\{\varphi_m\}$ *is a bounded sequence in* $K\{M_p\}$*), then* $\|\varphi_m\|_p \to 0$ *for any* $p \geqslant 1$, *i.e.,* $\varphi_m \to 0$ *in* $K\{M_p\}$.

Proof. Given p, choose p' as in (P). Then, for any $\epsilon > 0$, $x \in R_M^n$,

$$M_p(x) < \frac{\epsilon}{C_{p'}} M_{p'}(x)$$

if $|x| > N_1$ or if $M_p(x) > N_1$. We then have, for $|\alpha| \leqslant p$,

$$M_p(x)|D^\alpha \varphi_m(x)| \leqslant \frac{\epsilon}{C_{p'}} M_{p'}(x)|D^\alpha \varphi_m(x)| \leqslant \frac{\epsilon}{C_{p'}} \|\varphi_m\|_{p'} \leqslant \epsilon. \tag{2.9}$$

Since the set of points Γ in R_M^n where (2.9) is possibly not satisfied is contained in $|x| \leqslant N_1$ and $M_p(x) \leqslant N_1$ on that set, we can take m_0 sufficiently large so that if $m \geqslant m_0$,

$$M_p(x)|D^\alpha \varphi_m(x)| < \epsilon$$

holds for $x \in \Gamma$ and, hence, for all $x \in R_M^n$. We thus have $\|\varphi_m\|_p \leqslant \epsilon$ if $m \geqslant m_0$ and the proof is completed.

Corollary. *If* $\{\varphi_m\} \subset K\{M_p\}$, $\|\varphi_m\|_p \leqslant C_p$ *for all* $p \geqslant 1$ *and* $\varphi_m \to \varphi_0$ *properly, then* $\varphi_0 \in K\{M_p\}$ *and* $\varphi_m \to \varphi_0$ in $K\{M_p\}$.

Indeed, (2.2) is valid for any m with $\epsilon_m = 2C_p$. Taking $k \to \infty$ we obtain (2.3); hence (2.4). It follows that $\|\varphi_0\|_p < \infty$ for any $p \geqslant 1$. Hence $\varphi_0 \in \bigcap_p \Phi_p = K\{M_p\}$. Now apply Lemma 3 to $\{\varphi_m - \varphi_0\}$.

Theorem 3. *If the M_p satisfy the condition (P), then $K\{M_p\}$ is a perfect space.*

Proof. We have to show that bounded sets are relatively sequentially compact. Let A be a bounded set. Then $\|\varphi_p\| \leqslant C_p$ for all $\varphi \in A$, $1 \leqslant p < \infty$. Since $M_p(x) \geqslant 1$, for each α the family $\{D^\alpha\varphi; \varphi \in A\}$ is uniformly bounded. Using the lemma of Ascoli-Arzela and the standard diagonalization procedure, we can choose a sequence $\{\varphi_m\} \subset A$ which is properly convergent. Indeed, as a first step we choose a sequence which converges together with the first derivatives uniformly in $|x| \leqslant 1$; as a second step we choose a subsequence which converges together with the first and second derivatives uniformly for $|x| \leqslant 2$, etc. Arranging these sequences in rows, the φ_m are taken to be the elements of the main diagonal. Since $\{\varphi_m\}$ is properly convergent and bounded in $K\{M_p\}$, the corollary to Lemma 3 implies that $\{\varphi_m\}$ is a convergent sequence in $K\{M_p\}$. Hence, A is relatively sequentially compact.

It is clear that any $K\{M_p\}$ contains finite functions provided the interior of R^n_M is not empty. If the $M_p(x)$ are finite functions on R^n, then $K\{M_p\}$ contains C_c^∞. We shall now prove the following:

Theorem 4. *If the M_p satisfy the condition (P) then the finite functions of $K\{M_p\}$ form a dense linear subspace of $K\{M_p\}$.*

Proof. We introduce the function

$$h(x) = C \int_{-2}^{2} w(r - \rho)\, d\rho,$$

where $|x| = r$ and w is defined by

$$w(r) = \begin{cases} 0 & \text{if } |r| \geqslant 1, \\ \exp\,[1/(r^2 - 1)] & \text{if } |r| < 1. \end{cases}$$

w and, therefore, h are C^∞ functions, and if the constant C is appropriately chosen then $h(x) = 1$ if $|x| < 1$ and $h(x) = 0$ if $|x| \gtrsim 3$.

Now for any $\varphi \in K\{M_p\}$ define a sequence $\{\varphi_m\}$ by

$$\varphi_m(x) = \varphi(x)h\left(\frac{x}{m}\right). \tag{2.10}$$

Since $\varphi_m(x) = \varphi(x)$ for $|x| < m$, $\varphi_m \to \varphi$ properly. If we prove that

$$\|\varphi_m\|_p \leqslant C_p \qquad (1 \leqslant p < \infty) \tag{2.11}$$

then it follows, by the corollary to Lemma 3, that $\varphi_m \to \varphi$ in $K\{M_p\}$, and the proof of the theorem is completed.

To prove (2.11) let

$$|D^\alpha h(x)| \leqslant H_p \qquad \text{if } |\alpha| \leqslant p,\ x \in R^n,$$

where $\{H_p\}$ is a sequence of constants. Then also

$$\left| D^\alpha h\left(\frac{x}{m}\right)\right| \leqslant H_p.$$

Using Leibniz' rule,

$$D^\alpha(fg) = \sum_{\beta \leqslant \alpha} \binom{\alpha}{\beta} D^{\alpha-\beta}f \cdot D^\beta g, \tag{2.12}$$

we obtain

$$M_p(x)|D^\alpha \varphi_m(x)| \leqslant C'_p H_p \|\varphi\|_p,$$

and the proof of (2.11) with $C_p = C'_p H_p \|\varphi\|_p$ follows.

Two sequences $\{M_p(x)\}$, $\{M'_p(x)\}$ are said to be *equivalent* if the set R'_M where the $M'_p = \infty$ coincides with the set R_M where the $M_p = \infty$, and if for $x \in R^n_M$ (the complement of R_M)

$$0 < C(p) \leqslant \frac{M_p(x)}{M'_p(x)} \leqslant C'(p) < \infty. \tag{2.13}$$

It follows that

$$C(p)\|\varphi\|'_p \leqslant \|\varphi\|_p \leqslant C'(p)\|\varphi\|'_p \tag{2.14}$$

so that the two sequences of norms are equivalent.

We conclude this section by noting that both sequences (1.5) and (1.7) satisfy the condition (P). Hence $K(N) \equiv (D_N)$ (N compact) and S are perfect spaces. Furthermore, the set C^∞_c is dense in S.

3. The Spaces $Z\{M_p\}$

Let $\Psi \equiv Z\{M_p\}$ and denote by $\bar{\Psi}_p$ the set of all entire functions for which $\|\psi\|_p < \infty$. The method used in proving Lemma 1 can easily be modified to show that $\bar{\Psi}_p$ is a complete normed space (use is made of the inequalities $M_p(z) \geqslant C(y) > 0$). If we denote by Ψ_p the completion of Ψ with respect to the norm $\|\cdot\|_p$, then $\Psi \subset \Psi_p \subset \bar{\Psi}_p$. Since $\Psi = \bigcap_{p=1}^{\infty} \bar{\Psi}_p$, it follows that

$$\Psi \equiv Z\{M_p\} = \bigcap_{p=1}^{\infty} \Psi_p. \tag{3.1}$$

An analogue of Lemma 2 is valid for the present norms (1.9) and the proof is analogous to that of Lemma 2. We thus have the following:

Theorem 5. *The space $Z\{M_p\}$ is a complete countably normed space.*

To prove that $Z\{M_p\}$ is a perfect space we need a condition similar to the condition (P) (or, rather, similar to (2.7)), namely

(P_0) *To any integer* $p \geqslant 1$ *there corresponds an integer* $p' > p$ *such that*

$$\lim_{|z|\to\infty} \frac{M_p(z)}{M_{p'}(z)} = 0. \tag{3.2}$$

We can now prove an analogue of Lemma 3 and its corollary. In fact, we have the following:

Lemma 4. *If* $\{\psi_m\} \subset Z\{M_p\}$, $\|\psi_m\|_p \leqslant C_p$ *for all* $p \geqslant 1$, *and* $\psi_m \to \psi_0$ *pointwise in* R^n, *then* $\psi_0 \in Z\{M_p\}$ *and* $\psi_m \to \psi_0$ *in* $Z\{M_p\}$.

Proof. Since $M_p(z) \geqslant C(y) > 0$, in any bounded domain $G \subset C^n$ containing the origin the sequence $\{\psi_m\}$ is uniformly bounded. But then this family of functions is a normal family, i.e., from any sequence of the family one can pick a subsequence which converges uniformly in closed subsets of G. The limits of all the convergent subsequences must be the same analytic function, since any two such limits are analytic functions which coincide for $z = x$ in some neighborhood of the origin. Hence $\psi_0(x)$ can be extended as an entire function $\psi_0(z)$ and $\psi_m(z) \to \psi_0(z)$ uniformly on bounded sets of C^n. We can now proceed as in the proof of Lemma 3 and its corollary.

Remark. From the above proof we see that the lemma is still valid if the assumption that $\psi_m \to \psi_0$ pointwise in R^n is replaced by the assumption that $\psi_m(x) \to \psi_0(x)$ for each point x in some neighborhood of the origin.

Using Lemma 4 and the tool of normal families the proof of Theorem 3 can be modified to yield the following theorem.

Theorem 6. *If the* $M_p(z)$ *satisfy the condition* (P_0), *then* $Z\{M_p\}$ *is a perfect space.*

We conclude this section with two remarks. (1) The concept of equivalent sequences $\{M_p(z)\}$, $\{M'_p(z)\}$ can be introduced as for $K\{M_p\}$. The sequences of norms of equivalent sequences are equivalent. (2) Theorem 4 is obviously false for the spaces $Z\{M_p\}$.

4. Multiplication and the Derivatives of Generalized Functions

A function $f(x)$ is said to define a generalized function on Φ if $f(x)$ is measurable on R and if

$$f(\varphi) \equiv \int_R f(x)\varphi(x)\,dx$$

is a continuous linear functional on Φ. The following simple theorem is sometimes useful.

Theorem 7. *Let* $\{f_m\}$ *be a sequence of measurable functions on* R *which satisfy the following conditions:* (a) $|f_m(x)| \leqslant g(x)$ *where* $g(x)$ *defines a generalized function on* Φ *and furthermore,* $\int_R g(x)|\varphi(x)|\,dx$ *remains bounded when* φ *varies in bounded sets of* Φ; (b) $\lim f_m(x) = f_0(x)$ *almost everywhere on* R. *Then* $f_0(x)$ *also defines a generalized function on* Φ *and* $f_m \to f_0$ *in the weak topology of* Φ'.

Proof. Using a theorem of Lebesgue on bounded convergence, we have

$$\begin{aligned}\int_R |f_0(x)\varphi(x)|\,dx &= \lim_{m\to\infty} \int_R |f_m(x)\varphi(x)|\,dx \\ &\leqslant \int_R g(x)|\varphi(x)|\,dx.\end{aligned}$$

Hence, $f_0(x)$ defines a generalized function on Φ. Since

$$\lim_{m\to\infty} \int_R [f_0(x) - f_m(x)]\varphi(x)\,dx = 0,$$

$f_m \to f_0$ weakly in Φ'.

Definition. Let α be a multiplier in a test space Φ. For any generalized function $f \in \Phi'$ we then define αf (or $\alpha \cdot f$) by

$$(\alpha f, \varphi) = (f, \alpha\varphi). \tag{4.1}$$

The right side is easily seen to be a continuous linear functional of $\varphi \in \Phi$; hence αf is a generalized function. We call αf the *multiplicative product* of α with f. $f \to \alpha f$ is a continuous mapping from Φ' into Φ': if $f_m \to 0$ in the weak (or strong) topology then $\alpha f_m \to 0$ in the weak (or strong) topology. We say that α is also a multiplier in Φ'.

Using the rule (2.12) one can verify that if in a $K\{M_p\}$ space, for any $p \geqslant r \geqslant 1$ there exists a $q \geqslant p$ and a constant C_{pr} such that

$$M_p(x)M_r(x) \leqslant C_{pr}M_q(x), \tag{4.2}$$

then every C^∞ function $g(x)$ which satisfies

$$|D^\beta g(x)| \leqslant C_\beta M_{k(\beta)}(x) \qquad (0 \leqslant |\beta| < \infty) \tag{4.3}$$

(for some function $k(\beta)$) is a multiplier in $K\{M_p\}$. Details are left to the reader.

In particular, the C^∞ functions satisfying

$$|D^\beta g(x)| \leqslant C_\beta(1 + |x|)^{k(\beta)} \qquad (0 \leqslant |\beta| < \infty) \tag{4.4}$$

are multipliers in S. Any C^∞ function is a multiplier in $K(a)$.

In $Z(a)$, polynomials are multipliers.

Definition. Let Φ be a test space which admits differentiation. Then, for any $f \in \Phi'$ we define its *first derivatives* $D_i f \equiv \partial f/\partial x_i$ by

$$(D_i f, \varphi) = -(f, D\varphi_i). \tag{4.5}$$

$D_i f$ is easily seen to be a continuous linear functional on Φ. $D^\alpha f$ is defined similarly by

$$(D^\alpha f, \varphi) = (-1)^{|\alpha|}(f, D^\alpha \varphi) \tag{4.6}$$

and it also belongs to Φ'. Since the operator $A^* f = D^\alpha f$ is the conjugate of $A\varphi = (-1)^{|\alpha|} D^\alpha \varphi$, we conclude (see Chap. 1, Secs. 7, 8) that A^* is a continuous operator. Hence:

Theorem 8. *If $f_m \to f$ weakly (strongly) in Φ', then $D^\alpha f_m \to D^\alpha f$ weakly (strongly) in Φ'.*

The motivation for the definition (4.5) is the following: if f has first continuous derivatives on R^n and φ has a compact support then (with $D_i f$ being the usual derivation of the function $f(x)$)

$$\int_{R^n} D_i f(x) \cdot \varphi(x)\, dx = -\int_{R^n} f(x)\, D_i \varphi(x)\, dx, \tag{4.7}$$

which means that the generalized function defined by $D_i f(x)$ coincides with the D_i-derivative of the generalized function defined by $f(x)$. In the future, when there is a possibility of confusion, we denote the usual derivatives of functions by $[D_i f]$ or $[\partial f/\partial x_i]$. The $D_i f$ will then also be called *generalized derivatives.*

The spaces $K(N)$, S and in fact all the $K\{M_p\}$ spaces admit differentiation. The proof is left to the reader.

$Z(a)$ also admits differentiation. Indeed, if $\psi \in Z(a)$,

$$M_p(z) \left| \frac{\partial \psi(z)}{\partial z_j} \right| \leqslant \frac{1}{2\pi} \int_{|\zeta_j - z_j| = 1} \frac{M_p(z)}{M_p(z')} \frac{M_p(z')|\psi(z')|}{|\zeta_j - z_j|^2} |d\zeta_j| \leqslant C_p \|\psi\|_p,$$

where $z' = (z_1, \ldots, z_{j-1}, \zeta_j, z_{j+1}, \ldots, z_n)$ and C_p is a constant independent of ψ. We thus get

$$\left\| \frac{\partial \psi}{\partial z_j} \right\|_p \leqslant C_p \|\psi\|_p,$$

which proves that $\partial\psi/\partial z_j$ belongs to $Z\{M_p\}$ and that the mapping $\psi \to \partial\psi/\partial z_j$ is a continuous mapping of $Z(a)$ into itself.

The following theorem will be used in the next section.

Theorem 9. *Let f be a locally integrable function, absolutely continuous in x_1 for almost all $(x_2, \ldots, x_n)$, and assume that the derivative $[\partial f/\partial x_1]$ (which exists almost everywhere for almost all $(x_2, \ldots, x_n)$) is equal to a locally* *integrable function $g(x)$. If f and g are generalized functions over $K\{M_p\}$ and if the M_p satisfy the condition (P), then*

$$\partial f/\partial x_1 = g. \tag{4.8}$$

Proof. Let $\varphi \in C_c^\infty$. Then, using Fubini's theorem and integration by parts, we have

$$\begin{aligned}\frac{\partial f}{\partial x_1}(\varphi) &= \int_{R^n} f(x)\left(-\frac{\partial \varphi(x)}{\partial x_1}\right) dx \\ &= \int_{-\infty}^{\infty} \cdots \int_{-\infty}^{\infty} dx_2 \cdots dx_n \int_{-\infty}^{\infty} f(x)\left(-\frac{\partial \varphi(x)}{\partial x_1}\right) dx_1 \\ &= \int_{-\infty}^{\infty} \cdots \int_{-\infty}^{\infty} dx_2 \cdots dx_n \int_{-\infty}^{\infty} g(x)\varphi(x)\, dx_1 \\ &= \int_{R^n} g(x)\varphi(x)\, dx = g(\varphi).\end{aligned}$$

Thus, the generalized derivative $\partial f/\partial x_1$ coincides with g on the set of all C_c^∞ functions which belong to $K\{M_p\}$. By Theorem 4, Sec. 4, this set is dense in $K\{M_p\}$, hence (4.8) follows.

Let $P(D)$ be a power series in D and assume that Φ admits differentiation and that for every $\varphi \in \Phi$ the series $P(-D)\varphi$ is convergent in the Φ topology (and hence also pointwise). Denoting by $\{P_m(D)\}$ any sequence of partial sums of $P(D)$ we thus have: $P_m(-D)\varphi \to P(-D)\varphi$ in the Φ topology, as $m \to \infty$. For any $u \in \Phi'$ we define $P(D)u$ by

$$(P(D)u, \varphi) = \lim_{m\to\infty} (P_m(D)u, \varphi). \tag{4.9}$$

Since $(P_m(D)u, \varphi) = (u, P_m(-D)\varphi)$, the limit on the right exists and is independent of the particular sequence $\{P_m(D)\}$. Furthermore,

$$(P(D)u, \varphi) = (u, (P(-D)\varphi). \tag{4.10}$$

If $P(-D)\varphi$ is convergent uniformly with respect to φ in bounded sets of Φ, then $P(D)u$ belongs to Φ' and, in fact, $P(D)$ is a bounded operator in Φ'.

5. Structure of Generalized Functions on $\mathbf{K}\{\mathbf{M}_p\}$

One of the most interesting and important problems in the theory of generalized functions is the problem of finding the structure of generalized functions by expressing them in terms of differential operators acting on functions or on measures. In this section we shall obtain such theorems for the spaces $K\{M_p\}$.

Let f be a generalized function on $K\{M_p\}$. Then, by Chap. 1, Sec. 4, f is a continuous linear functional on some Φ_p.

To each $\varphi \in \Phi_p$ there corresponds in a one-to-one way a vector-function $\{\psi\}$ of length ν whose components

$$(5.1) \qquad \psi_\alpha(x) = M_p(x)\, D^\alpha\varphi(x) \qquad (0 \leqslant |\alpha| \leqslant p)$$

are defined for $x \in R^n_M$. Let Γ^ν be the (Cartesian) product of ν identical spaces Γ, each Γ consisting of all the continuous functions $h(x)$ on R^n_M having a finite norm

$$\|h\| = \sup_{x \in R^n{}_M} |h(x)|.$$

The norm in Γ^ν is taken to be, say, the sum of the norms of the components. The vector-functions $\{\psi\}$ form a linear subspace Δ^ν of Γ^ν and, since $f \in \Phi'_p$, the functional $L(\{\psi\}) = f(\varphi)$ is a continuous linear functional on Δ^ν provided with the norm of Γ^ν. By Theorem 8a, Sec. 4, Chap. 1, L can be extended into a continuous linear functional $\tilde{L}$ on Γ^ν with norm (on Γ^ν) equal to the norm of L on Δ^ν.

By a theorem of F. Riesz, the continuous linear functionals $\tilde{L}$ on Γ^ν have the form

$$(5.2) \qquad \tilde{L}(\theta) = \sum_{|\alpha| \leqslant p} \int_{R^n{}_M} \theta_\alpha(x)\, d\sigma_\alpha(x) \qquad (\theta = \{\theta_\alpha\}),$$

where the σ_α are measures, and

$$(5.3) \qquad \|\tilde{L}\| = \sum_{|\alpha| \leqslant p} \int_{R^n{}_M} |d\sigma_\alpha(x)|.$$

Applying (5.2) to the elements $\theta = \{\psi\}$ of Δ^ν we obtain

$$(5.4) \qquad f(\varphi) = \sum_{|\alpha| \leqslant p} \int_{R^n{}_M} M_p(x)\, D^\alpha\varphi(x)\, d\sigma_\alpha(x).$$

We have thus proved the following:

Theorem 10. *Any generalized function f over $K\{M_p\}$ has the form (5.4), for some $p \geqslant 1$ and measures $\sigma_\alpha(x)$ on R^n_M, and the norm of f on Φ_p is equal to*

$$(5.5) \qquad \sum_{|\alpha| \leqslant p} \int_{R^n{}_M} |d\sigma_\alpha(x)|.$$

To get a more refined representation we make the following assumptions on the M_p:

(N_1) *For every integer $p \geqslant 1$ there exists an integer $p' \geqslant p$ such that*

$$(5.6) \qquad m_{pp'}(x) \equiv \frac{M_p(x)}{M_{p'}(x)}$$

is integrable on R^n_M.

(N_2) *For every integer* $p \geqslant 1$ *there exists an integer* $p'' \geqslant p$ *such that for any* $x \in R^n_M$, $|x' - x| \leqslant 1$,

$$\frac{M_p(x)}{M_{p''}(x')} \leqslant C_p \qquad (C_p \text{ a constant}). \tag{5.7}$$

We shall prove that under the assumptions (N_1), (N_2), the norms $\|\varphi\|_p$ defined in (1.4) are equivalent to the norms

$$\|\varphi\|'_p = \sup_{|\alpha| \leqslant p} \int_{R^n_M} M_p(x)|D^\alpha\varphi(x)|\,dx. \tag{5.8}$$

We have, if $x \in R^n_M$,

$$M_p(x)|D^\alpha\varphi(x)| = m_{pp'}(x)M_{p'}(x)|D^\alpha\varphi(x)| \leqslant m_{pp'}(x)\|\varphi\|_{p'}.$$

Hence, by integration,

$$\|\varphi\|'_p \leqslant B_p\|\varphi\|_{p'} \qquad \left(B_p = \int_{R^n_M} m_{pp'}(x)\,dx\right). \tag{5.9}$$

We next need the elementary inequality, for any $x \in R^n$,

$$|\varphi(x)| \leqslant A \sum_{|\beta| \leqslant q} \int_{|\xi - x| < 1} |D^\beta\varphi(\xi)|\,d\xi, \tag{5.10}$$

where A and q are independent of φ. Such an inequality holds with $q = n$. Indeed, let $\gamma(t)$ (t real) be a continuously differentiable function which equals 1 at $t = 0$ and 0 for $|t| \geqslant \epsilon$, where ϵ is a fixed positive number. Since

$$\begin{aligned}\varphi(x) &= \gamma(0)\varphi(x) - \gamma(-\epsilon)\varphi(x_1 - \epsilon, x_2, \ldots, x_n)\\ &= \int_{x_1-\epsilon}^{x_1} \frac{\partial}{\partial \xi_1}[\gamma(\xi_1 - x_1)\varphi(\xi_1, x_2, \ldots, x_n)]\,d\xi_1\\ &= \int_{x_1-\epsilon}^{x_1} \left(\frac{\partial\gamma}{\partial\xi_1}\varphi + \gamma\frac{\partial\varphi}{\partial\xi_1}\right) d\xi_1,\end{aligned}$$

we have

$$|\varphi(x)| \leqslant A_0 \int_{x_1-\epsilon}^{x_1} |\varphi(\xi_1, x_2, \ldots, x_n)|\,d\xi_1 + A_0 \int_{x_1-\epsilon}^{x_1} \left|\frac{\partial}{\partial\xi_1}\varphi(\xi_1, x_2, \ldots, x_n)\right| d\xi_1.$$

Applying the same argument to the integrands

$$\varphi(\xi_1, x_2, \ldots, x_n), \qquad \frac{\partial\varphi(\xi_1, x_2, \ldots, x_n)}{\partial\xi_1},$$

but with x_1, $\gamma(x_1)$ replaced by x_2, $\gamma(x_2)$, and proceeding in this way step by step, we arrive at the inequality (5.10) provided we take $\epsilon\sqrt{n} < 1$.

Using (5.10) with φ replaced by $D^\alpha\varphi$, and assuming, as we may, that $p'' \geqslant p + q$, we have (if $x \in R^n_M$)

$$M_p(x)|D^\alpha\varphi(x)| \leqslant A \sum_{|\beta| \leqslant q} \int_{\substack{|\xi - x| < 1 \\ \xi \in R^n{}_M}} \frac{M_p(x)}{M_{p''}(\xi)} M_{p''}(\xi)|D^{\alpha+\beta}\varphi(\xi)| \, d\xi$$

$$\leqslant A'C_p\|\varphi\|'_{p''} \qquad (A' \text{ a constant}).$$

Hence,

$$\|\varphi\|_p \leqslant A'C_p\|\varphi\|'_{p''}. \tag{5.11}$$

Equations (5.9) and (5.11) show that the sequences of norms $\|\cdot\|_p$, $\|\cdot\|'_p$ are equivalent. Given a generalized function f we can therefore proceed as in the proof of Theorem 10 but with the norms $\|\cdot\|_p$ replaced by $\|\cdot\|'_p$. On the space Γ^ν we take the norm

$$\|\theta\| = \sum_{j=1}^{\nu} \int_{R^n{}_M} |\theta_j(x)| \, dx, \tag{5.12}$$

where $\theta = (\theta_1, \ldots, \theta_\nu) \in \Gamma^\nu$. $L(\{\psi\}) = f(\varphi)$ is then a continuous functional on the subspace Δ^ν. Since every continuous linear functional T on Γ^ν normed by (5.12) has the form

$$T(\theta) = \sum_{j=1}^{\nu} \int_{R^n{}_M} \theta_j(x) g_j(x) \, dx,$$

where the $g_j(x)$ are measurable and essentially bounded functions, and

$$\|T\| = \sum_{j=1}^{\nu} \underset{x \in R^n{}_M}{\text{essen. sup}} |g_j(x)|,$$

we obtain the following:

Theorem 11. *Let the M_p satisfy the conditions (N_1), (N_2). Then any generalized function on $K\{M_p\}$ has the form*

$$(f, \varphi) = \sum_{|\alpha| \leqslant p} \int_{R^n{}_M} M_p(x) f_\alpha(x) D^\alpha\varphi(x) \, dx \tag{5.13}$$

for some integer $p \geqslant 1$ and for some measurable and essentially bounded functions $f_\alpha(x)$, and the norm of f on Φ, normed by $\|\cdot\|'_p$, is equal to

$$\sum_{|\alpha| \leqslant p} \underset{x \in R^n{}_M}{\text{essen. sup}} |f_\alpha(x)|. \tag{5.14}$$

Consider the space S of fast-decreasing C^∞ functions. The conditions (N_1), (N_2) are clearly satisfied. $R^n_M = R^n$ and each term on the right side of (5.13) can be written as

$$(-1)^{|\alpha|} D^\alpha(M_p f_\alpha) \cdot \varphi, \tag{5.15}$$

where $M_p f_\alpha$ is a functional on S of function type. To each integral

$$I_i = \int_0^{x_i} M_p f_\alpha \, dx_i \qquad (i = 1, \ldots, n)$$

we can apply Theorem 9, Sec. 4, and thus conclude that

$$D_i I_i = M_p f_\alpha$$

as elements of S'. The same holds for any multiple integral. Hence, each functional in (5.15) can be written in the form $D^\beta g_\alpha$, with the same β for all the α, and the g_α are continuous functions on R^n satisfying

$$|g_\alpha(x)| \leqslant A_\alpha(1 + |x|)^{m_\alpha}.$$

We have thus proved the following:

Theorem 12. *Every generalized function f over S has the form*

$$f = D^\beta[(1 + |x|^2)^m F(x)] \tag{5.16}$$

where $F(x)$ is a continuous bounded function on R^n.

Suppose f varies in a bounded set B' of S'. By Theorem 13, Sec. 5, Chap. 1, B' is contained in some Φ_p' and it is bounded in the norm $\|\cdot\|_p$ of Φ_p'. Hence the representation asserted in Theorem 11 is valid for all $f \in B'$ with the same p, and the set of numbers (5.14) is also bounded. Proceeding similarly to the proof of Theorem 12, we obtain

Theorem 13. *If B' is a bounded set in S', then for every $f \in B'$ there exists a representation of the form* (5.16) *with the same β and m. Furthermore, the family of continuous functions $\{F(x); f \in B'\}$ is uniformly bounded on R^n.*

The functionals over $K(N)$ will be considered in detail in the next chapter.

PROBLEMS

1. Prove: if, for some q, $|k(x)|/M_q(x)$ is integrable over R^n, then the right side of (1.12) is a generalized function over $K\{M_p\}$.

2. Show that each of the sequences

$$M_p'(x) = \prod_{j=1}^{n} (1 + |x_j|)^p, \qquad M_p''(x) = (1 + |x|^2)^{p/2}$$

is equivalent to the sequence (1.7).

3. Show that each of the sequences

$$M'_p(z) = e^{-a|y|} \prod_{j=1}^{n} (1 + |z_j|)^p, \qquad M''_p(z) = e^{-a|y|}(1 + |z|^2)^{p/2}$$

is equivalent to the sequence (1.10).

4. Let $f(z)$ be an entire function satisfying

$$|f(z)| \leqslant C(1 + |z|^m)e^{b|y|} \qquad (m \geqslant 0, b \geqslant 0).$$

Prove that the operator $A\varphi = f\varphi$ is a continuous operator from $Z(a)$ into $Z(a + b)$, for any $a > 0$.

5. Prove the converse of Theorem 10, namely, that every functional of the form (5.4), with σ_α for which (5.5) is finite, is a generalized function over $K\{M_p\}$. Prove, similarly, the converse of Theorem 11 and of Theorem 12.

CHAPTER 3

THEORY OF DISTRIBUTIONS

In this chapter we present the theory of distributions of L. Schwartz, i.e., the theory of generalized functions over $(D) \equiv K$. Some concepts and results that are valid for distributions, such as the theory of Fourier transforms, can be extended to all generalized functions. Many other theorems which will be proved for distributions, such as the structure theorems, do not have analogues for arbitrary generalized functions, but still can be extended to some classes of generalized functions (such as $(K\{M_p\})'$). Various applications of distributions to differential equations will be given both in this chapter and in subsequent ones.

1. Spaces of Functions

$x = (x_1, \ldots, x_n)$ is a variable point in the real n-dimensional Euclidean space R^n and $|x| = (\sum_{i=1}^{n} x_i^2)^{1/2}$ is the norm of x. For $\alpha = (\alpha_1, \ldots, \alpha_n)$ we set $D^\alpha = D_1^{\alpha_1} \ldots D_n^{\alpha_n}$ where $D_i = \partial/\partial x_i$, $x^\alpha = x_1^{\alpha_1} \ldots x_n^{\alpha_n}$, $|\alpha| = \alpha_1 + \cdots + \alpha_n$, and $\alpha! = \alpha_1! \cdots \alpha_n!$. (There will be no confusion between the notations $|x|$, $|\alpha|$.) The class of m-times continuously differentiable complex-valued functions in R^n is denoted by C^m $(0 \leqslant m \leqslant \infty)$ and the subclass consisting of those functions which have compact support is denoted by C_c^m. A sequence $\{\varphi_j\} \subset C_c^m$ is said to *converge* to zero if (a) there exists a bounded set N containing the supports of all the φ_j, and (b) for any α, $|\alpha| \leqslant m$, $D^\alpha \varphi_j(x) \to 0$ as $j \to \infty$, uniformly with respect to x.

Let $\Omega = \{\Omega_0 = \phi, \Omega_1, \Omega_2, \ldots\}$, where ϕ is the empty set and $\Omega_j = \{x; |x| < j\}$; $\{\epsilon\} = \{\epsilon_0, \epsilon_1, \epsilon_2, \ldots\}$, where the ϵ_i are arbitrary positive numbers decreasing monotonically to 0; and $\{m\} = \{m_0, m_1, m_2, \ldots\}$, where the m_i are arbitrary non-negative integers increasing monotonically to ∞. Denote by $V(\{m\}, \{\epsilon\})$ the set of all functions $\varphi \in C_c^\infty$ satisfying

$$|D^\alpha \varphi(x)| < \epsilon_j \qquad \text{if } |\alpha| \leqslant m_j,\ x \notin \Omega_j,\ j = 0, 1, 2, \ldots.$$

Taking all the sets $V(\{m\}, \{\epsilon\})$ as a basis of neighborhoods of 0 in C_c^∞, the linear space C_c^∞ becomes a linear topological space (in fact, locally convex), which we shall denote by (D).

Theorem 1. *A sequence $\{\varphi_j\} \subset C_c^\infty$ is convergent to 0 if and only if it converges to 0 in the topology of (D).*

Proof. Let $\varphi_j \to 0$. Then the supports of the φ_j lie in some Ω_k. In Ω_k each sequence $\{D^\alpha \varphi_j\}$ is uniformly convergent to 0. Hence, given any neighborhood $V(\{m\}, \{\epsilon\})$, the φ_j belong to it if j is sufficiently large. Suppose conversely that $\varphi_j \to 0$ in the topology of (D). Then, the supports of the φ_j must lie in some bounded set. Indeed, in the contrary case we have $\varphi_{j_m}(x^m) \neq 0$ for some sequence $\{j_m\}$ and for some $|x^m| \geqslant i_m$ where the i_m increases monotonically to ∞. Defining $V(\{m\}, \{\epsilon\})$ by

$$\epsilon_{i_m} = \tfrac{1}{2} |\varphi_{j_m}(x^m)1 \qquad (m = 0, 1, 2, \ldots)$$

and by making an arbitrary choice of the remaining ϵ's and of the m's, it is clear that $\varphi_i \notin V(\{m\}, \{\epsilon\})$ for the subsequence $i = j_m$, thus contradicting our assumption that $\varphi_i \to 0$ in (D). The uniform convergence of $\{D^\alpha \varphi_j(x)\}$ to zero, for any α, follows by taking neighborhoods $V(\{m\}, \{\epsilon\})$ with $m_0 = |\alpha|$ and ϵ_0 arbitrarily small, and recalling that the φ_j must belong to each such neighborhood for j sufficiently large.

We denote by (D_G) the set of all $\varphi \in (D)$ with support in G. If $G = N$ is a compact set in R^n, we introduce in (D_N) the topology induced by (D). As is easily seen, a basis of neighborhoods at 0 in (D_N) is given by

$$V(m, \epsilon, N) = \{\varphi; \varphi \in (D_N), |D^\alpha \varphi(x)| < \epsilon \qquad \text{for } |\alpha| \leqslant m, x \in N\}.$$

(D_N) is a perfect space (see Chap. 1, Sec. 6).

Let N be a compact set in R^n and let $\{M\} = \{M_0, M_1, M_2, \ldots\}$ be a sequence of positive numbers which is monotonically increasing. We denote by $B(\{M\}, N)$ the set of all $\varphi \in (D_N)$ satisfying

$$|D^\alpha \varphi(x)| \leqslant M_m \qquad \text{for } |\alpha| \leqslant m, x \in R^n, m = 0, 1, 2, \ldots.$$

Theorem 2. *A set $W \subset (D)$ is bounded in (D) if and only if it is contained in some $B(\{M\}, N)$, i.e., if and only if the supports of the $\varphi \in W$ lie in some compact set N and, for each α, the set $\{D^\alpha \varphi; \varphi \in W\}$ is uniformly bounded.*

Proof. We recall that a set W in a linear topological space (D) is said to be bounded if for any neighborhood V of 0 in (D) there exists a $\lambda > 0$ such that $W \subset \lambda V$. It is easily seen that each $B(\{M\}, N)$ is a bounded set. Hence if W is contained in such a set, it is also bounded. Conversely, let W be a bounded set. We claim that the supports of the

$\varphi \in W$ must be contained in some bounded set. Indeed, if this is not the case, then there exist sequences $\{\varphi_j\} \subset W$, $\{x^j\} \subset R^n$ such that $\varphi_j(x^j) \neq 0$ and $|x_j| \geqslant j$. Choose $\epsilon_j = |\varphi_j(x^j)|/j$ and an arbitrary $\{m\}$. Then W is not contained in $\lambda V(\{m\}, \{\epsilon\})$ for any $\lambda > 0$. This contradicts the assumption that W is a bounded set. We have thus proved that $W \subset (D_N)$ for some compact set N. Since δW is contained in any neighborhood of 0 in (D) if δ is sufficiently small, the set $\{D^\alpha \varphi; \varphi \in W\}$ is uniformly bounded, say by M_α. Hence $W \subset B(\{M\}, N)$ where $M_m = \max_{|\alpha| \leqslant m} M_\alpha$.

The space C_c^h can be topologized by introducing the system of neighborhoods $V(\{m\}, \{\epsilon\}) \equiv V(\{\epsilon\})$ with $\{m\} = \{h, h, h, \ldots\}$. The linear topological space thus obtained is denoted by (D^h). As in Theorem 1, $\varphi_j \to 0$ in the sense of (D^h) if and only if their supports are uniformly bounded and $D^\alpha \varphi_j \to 0$ for $|\alpha| \leqslant h$ uniformly with respect to $x \in R^n$. (D_G^m) and the topology on (D_N^m) (N compact) are defined analogously to (D_G) and the topology on (D_N).

2. Partition of Unity

Theorem 3. *Let $\{\Omega_i\}$ be a countable covering of an open set $\Omega \subset R^n$ by open sets Ω_i (i ranges over a denumerable set). Then there exist functions α_i having the following properties:*

(a) $\alpha_i \geqslant 0$, $\Sigma\, \alpha_i = 1$ in Ω;
(b) *α_i is C^∞ and its support lies in Ω_i;*
(c) *Every compact set in Ω intersects only a finite number of supports of the α_i.*

The α_i are said to constitute a *partition of unity, subordinate to* the covering $\{\Omega_i\}$.

The proof of the theorem relies upon the following lemma.

Lemma 1. *Let A be a compact set in R^n, and let B be an open set in R^n which contains A. Then there exists a C^∞ function β having the following properties:* (i) *the support of β lies in B;* (ii) *$\beta(x) = 1$ if $x \in A$, and* (iii) *$0 \leqslant \beta(x) \leqslant 1$ for all x.*

Proof of Lemma 1. For any $\epsilon > 0$, we introduce the function

$$\rho_\epsilon(x) = \begin{cases} 0 & \text{if } |x| \geqslant \epsilon, \\ \dfrac{k}{\epsilon^n} \exp\left(\dfrac{-\epsilon^2}{\epsilon^2 - x^2}\right) & \text{if } |x| < \epsilon, \end{cases} \tag{2.1}$$

which belongs to (D) and satisfies the following properties: $\rho_\epsilon(x) \geqslant 0$ in R^n, $\rho_\epsilon(x) > 0$ if $|x| < \epsilon$, $\rho_\epsilon(x) = 0$ if $|x| > \epsilon$, and $\int \rho_\epsilon(x)\, dx = 1$ if k is determined by

$$k \int_{|x|\leq 1} \exp\left(\frac{1}{|x|^2 - 1}\right) dx = 1.$$

Here we have used the obvious notation $dx = dx_1 \cdots dx_n$.

Let G be a bounded open set such that $A \subset G \subset \bar{G} \subset B$ and denote by $\chi(x)$ the characteristic function of G. If ϵ is sufficiently small then

$$\beta(x) = \int_{R^n} \chi(\xi)\rho_\epsilon(x - \xi)\, d\xi$$

satisfies all the requirements of the lemma.

Proof of Theorem 3. We say that a covering $\{U_i\}$ of Ω is *subordinate to* the covering $\{\Omega_i\}$ if for each i there exists a j such that $\bar{U}_i \subset \Omega_j$. A covering $\{U_i\}$ of Ω is said to be *locally finite* if every compact set in Ω intersects only a finite number of the U_i. The main effort in proving the theorem is in the construction of a locally finite covering $\{G_i\}$ of Ω (each G_i an open set) which is subordinate to $\{\Omega_i\}$. We proceed to describe this construction.

We first construct a subordinate covering $\{\Omega_i'\}$. For each point $x \in \Omega$ there exists a bounded open neighborhood whose closure lies in Ω and in some Ω_j. Choosing a denumerable set of such neighborhoods, denote it by $\{\Omega_i'\}$, we thus obtain a subordinate covering to $\{\Omega_i\}$.

We next define, recursively, a sequence of compact sets K_m and a sequence of integers j_m: $K_1 = \Omega_1'$, $j_1 = 1$. Suppose K_{m-1} and j_{m-1} have been defined; we then let j_m be the smallest integer $> j_{m-1}$ such that $K_{m-1} \subset \bigcup_{i=1}^{j_m} \Omega_i'$ and take $K_m = \bigcup_{i=1}^{j_m} \overline{\Omega_i'}$.

K_{m-1} thus lies in the interior of K_m and every compact set of Ω is contained in some K_m. Set

$$G_i = \Omega_i' \qquad \text{if } i \leqslant j_2,$$

$$G_i = \Omega_i' \cap \hat{K}_{m-1} \qquad \text{if } j_m < i \leqslant j_{m+1} \qquad (\text{for } m > 1),$$

where $\hat{K}_{m-1}$ is the complement of K_{m-1} in R^n. We can prove by induction that

$$\bigcup_{i=1}^{j_{m+1}} G_i = \bigcup_{i=1}^{j_{m+1}} \Omega_i'.$$

Indeed, suppose this has been proved for j_m and set

$$A = \bigcup_{i=1}^{j_m} G_i = \bigcup_{i=1}^{j_m} \Omega_i'.$$

Noting that $K_{m-1} \subset A$ it is clear that $G_i \cup A = \Omega_i' \cup A$ if $j_m < i \leqslant j_{m+1}$, from which the conclusion follows. The G_i thus form a covering of Ω subordinate to the covering $\{\Omega_i\}$. This covering is locally finite since each compact set N of Ω is contained in some K_m, thus implying that $N \cap G_i = \emptyset$ if $i > j_{m+1}$.

Let $\{G'_i\}$ be a covering of Ω such that $\overline{G'_i} \subset G_i$. It can be constructed inductively as follows: suppose that G'_i have been constructed for $i < k$ in such a way that the G'_i and the G_j, $j \geqslant k$ form a covering of Ω. Then

$$B = G_k - \bigcup_{i<k} (G_k \cap G'_i) - \bigcup_{j>k} (G_k \cap G_j)$$

is a subset of G_k which is closed since it is equal to

$$\bar{G}_k - \bigcup_{i<k} (\bar{G}_k \cap G'_i) - \bigcup_{j>k} (\bar{G}_k \cap G_j)$$

(recall that $\bar{G}_k \subset \Omega$ and hence a neighborhood of the boundary of G_k is contained in $(\bigcup_{i<k} G'_i) \cup (\bigcup_{j>k} G_j)$). Take G'_k to be any set containing B and such that $\overline{G'_k} \subset G_k$.

Since $\overline{G'_i} \subset G_i$, by Lemma 1 there exists a C^∞ function ψ_i which equals 1 on G'_i, has its support in G_i, and is such that $0 \leqslant \psi_i \leqslant 1$. The sum $\psi = \Sigma\, \psi_i$ is a C^∞ function since, for every compact set in Ω, only a finite number of ψ_i do not vanish identically on that set. Since $\psi \geqslant 1$, the functions $\varphi_i = \psi_i/\psi$ are C^∞ and, in addition, $\Sigma\, \varphi_i = 1$. We now define $\alpha_1 = \Sigma'\, \varphi_i$ where the sum ranges over those i's for which $G_i \subset U_1$. Next, $\alpha_2 = \Sigma''\, \varphi_i$ where the sum ranges over the remaining i's for which $G_i \subset U_2$, etc. The α_i clearly satisfy all the required properties.

Remark. For any $\varphi \in C_c^m$, define

$$\varphi_\epsilon(x) = \int_{R^n} \varphi(\xi)\rho_\epsilon(x - \xi)\, d\xi.$$

Then $\varphi_\epsilon \in C_c^\infty$ and, as is easily seen, $\varphi_\epsilon \to \varphi$ in the sense of (D^m). We have thus proved that the elements of (D) form a dense subset in the topological space (D^m).

3. Definition and Some Properties of Distributions

The conjugate space of (D) $((D^m))$ is denoted by (D') $((D'^m))$. The elements of (D') are called *distributions*. Thus, distributions are continuous linear functionals on (D). The application of a distribution T to $\varphi \in (D)$ is denoted by $T(\varphi)$ or by $T \cdot \varphi$. A weak topology is defined on (D') by giving the basis of neighborhoods

$$V(\varphi_1, \ldots, \varphi_m, \epsilon) = \{T; |T(\varphi_1)| < \epsilon, \ldots, |T(\varphi_m)| < \epsilon\}$$

of 0. A sequence $\{T_j\} \subset (D')$ is said to be *weakly convergent* to 0 if $T_j(\varphi) \to 0$ for any $\varphi \in (D)$. A sequence $\{T_j\} \subset (D')$ is said to be *strongly convergent*, or, simply, *convergent* to 0 (we write $T_j \to 0$) if $T_j(\varphi) \to 0$ for any $\varphi \in (D)$ and uniformly so for φ in bounded sets of (D).

Note that if B is a bounded set of (D) then, by definition, it is contained in λV for any neighborhood V of 0 in (D), provided $\lambda > 0$ is sufficiently large. Now, since T is continuous, there exist neighborhoods V of 0 in (D) such that $\{|T(\varphi)|; \varphi \in V\}$ is a bounded set. It follows then that if $T \in (D')$,

$$T(B) \equiv \sup_{\varphi \in B} |T(\varphi)|$$

is finite for any bounded set B. (The converse is also true; see Theorem 5′.) We can therefore rephrase the definition of convergence: $T_j \to 0$ if and only if $T_j(B) \to 0$ as $j \to \infty$, for any bounded set B.

A topology (the "strong topology") is introduced in (D') by giving a basis of neighborhoods $V(B, \epsilon)$ of 0 in (D'), where B is an arbitrary bounded set of (D) and $\epsilon > 0$. $V(B, \epsilon)$ is the set of all $T \in (D')$ satisfying $T(B) < \epsilon$. In the future, when we refer to the topology of (D') without further specifications we shall understand it to be the (strong) topology determined by the $V(B, \epsilon)$. It is clear that $T_j \to 0$ if and only if $T_j \to 0$ in the topology of (D'). Note also that B' is a bounded set in (D') if and only if

$$B'(B) = \sup_{T \in B'} T(B) < \infty \tag{3.1}$$

for any bounded set $B \subset (D)$. The above definitions of strong and weak topologies fit in with the general terminology of Chap. 1, Sec. 5.

The following theorem relates the topology of (D') to the weak topology of (D'). It will not be used in the future and the proof will not be given here.

Theorem 4. *If $T_j(\varphi) \to T(\varphi)$ for every $\varphi \in (D)$, where T_j are distributions, then T is a distribution and $T_j - T \to 0$,* i.e., *weak convergence implies convergence for any sequence of distributions. Moreover, in bounded sets of (D') the topology of (D') is equivalent to the weak topology of (D').*

The following theorem is useful in proving the continuity of linear functionals on (D).

Theorem 5. *A linear functional T on (D) is continuous if and only if, for any compact set N, the restriction of T to (D_N) is continuous.*

Using Theorem 5, Sec. 4, Chap. 1 we can restate Theorem 5 as follows:

Theorem 5′. *A linear functional T on (D) is continuous if and only if, for any bounded set $B \subset (D)$, $T(B)$ is finite.*

The proof of Theorem 5 is based upon the following theorem.

Theorem 6. *A convex set W is a neighborhood of* 0 *in* (D) *if and only if, for any compact set $N \subset R^n$, the intersection of W with (D_N) is a convex neighborhood of* 0 *in* (D_N).

In other words: (D) is the inductive limit of the (D_N).

These theorems are true also for (D^m).

Proof of Theorem 6. It is clear that $V(\{m\}, \{\epsilon\})$ intersects each (D_N) in a neighborhood of (D_N). Since any neighborhood of 0 in (D) contains some $V(\{m\}, \{\epsilon\})$, one part of the theorem is proved.

Suppose now that W is a convex set in (D) which intersects each (D_N) in a neighborhood of 0 in (D_N). We shall prove that W is a neighborhood of 0 in (D). Taking $N = \{x; |x| \leqslant h + 2\}$ we conclude that for any $h \geqslant 0$ there exists an integer $m_h \geqslant 0$ and a positive number η_h such that if $\varphi \in (D)$ and its support lies in $\{x; |x| \leqslant h + 2\}$, and if

$$|D^\beta \varphi(x)| \leqslant \eta_h \qquad \text{for } |\beta| \leqslant m_h, \tag{3.2}$$

then $\varphi \in W$. Without loss of generality we may take the m_h to be monotone increasing to ∞.

By Theorem 3 there exist C^∞ functions α_h such that $\alpha_h \geqslant 0$, $\Sigma\, \alpha_h = 1$, and the support of α_h lies in $h \leqslant |x| \leqslant h + 2$. We can write every $\varphi \in (D)$ in the form

$$\varphi = \sum_{h=0}^{\infty} \frac{1}{2^{h+1}} (2^{h+1}\alpha_h \varphi).$$

Since W is convex, if $2^{h+1}\, \alpha_h\, \varphi \in W$ for every h, then also $\varphi \in W$.

Observe next that if

$$|D^\beta \varphi(x)| \leqslant \epsilon_h \qquad \text{for } |\beta| \leqslant m_h,\ |x| \geqslant h,$$

then $|2^{h+1} D^\beta(\alpha_h \varphi)| \leqslant k_h \epsilon_h$ for all $x \in R^n$, where the k_h are constants independent of the particular $\varphi \in (D)$ and of ϵ_h. Hence, if we take a monotone sequence $\{\epsilon_h\}$ decreasing to 0 and satisfying $k_h \epsilon_h < \eta_h$, then $\varphi \in V(\{m\}, \{\epsilon\})$ implies (see Eq. (3.2)) that $2^{h+1}\alpha_h \varphi \in W$, and, consequently $\varphi \in W$. Thus W is neighborhood of 0 in (D).

Theorem 5 is a special case of Theorem 33, Sec. 8, Chap. 1.

Setting $N_m = \{x; |x_i| \leqslant m,\ i = 1, \ldots, n\}$ we conclude from the proof of Theorem 6:

Corollary 1 (*to Theorem* 6). *(D) is the strict inductive limit of the spaces* $(D_{N_m}) \equiv K(m)$ $(m = 1, 2, \ldots)$.

Thus, (D) is a fundamental space (or a test space) in the terminology introduced in Chap. 2, Sec. 1.

Using this corollary, we see that Theorem 2 then follows from Theorem 35, Sec. 8, Chap. 1. Using Theorem 36, Sec. 8, Chap. 1 we also conclude:

Corollary 2 (*to Theorem* 6). (D) *is a complete space (in the sense of Cauchy sequences).*

We remark that Theorem 4 follows from general theorems on topological spaces. In fact, the inductive limit of Fréchet spaces is a space possessing the property asserted in Theorem 2, Sec. 3, Chap. 1; thus (D) possesses this property. In addition, bounded sets in (D) are easily seen to be relatively compact. Hence (D) is a Montel space, and for Montel spaces the weak and strong topologies, in bounded sets of the conjugate space, are equivalent topologies. We also mention without proof that every Montel space is reflexive and that bounded sets in the conjugate space are relatively compact. We shall not make use of these facts in the future.

We finally mention that, in view of Theorem 6, Sec. 4, Chap. 1, Theorem 5′ is equivalent to the following theorem:

Theorem 5″. *A linear functional* T *on* (D) *is continuous if and only if* $T(\varphi_m) \to 0$ *whenever* $\varphi_m \to 0$ *in* (D).

There exists a one-to-one correspondence between locally (Lebesgue) integrable functions $f(x)$ and a linear subspace of (D'); namely, we correspond to $f(x)$ the functional

$$f(\varphi) = \int_{R^n} f(x)\varphi(x)\,dx. \tag{3.3}$$

In fact, if $f(\varphi) = f'(\varphi)$ for all $\varphi \in (D)$ then $f(\varphi) = f'(\varphi)$ also for $\varphi \in (D^0)$ (by the remark at the end of Sec. 2), from which it follows that $f(x) = f'(x)$ almost everywhere.

Let L_N^p $(1 \leqslant p < \infty)$ be the set of functions which are defined on a compact set N and whose pth power is integrable. L_N^p is a Banach space if we define the norm by

$$\|f\|_{p,N} = \left\{\int_N |f(x)|^p\,dx\right\}^{1/p}.$$

L_N^∞ is the Banach space of measurable and essentially bounded functions with the norm $\|f\|_{\infty,N} = \underset{x\in N}{\text{essen. sup}}\,|f(x)|$. We shall prove the following:

Theorem 7. *If a sequence* $\{f_j\}$ *is convergent to* 0 *in the* L_N^1 *sense, for any compact set* N, *then* $f_j \to 0$ *in the sense of* (D').

Proof. We have to show that for any bounded set $B \subset (D)$, $f_j(B) \to 0$

as $j \to \infty$. By Theorem 2, B is contained in some $B(\{M\}, N)$. Hence

$$\begin{aligned} f_j(B) &= \sup_{\varphi \in B} |f_j(\varphi)| = \sup_{\varphi \in B} \left| \int_N f_j(x)\varphi(x)\,dx \right| \leqslant M_0 \int_N |f_j(x)|\,dx \\ &= M_0 \|f_j\|_{1,N} \to 0 \qquad \text{as } j \to \infty. \end{aligned}$$

A measure μ is a totally additive complex-valued function defined on the bounded Borel sets of R^n, i.e., on all the bounded sets which are obtained from the open sets of R^n by taking countable unions of finite intersections. Total additivity means that if $A = \bigcup_j A_j$ where $A_i \cap A_j = \emptyset$ for $i \neq j$ and the A_j are Borel sets, and if A is bounded, then $\mu(A) = \sum_j \mu(A_j)$ and the series is absolutely convergent. There exists a one-to-one correspondence between measures and a linear subset of (D'), given by

$$\mu(\varphi) = \int_{R^n} \varphi(x)\,d\mu(x). \tag{3.4}$$

In fact, by the remark at the end of Sec. 2, (D) is dense in (D^0) and hence if μ and μ' coincide on (D) they coincide on (D^0) and, therefore, are identical measures.

The functional $\mu(\varphi)$ is continuous in (D) provided with the topology of (D^0). Conversely, if $\mu(\varphi)$ is a continuous linear functional on (D) provided with the topology of (D^0), then we can extend it, by continuity (Theorem 7 of Chap. 1 holds for (D^0)), into a continuous functional on (D^0). A theorem of F. Riesz asserts that any continuous linear functional of (D^0) can be represented in the form (3.4) where $\mu(x)$ is a measure. Hence, measures can be identified with those distributions which are continuous on (D) provided with the topology of (D^0).

A distribution T is *real* if $T(\varphi)$ is real for any real $\varphi \in (D)$. It is *positive* (we then write $T \geqslant 0$) if $T(\varphi) \geqslant 0$ for any $\varphi \geqslant 0$ in (D). Setting $T_1(\varphi) = \operatorname{Re} T(\varphi)$, $T_2(\varphi) = \operatorname{Im} T(\varphi)$, we can write $T = T_1 + iT_2$ and T_1, T_2 are real distributions. If T is a measure T_j $(j = 1,2)$ can be further decomposed as a difference of two positive measures. If T is not a measure, T_j does not possess such a decomposition in view of the following theorem.

Theorem 8. *A positive distribution is a positive measure.*

Proof. We have to prove that T is continuous on (D) provided with the topology of (D^0). By the analogue of Theorem 5 with (D^0) instead of (D), it suffices to prove that T is continuous on (D_N) provided with the topology of (D_N^0) for any compact N. Thus, given $\delta > 0$, we have to find a neighborhood

$$V = \{\varphi; \varphi \in (D_N), |\varphi(x)| < \epsilon\}$$

such that $|T(\varphi)| < \delta$ if $\varphi \in V$. Let $\psi \in C_c^\infty$ be a non-negative function which is equal to 1 on N. (Its existence follows from Lemma 1, Sec. 2.) Writing $\varphi = \varphi_1 + i\varphi_2$ and noting that $|\varphi(x)| < \epsilon\psi(x)$, for all x, we obtain

$$(3.5) \qquad -\epsilon\psi < \varphi_1 < \epsilon\psi, \qquad -\epsilon\psi < \varphi_2 < \epsilon\psi.$$

Since $T \geqslant 0$, $T(\alpha) \leqslant T(\beta)$ if $\alpha \leqslant \beta$. Using this, we obtain from (3.5)

$$|T(\varphi_1)| \leqslant \epsilon T(\psi), \qquad |T(\varphi_2)| \leqslant \epsilon T(\psi).$$

Hence $|T(\varphi)| < \delta$ for all $\varphi \in V$, provided $2\epsilon T(\psi) < \delta$.

We conclude this section by saying a few words about the space (D_G) where G is an *open* set in R^n. We define topology in a similar way as for (D), taking

$$\Omega = \{\emptyset, \Omega_1, \Omega_2, \ldots\},$$

where $\Omega_j \subset \Omega_{j+1}$, $\Omega_j \subset G$ and every compact set of G is contained in Ω_j for some j sufficiently large. (D'_G) is the conjugate of (D_G). Results proved for (D) and (D') can easily be extended to (D_G) and (D'_G). The elements of (D'_G) are called *distributions defined on* G. It is clear that if a distribution is defined on G_1 then it is also defined on G_2 if $G_2 \subset G_1$. By an analogue of Theorem 5″, $T \in (D'_G)$ if and only if $T(\varphi_m) \to 0$ whenever $\varphi_m \to 0$ in (D_G) (i.e., the supports of the φ_m are contained in some compact set $N \subset G$ and, for any α, $D^\alpha\varphi_m \to 0$ uniformly in G). In this chapter, except for Theorem 15, Sec. 5, we shall not consider distributions defined on G (i.e., elements of (D'_G)) but we shall consider distributions $T \in (D')$ on G, i.e., restricted to the linear subspace (D_G) of (D) (see, for instance, Theorems 17–19, Sec. 5).

4. Derivatives of Distributions

If f is a continuously differentiable function then

$$\int_{R^n} D_i f(x) \cdot \varphi(x)\, dx = -\int_{R^n} f(x)\, D_i\varphi(x)\, dx \qquad \left(D_i = \frac{\partial}{\partial x_i}\right)$$

for any $\varphi \in (D)$. Using the notation (3.3) we can write this equality in the form

$$(4.1) \qquad D_i f(\varphi) = -f(D_i\varphi).$$

This equality motivates the following definition:

The derivative D_iT (or $\partial T/\partial x_i$) of a distribution T is defined by

$$(4.2) \qquad D_iT(\varphi) = -T(D_i\varphi).$$

(This definition fits in with the definition of a derivative for generalized functions.)

It is easily verified that T is a distribution. Similarly we define $D^\alpha T$ (or $\partial^{|\alpha|}T/\partial x_1^{\alpha_1} \cdots \partial x_n^{\alpha_n}$, $|\alpha| = \alpha_1 + \cdots + \alpha_n$) by

$$D^\alpha T(\varphi) = (-1)^{|\alpha|} T(D^\alpha \varphi). \tag{4.3}$$

The notation $T^{(\alpha)}$ for $D^\alpha T$ is also used occasionally.

When there may be a confusion between the distribution-derivative and the classical-derivative of a function f, we shall use $[D^\alpha f]$ for the classical-derivative. In view of Eq. (4.1), $D_i f = [D_i f]$ if $f \in C'$. Similarly, $D^\alpha f = [D^\alpha f]$ if $f \in C^{|\alpha|}$. If f is a continuous function and if its first derivatives are piecewise continuous (the discontinuity occurring on hypersurfaces), then $D_i f$ and $[D_i f]$ are different distributions; see Examples 2 and 3 below.

Example 1. Heaviside's function $Y(x)$ $(n = 1)$ is defined as 0 for $x < 0$, as 1 for $x > 0$, and (usually) as $\frac{1}{2}$ for $x = 0$. We then have

$$Y(\varphi) = \int_0^\infty \varphi(x)\, dx.$$

Hence,

$$Y'(\varphi) = -Y(\varphi') = -\int_0^\infty \varphi'(x)\, dx = \varphi(0).$$

The distribution $\delta(\varphi) = \varphi(0)$ is a measure and is called the *Dirac measure*. We have: $Y' = \delta$. One can also verify that

$$\delta^{(q)}(\varphi) = (-1)^q \varphi^{(q)}(0). \tag{4.4}$$

Example 2. Let $f(x)$ $(n = 1)$ have m uniformly continuous derivatives in each of the intervals (x_j, x_{j+1}) $(j = 0, \pm 1, \pm 2, \ldots)$ where $x_j \to \pm\infty$ as $j \to \pm\infty$. Let $f_j^{(k)} = f^{(k)}(x_j + 0) - f^{(k)}(x_j - 0)$ be the jump of $f^{(k)}$ at x_j. Since

$$f'(\varphi) = -f(\varphi') = -\int_{-\infty}^\infty f(x)\varphi'(x)\, dx = \sum_j \varphi(x_j) f_j^0 + \int_{-\infty}^\infty [f'(x)]\varphi(x)\, dx,$$

we have

$$f' = [f'] + \sum_j f_j^0 \delta_{(x_j)}, \tag{4.5}$$

where $\delta_{(a)}$ is defined by $\delta_{(a)}(\varphi) = \varphi(a)$.

Similarly one can prove, for any $k \leqslant m$, that

$$f^{(k)} = [f^{(k)}] + \sum_j f_j^{(k-1)} \delta_{(x_j)} + \sum_j f_j^{(k-2)} \delta'_{(x_j)} + \cdots + \sum_j f_j^{(0)} \delta_{(x_j)}^{(k-1)}. \tag{4.6}$$

Example 3. Let f be a continuously differentiable function in a closed domain $G \subset R^n$ having a smooth boundary S. Define f to be 0 outside G.

Using Green's formula, we have

$$D_i f(\varphi) = -f(D_i\varphi) = -\int_G f(x)\, D_i\varphi(x)\, dx$$
$$= -\int_S f(x)\varphi(x) \cos(\nu, x_i)\, dS + \int_G [D_i f](x)\varphi(x)\, dx,$$

where ν is the outward normal to S and dS is the surface element. We conclude that $D_i f$ differs from $[D_i f]$ by a distribution T_0,

$$T_0(\varphi) = \int_S f(x) \cos(\nu, x_i)\varphi(x)\, dS. \tag{4.7}$$

The derivative of a function $f \in C^m$ is a function of class C^{m-1}. This can be extended to distributions as follows.

We say that a distribution T is of order $\leqslant m$ if it is continuous in (D) provided with the topology of (D^m). Since (D) is a dense subspace of (D^m), T can also be identified with an element of (D'^m). T is said to be of order m if it is of order $\leqslant m$ but it is not of order $\leqslant m - 1$. Measures, for instance, are the distributions of order 0. We can now assert that if T is of order $\leqslant m$ then $D_i T$ (for $i = 1, \ldots, n$) is of the order $\leqslant m + 1$. The proof is obvious.

The converse of the last assertion is true if $n = 1$, but for $n > 1$ it is unknown.

Theorem 9. *For any α, the operator $T \to D^\alpha T$ is a continuous operator from (D') into (D').*

Proof. This follows from Chap. 2, Sec. 4, but we give here an independent proof. Recall that the topology of (D') is given in terms of neighborhoods $V(B, \epsilon)$. Now, for any bounded set $B \subset (D)$ the formula $D^\alpha T(\varphi) = (-1)^{|\alpha|}T(D^\alpha\varphi)$ implies that $D^\alpha T \in V(B, \epsilon)$ if and only if $T \in V(D^\alpha B, \epsilon)$ where $D^\alpha B = \{D^\alpha\varphi; \varphi \in B\}$. Since $D^\alpha B$ is a bounded set, as verified by using Theorem 2, Sec. 1, $V(D^\alpha B, \epsilon)$ is a neighborhood of 0 in (D') and the proof is completed.

Corollary 1. *If $T_j \to 0$ in (D') then $D^\alpha T_j \to 0$ in (D').*

From Theorem 7 and Corollary 1 we also get:

Corollary 2. *If $f_j \to 0$ in L^1_N for any compact set $N \subset R^n$, and if $L = \Sigma\, a_\alpha D^\alpha$ where the a_α are constants and the sum consists only of a finite number of terms, then $Lf_j \to 0$ in (D').*

Let $h \in R^n$. A *translation* τ_h is an operation defined by

$$\begin{aligned} \tau_h f(x) &= f(x + h) \qquad \text{for functions;} \\ \tau_h T(\varphi) &= T(\tau_{-h}\varphi) \qquad \text{for distributions.} \end{aligned} \tag{4.8}$$

These definitions are consistent for functions. Using Theorem 5′, Sec. 3, one easily verifies that $\tau_h T$ is a distribution.

Theorem 10. *If* $h = (0, \ldots, h_k, 0, \ldots, 0)$ *then, as* $h \to 0$,

$$\frac{\tau_h T - T}{h_k} \to \frac{\partial T}{\partial x_k} \qquad \textit{in } (D').$$

Proof. Setting

$$S_h = \frac{\partial T}{\partial x_k} - \frac{\tau_h T - T}{h_k},$$

$$\psi_h = -\frac{\partial \varphi}{\partial x_k} - \frac{\tau_{-h}\varphi - \varphi}{h_k},$$

we have $S_h(\varphi) = T(\psi_h)$. For any bounded set $B \subset (D)$ we introduce the set $B_h = \{\psi_h; \varphi \in B\}$. Then we also have $S_h(B) = T(B_h)$. From the characterization of bounded sets in (D) given in Theorem 2, Sec. 1, it follows that

$$|D^\alpha \psi| \leqslant M_\alpha(h) \qquad \text{for all } \psi \in B_h,$$

where $M_\alpha(h) \to 0$ if $h \to 0$. The supports of the $\psi \in B_h$ are contained in a bounded set independent of h if, say, $|h| \leqslant 1$. Hence, given any neighborhood $V(\{m\}, \{\epsilon\})$ of 0 in (D), B_h belongs to this neighborhood if $|h|$ is sufficiently small. Since, on the other hand, for any $\eta > 0$ there exists a neighborhood $V = V(\{m\}, \{\epsilon\})$ such that $T(V) < \eta$, we have $T(B_h) < \eta$ if $|h|$ is sufficiently small; hence also $S_h(B) < \eta$. We have thus proved that $S_h(B) \to 0$ as $|h| \to 0$, where B is any bounded set of (D), i.e., $S_h \to 0$ in (D').

From Theorem 10 it follows that the mapping $h \to \tau_h T$ is a continuous mapping from R^n into (D').

From the proof of Theorem 9, Sec. 4, Chap. 2 we obtain:

Theorem 11. *If f is a locally integrable function, absolutely continuous in x_1 for almost all $(x_2, \ldots, x_n)$, and if its derivative $[\partial f/\partial x_1]$ is equal almost everywhere to a locally integrable function $g(x)$, then $\partial f/\partial x_1 = g$ in the sense of (D').*

T is called a *primitive* of S if $D_i T = S$ for some i. The remainder of this section is devoted to proving the existence of primitives. We begin with the case $n = 1$.

Theorem 12. *Let $n = 1$. For any distribution S there exists a primitive distribution T. The difference between any two primitives of S is a constant.*

Proof. If T exists then

$$T(d\psi/dx) = -S(\psi) \qquad \text{for any } \psi \in (D). \tag{4.9}$$

Let (H) be the subspace of (D) consisting of all functions χ satisfying

$$\int_{-\infty}^{\infty} \chi(t)\,dt = 0.$$

If $\chi \in (H)$, then its definite integral

$$\psi(x) = \int_{-\infty}^{x} \chi(t)\,dt$$

has a compact support, and conversely.

Let φ_0 be a fixed function of (D) such that

$$\int_{-\infty}^{\infty} \varphi_0(t)\,dt = 1.$$

Then, any $\varphi \in (D)$ can be decomposed in the form $\varphi = \lambda\varphi_0 + \chi$, where

$$\lambda = \int_{-\infty}^{\infty} \varphi(t)\,dt \tag{4.10}$$

and $\chi = \varphi - \lambda\varphi_0$ belongs to (H). Writing $\chi = d\psi/dx$, where $\psi \in (D)$, we are led, in view of Eq. (4.9), to the equation

$$T(\varphi) = \lambda T(\varphi_0) - S(\psi), \tag{4.11}$$

provided T exists.

From (4.11) it follows that if T_1 and T_2 are two primitives of S, then

$$T_1(\varphi) - T_2(\varphi) = \lambda C = \int_{-\infty}^{\infty} C\varphi(t)\,dt,$$

where $C = T_1(\varphi_0) - T_2(\varphi_0)$. Hence, $T_1 - T_2$ is the constant C.

To prove the existence of a primitive distribution T we define

$$T(\varphi) = \gamma\lambda - S(\psi),$$

where γ is a constant and λ is defined by (4.10). T is a linear functional satisfying (4.9). To prove that T is also continuous it suffices to show (in view of Theorem 5′, Sec. 3) that T is bounded on bounded sets $B \subset (D)$. Now, if φ varies in B, λ varies in a bounded set N of complex numbers. Hence $\chi = \varphi - \lambda\varphi_0$ also varies in a bounded set of (D). Its definite integral ψ will vary in a bounded set B_1 of (D) (the support of ψ is contained in any interval containing the support of χ). Since $S(B_1) < \infty$ we conclude from (4.11) that $T(B) < \infty$.

From Theorem 12 it follows that if $DT = 0$ then $T = C$ (a constant). Applying this remark to DT, when T satisfies the equation

$$d^2T/dx^2 = 0, \tag{4.12}$$

we conclude that $DT = C$. One solution of this equation is $T = x$. In view of the uniqueness assertion of Theorem 12 it follows that if T

satisfies (4.12) then $T = x + C$. Proceeding by induction we arrive at the following:

Corollary. *Let $n = 1$. If $D^pT = 0$ then T is a polynomial of degree $< p$.*

We now turn to the general case, $n \geqslant 1$.

Theorem 13. *For any distribution S there exists a distribution T such that*

$$\partial T/\partial x_1 = S.$$

Proof. Let (H_1) be the linear space of all the functions $\chi_1 \in (D)$ satisfying

$$\int_{-\infty}^{\infty} \chi_1(t_1, x_2, \ldots, x_n)\, dt_1 = 0.$$

χ_1 belongs to (H_1) if and only if its definite integral

$$\psi_1(x) = \int_{-\infty}^{x_1} \chi_1(t_1, x_2, \ldots, x_n)\, dt_1$$

belongs to (D). With φ_0 being the same function as in the proof of Theorem 12, we can decompose any $\varphi \in (D)$ in the form

$$\varphi(x) = \lambda_1(x_2, \ldots, x_n)\varphi_0(x_1) + \chi_1(x),$$

where $$\lambda_1(x_2, \ldots, x_n) = \int_{-\infty}^{\infty} \varphi(t_1, x_2, \ldots, x_n)\, dt_1$$

and, hence, $\chi_1 \in (H_1)$. We are then led to define

$$T(\varphi) = \Sigma(\lambda_1) - S_1(\psi_1), \tag{4.13}$$

where Σ is an arbitrary distribution on $(D)_{x_2,\ldots,x_n}$. T thus defined is a linear functional satisfying

$$T(\partial\psi/\partial x_1) = -S(\psi) \qquad \text{for any } \psi \in (D).$$

The proof that T is continuous is analogous to the proof of the corresponding fact in Theorem 12.

Theorem 14. *Consider the system of equations*

$$\frac{\partial T}{\partial x_1} = S_1, \qquad \frac{\partial T}{\partial x_2} = S_2, \ldots, \qquad \frac{\partial T}{\partial x_k} = S_k, \tag{4.14}$$

where $S_1, \ldots, S_k$ are distributions. If there exists a solution then

$$\frac{\partial S_i}{\partial x_j} = \frac{\partial S_j}{\partial x_i} \qquad \text{for } i, j = 1, \ldots, k. \tag{4.15}$$

Conversely, if (4.15) is satisfied then there exists a solution of (4.14). The difference between two solutions is a distribution of the form $U(\varphi) = \sum_k (\lambda_k)$ *where* $\sum_k$ *is an arbitrary distribution over* $(D)_{x_{k+1},\ldots,x_n}$ *and*

$$(4.16) \quad \lambda_k(x_{k+1}, \ldots, x_n) = \int_{-\infty}^{\infty} \cdots \int_{-\infty}^{\infty} \varphi(t_1, \ldots, t_k, x_{k+1}, \ldots, x_n)\, dt_1 \cdots dt_k.$$

Proof. The first part of the theorem is obvious. Assuming next that (4.15) is satisfied, we shall prove the existence of solutions as asserted. For $k = 1$ this follows from Theorem 13 and its proof. Proceeding by induction, we assume that there exists a solution T_1 for the first $k - 1$ equations of (4.14), for *any* $S_1, \ldots, S_{k-1}$ satisfying (4.15), and that the general solution T is

$$(4.17) \qquad T(\varphi) = T_1(\varphi) + \sum_{k-1} (\lambda_{k-1}),$$

where $\sum_{k-1}$ is an arbitrary distribution over $(D)_{x_k,\ldots,x_n}$. The last equation in (4.14) becomes

$$(4.18) \qquad \frac{\partial \sum_{k-1} (\lambda_{k-1})}{\partial x_k} = S_k(\varphi) - \frac{\partial T_1(\varphi)}{\partial x_k}.$$

Since, for $i < k$,

$$\frac{\partial}{\partial x_i}\left(S_k - \frac{\partial T_1}{\partial x_k}\right) = \frac{\partial S_k}{\partial x_i} - \frac{\partial}{\partial x_i}\frac{\partial T_1}{\partial x_k} = \frac{\partial}{\partial x_k}\left(S_i - \frac{\partial T_1}{\partial x_i}\right) = 0,$$

it follows, upon using the inductive assumption, that

$$S_k(\varphi) - \frac{\partial T_1(\varphi)}{\partial x_k} = \overset{0}{\sum_{k-1}} (\lambda_{k-1}),$$

where $\overset{0}{\sum_{k-1}}$ is some distribution over $(D)_{x_k,\ldots,x_n}$. Substituting this into (4.18) we obtain

$$\frac{\partial \sum_{k-1} (\lambda_{k-1})}{\partial x_k} = \overset{0}{\sum_{k-1}} (\lambda_{k-1}).$$

The general solution of this equation is

$$(4.19) \qquad \sum_{k-1} (\lambda_{k-1}) = \overset{1}{\sum_{k-1}} (\lambda_{k-1}) + \sum_k (\lambda_k),$$

where $\overset{1}{\sum_{k-1}}$ is a particular solution. Substituting (4.19) into (4.17), the proof is completed.

If $k = n$ then the difference between any two solutions of (4.14) is a constant. This remark can be used to establish the following:

Corollary. *If* $D^p T = 0$ *for all* $|p| \leqslant m$, *then* T *is a polynomial of degree* $< m$.

5. Structure of Distributions

We say that T is equal to 0 on an open set Ω if $T(\varphi) = 0$ for all $\varphi \in (D_\Omega)$. $T_1 = T_2$ on Ω if $T_1 - T_2 = 0$ on Ω.

Theorem 15 (*Principle of Localization*). *Let* $\{\Omega_i\}$ *be a countable covering of an open set* Ω, *where the* Ω_i *are open sets, and let* $\{T_i\}$ *be a family of distributions, each* T_i *being defined on* Ω_i. *Suppose further that if* $\Omega_i \cap \Omega_j \neq \emptyset$ *then* $T_i = T_j$ *on the intersection set. Then there exists one and only one distribution* T, *defined on* Ω, *which coincides on each* Ω_i *with* T_i.

Proof. Let $\{\alpha_i\}$ be a partition of unity subordinate to $\{\Omega_i\}$. If $\varphi \in (D_\Omega)$ then

$$\varphi = \sum_i \alpha_i \varphi \qquad \text{for } x \in R^n. \tag{5.1}$$

Since the support of φ intersects only a finite number of the supports of the α_i, the series in (5.1) consists of a finite number of terms. Hence, if T exists, then

$$T(\varphi) = \sum_i T(\alpha_i \varphi) = \sum_i T_i(\alpha_i \varphi), \tag{5.2}$$

which proves that T is uniquely determined.

To prove the existence of T with the required properties, we define $T(\varphi)$, for $\varphi \in (D_\Omega)$, by the right side of (5.2). T is then linear. To prove that T is continuous it suffices to show (in view of Theorem 5, Sec. 3) that T is continuous on (D_N) for any compact set $N \subset \Omega$. But if $\varphi \in (D_N)$ the series $\Sigma\, T_i(\alpha_i\varphi)$ consists only of that finite number of terms for which the support of α_i intersects N. Since each T_i is continuous, we conclude that T is also continuous.

It remains to prove that $T = T_j$ on Ω_j. Let $\varphi \in (D_{\Omega_j})$. The support of $\alpha_i\varphi$ lies in $\Omega_i \cap \Omega_j$ and, hence, if $\alpha_i\varphi \not\equiv 0$ then $\Omega_i \cap \Omega_j \neq \emptyset$ and, by assumption, $T_i(\alpha_i\varphi) = T_j(\alpha_i\varphi)$. Summing up with respect to i, we obtain

$$T_j(\varphi) = \sum_i T_j(\alpha_i\varphi) = \sum_i T_i(\alpha_i\varphi) = T(\varphi).$$

Corollary. *If* $T = 0$ *on each open set* Ω_i, *then* $T = 0$ *on the union* $\Omega^* = \bigcup_i \Omega_i$. *If* $T \geqslant 0$ *on each* Ω_i, *then* $T \geqslant 0$ *on* Ω^*. *If a sequence* $T_m \to 0$ *on each* Ω_i, *then* $T_m \to 0$ *on* Ω^*.

Consider the union of all open sets on which T vanishes. By the preceding corollary, this union is the largest open set on which T vanishes. The complement of this set in R^n is called the *support* of T. This definition is consistent, for functions and for measures, with the standard definitions.

From the definition it follows that $T(\varphi) = 0$ if the support of φ does not intersect the support of T.

As an example we mention that the support of the distribution T_0 in (4.7) is contained in S.

In the remainder of this section we shall study the structure of distributions on open bounded sets Ω by representing them in terms of derivatives of functions. In the next section we shall extend the results to distributions on R^n.

Theorem 16. *If T is a distribution and N a compact set in R^n, then there exists an integer $m \geqslant 0$ such that if $\varphi_j \in (D_N)$ and $D^\alpha \varphi_j \to 0$ uniformly in R^n, for $|\alpha| \leqslant m$, then $T(\varphi_j) \to 0$.*

Proof. T is a continuous functional on (D_N). Hence, for any $\epsilon > 0$ there exists a neighborhood of 0 in (D_N), say $V(m, \eta, N)$ such that if $\varphi \in V(m, \eta, N)$ then $|T(\varphi)| \leqslant \epsilon$. Take ϵ fixed. Then, for any $\lambda > 0$, $\varphi \in V(m, \lambda\eta, N)$ implies $|T(\varphi)| \leqslant \lambda\epsilon$, and the assertion follows by taking λ arbitrarily small.

Theorem 17. *Let T be a distribution and let Ω be a bounded open set in R^n. Then, on Ω, $T = D^p g$ for some p, g, where g is a continuous function whose support lies in an arbitrary neighborhood of $\bar{\Omega}$.*

Proof. Set $N = \bar{\Omega}$. By the proof of Theorem 16 there exists an integer $m \geqslant 0$ and $\eta > 0$ such that if $\varphi \in (D_N)$ and if

$$|D^\alpha \varphi| < \eta \qquad \text{for } |\alpha| \leqslant m, \tag{5.3}$$

then $|T(\varphi)| \leqslant \epsilon$.

Let $\theta \in (D_N)$. Each derivative $D^\alpha \theta$ ($|\alpha| \leqslant m$) can be represented as an iterated integral:

$$D^\alpha \theta(x) = \underbrace{\int_{-\infty}^{x_n} \int_{-\infty}^{\zeta_n} \cdots \int_{-\infty}^{\xi_n}}_{\gamma_n \text{ times}} \cdots \underbrace{\int_{-\infty}^{x_1} \int_{-\infty}^{\zeta_1} \cdots \int_{-\infty}^{\xi_1}}_{\gamma_1 \text{ times}}$$
$$D_1^{m+1} \cdots D_n^{m+1} \theta(\eta_1, \ldots, \eta_n)\, d\eta_1\, d\xi_1 \cdots d\zeta_1 \cdots d\eta_n\, d\xi_n \cdots d\zeta_n,$$

where $\gamma_i = m + 1 - \alpha_i \geqslant 1$. Letting A be a number $\geqslant 1$ and $\geqslant$ diameter of Ω, we conclude that if

$$\int_{R^n} |D_1^{m+1} \cdots D_n^{m+1} \varphi(x)|\, dx < \frac{\eta}{A^{mn}} \tag{5.4}$$

then φ satisfies (5.3).

Consider the linear subspace

$$(\Delta) = \{\psi; \psi = D_1^{m+1} \cdots D_n^{m+1}\varphi, \varphi \in (D_N)\}$$

of (D_N). On (Δ) we define a linear functional $L(\psi)$ by

$$L(\psi) = T(\varphi) \qquad \text{if } \psi = D_1^{m+1} \cdots D_n^{m+1}\varphi.$$

This definition is unique since if $\psi = 0$ and $\varphi \in (D)$, then $\varphi = 0$. We claim that L is continuous on (Δ) as a subspace of L_N^1. Indeed, since (5.4) implies $|L(\psi)| = |T(\varphi)| \leqslant \epsilon$, we have

$$|L(\psi)| \leqslant \frac{\epsilon A^{mn}}{\eta} \|\psi\|_{1,N}. \tag{5.5}$$

By the Hahn-Banach theorem (Theorem 8a, Sec. 4, Chap. 1), L can be extended into a continuous linear functional on L_N^1, having the same norm as L on (Δ). Denoting the extended functional again by L we can represent L in the form

$$L(\psi) = \int_N F(x)\psi(x)\,dx, \tag{5.6}$$

where F is an essentially bounded function on N, the essential supremum being the norm of L.

Applying (5.6) to those $\psi \in (\Delta)$ for which the corresponding φ belongs to (D_Ω), we get

$$T(\varphi) = F(D_1^{m+1} \cdots D_n^{m+1}\varphi) \qquad \text{for } \varphi \in (D_\Omega).$$

Hence, $T = (-1)^{(m+1)n}D_1^{m+1} \cdots D_n^{m+1}F$.

Extend the definition of F outside N by defining it to be 0 and consider the function

$$f(x) = \int_{-\infty}^{x} F(\xi)\,d\xi.$$

$f(x)$ is clearly a continuous function and it satisfies all the assumptions of Theorem 11. Hence, $D_1 f = [D_1 f]$. Applying Theorem 11 to $[D_1 f]$ we obtain $D_1 D_2 f = [D_2[D_1 f]]$. Proceeding step by step in this manner we arrive at $D_1 D_2 \cdots D_n f = F$. Hence,

$$T = (-1)^{(m+1)n} D_1^{m+2} D_2^{m+2} \cdots D_n^{m+2} f.$$

To complete the proof of the theorem, take $g = (-1)^{(m+1)n}\alpha f$, where $\alpha = 1$ and $\bar{\Omega}$ and $\alpha \in (D_U)$, U being an arbitrarily given neighborhood of $\bar{\Omega}$.

Corollary. *On (D_Ω) (Ω a bounded open set), every distribution is of finite order.*

The next two theorems consider a simultaneous representation of a family of distributions.

Theorem 18. *If B' is a bounded set of distributions and if Ω is a bounded open set in R^n then there exists a p such that, for any $T \in B'$, $T = D^p g$ on Ω and $\{g\}$ is a family of continuous and uniformly bounded functions whose supports lie in an arbitrarily given neighborhood U of $\bar{\Omega}$.*

Theorem 19. *Let $T_j \to 0$ in (D') and let Ω be a bounded open set in R^n. Then there exists a p such that, for each j, $T_j = D^p g_j$ on Ω where $\{g_j\}$ is a sequence of continuous functions vanishing outside any given neighborhood of $\bar{\Omega}$, and $g_j(x) \to 0$ uniformly in R^n.*

The converse of Theorem 19 follows from Corollary 2 to Theorem 9, Sec. 4.

The proofs of both theorems rely upon the following theorem.

Theorem 20. *Let B' be a bounded set in (D'). Then, for any compact set N there exists a neighborhood V of 0 in (D_N) such that*

$$B'(V) \equiv \sup_{T \in B'} T(V) < \infty .$$

This theorem is a special case of Theorem 12, Sec. 5, Chap. 1, since (D_N) satisfies the first countability axiom.

Proof of Theorem 18. We repeat, word by word, the proof of Theorem 17, using for m, η, ϵ in (5.3) the numbers corresponding to the neighborhood V of Theorem 20. We then obtain (5.5) with ϵ, A, η independent of the particular T in B'. By the Hahn-Banach extension the norm of L is unchanged, hence

$$\text{essen. sup } |F(x)| \leqslant A_0,$$

where A_0 is independent of $T \in B'$. The proof is thereby completed.

Proof of Theorem 19. Since $T_j \to 0$, $\{T_j\}$ is a bounded set in (D') and therefore there exist m, η, ϵ, independent of j such that (5.3) implies $|T_j(\varphi)| \leqslant \epsilon$. We have to modify the proof of Theorem 17 by considering (Δ) as a subspace of L_N^2 instead of L_N^1. Since $\|f\|_{1,N} \leqslant$ const. $\|f\|_{2,N}$, we have, in view of (5.5),

$$|L_j(\psi)| \leqslant A_1 \|\psi\|_{2,N} \tag{5.7}$$

for all $\psi \in (\Delta)$, where A_1 is independent of j. We can therefore extend each L_j to $\overline{(\Delta)}$ by continuity. Let (Γ) be the orthogonal complement of (Δ) in L_N^2 and define $L_j(\psi) = 0$ if $\psi \in (\Gamma)$. Then extend L_j to the whole space L_N^2 by linearity. If we represent the extended functional, which is again denoted by L_j, in the form

$$L_j(\psi) = \int_N F_j(x)\psi(x)\,dx \qquad (\psi \in L_N^2),$$

then we have

(a) $\|F_j\|_{2,N} \leqslant A_1$,
(b) $\int_N F_j(x)\psi(x)\,dx \to 0$ for any $\psi \in L_N^2$.

Consider the function

$$K_x(\xi) = \begin{cases} 1 & \text{if } \xi_i \leqslant x_i \ (i = 1, \ldots, n) \text{ and } \xi \in N, \\ 0 & \text{otherwise.} \end{cases}$$

Since $K_x(\xi) \in L_N^2$ for fixed x, we conclude, using (b) and setting $F_j(x) = 0$ if $x \notin N$, that

$$f_j(x) = \int_{-\infty}^{x} F_j(\xi)\,d\xi = \int_N K_x(\xi)F_j(\xi)\,d\xi \to 0$$

as $j \to \infty$. Since, by (a), the sequence $\{f_j\}$ is equicontinuous, $f_j \to 0$ uniformly with respect to x, and the proof is now easily completed by taking $g_j = (-1)^{(m+1)n}\alpha f_j$ as in the proof of Theorem 17.

Note that Theorem 17 can also be proved if L_N^1 is replaced by L_N^2. In proving Theorem 19, however, the use of the L_N^2 norm is essential to the proof (it is used in establishing (b)).

6. Structure of Distributions (Continued)

In Sec. 5 we studied the structure of distributions on bounded open sets $\Omega \subset R^n$. In this section we study the structure of distributions on R^n. We begin with distributions having a compact support.

Let $T \in (D')$ and let $\varphi \in C^\infty$. Assume that the support of φ intersects the support of T in a compact set N_0. Let α be any function in (D) which is equal to 1 in some neighborhood of N_0. We define

$$T(\varphi) = T(\alpha\varphi). \tag{6.1}$$

This definition is independent of the particular choice of α since if β is another function of (D), $\beta = 1$ in some neighborhood of N_0, then the support of $\alpha - \beta$ does not intersect the support of T and hence

$$T(\alpha\varphi) - T(\beta\varphi) = T((\alpha - \beta)\varphi) = 0.$$

One easily verifies that $DT(\varphi) = -T(D\varphi)$. Clearly, the definition (6.1) is consistent with the definition of $T(\varphi)$ when $\varphi \in (D)$.

Applying Theorem 16 to $\alpha\varphi_j$ instead of φ_j we obtain the following:

Theorem 21. *Let T be a distribution with a compact support N_0. Then there exists an integer $m \geqslant 0$ such that if $\{\varphi_j\}$ is any sequence of C^∞ functions* *which converges to* 0 *together with their first m derivatives, uniformly in some neighborhood of* N_0, *then* $T(\varphi_j) \to 0$.

In the future we shall use the notation $q \leqslant p$ to mean $q_i \leqslant p_i$ for $i = 1, \ldots, n$ and $\binom{p}{q}$ to mean $\binom{p_1}{q_1} \cdots \binom{p_n}{q_n}$.

Theorem 22. *Any distribution* $T \in (D')$ *with a compact support* N_0 *can be written in the form*

$$T = \sum_{q \leqslant p} D^q G_q,$$

where the G_q *are continuous functions which vanish outside any given neighborhood* U *of* N_0.

Proof. Let Ω be a bounded open set satisfying $N_0 \subset \Omega \subset \bar{\Omega} \subset U$. By Theorem 17, $T = D^p g$ in (D_Ω), i.e., if $\varphi \in (D_\Omega)$ then

$$T(\varphi) = (-1)^{|p|} \int_{R^n} g(x)\, D^p\varphi(x)\, dx,$$

and g is a continuous function whose support lies in some neighborhood of Ω.

Let $\alpha \in (D)$, $\alpha = 1$ in some neighborhood of N_0 lying in Ω and let $\varphi \in (D)$. Then

$$T(\varphi) = T(\alpha\varphi) = (-1)^{|p|} \int_{R^n} g\, D^p(\alpha\varphi)\, dx.$$

Using Leibniz' rule, written briefly in the form

$$D^p(\alpha\varphi) = \sum_{q \leqslant p} \binom{p}{q} D^{p-q}\alpha \cdot D^q\varphi,$$

we obtain

$$T(\varphi) = (-1)^{|p|} \sum_{q \leqslant p} \int \left[\binom{p}{q} g\, D^{p-q}\alpha\right] D^q\varphi\, dx.$$

Hence, $T = \sum_{q \leqslant p} D^q G_q$, where $G_q = (-1)^{|p+q|} \binom{p}{q} g\, D^{p-q}\alpha$. Since the support of G_q lies in U, the proof is completed.

Theorem 22 shows that if $T \in (D')$ has a compact support then it is of finite order, i.e., it belongs to some (D'^m). We can obtain representations of T in which the derivatives D^q are of order $\leqslant m$; more precisely,

Theorem 23. *If* T *is a distribution with a compact support* N_0 *and of order* $\leqslant m$ *then*

$$T = \sum_{|q| \leqslant m} D^q \mu_q,$$

where the μ_q are measures whose supports are contained in any given neighborhood U of N_0.

The converse is trivial.

Proof. The proof is similar to the proof of Theorem 10, Sec. 5, Chap. 2. Let Ω be an open set such that $N_0 \subset \Omega \subset \bar{\Omega} \subset U$. To every $\varphi \in (D)$ there corresponds a vector of length, say, ν,

$$\{\varphi_q\} = \{D^q\varphi\} \qquad \text{where } |q| \leqslant m.$$

The correspondence $\varphi \to \{\varphi_q\}$ is one-to-one. Let Δ^ν denote the subspace $\{\varphi_q\}$ in the linear space Γ^ν consisting of all continuous vector-functions of length ν over $\bar{\Omega}$. Γ^ν is taken with the usual norm of l.u.b. We introduce on Δ^ν the linear functional

$$L(\{\varphi_q\}) = T(\varphi)$$

which is continuous since T is of order $\leqslant m$. By the Hahn-Banach theorem (Theorem 8a, Sec. 4, Chap. 1), L can be extended into a continuous linear functional on Γ^ν, and by a theorem of F. Riesz,

$$L(\{\varphi_q\}) = \sum_{|q| \leqslant m} \nu_q(\varphi_q) = \sum_{|q| \leqslant m} \nu_q(D^q\varphi),$$

where the ν_q are measures with support in $\bar{\Omega}$. Taking $\mu_q = (-1)^{|q|}\nu_q$ we conclude that

$$T(\varphi) = \sum_{|q| \leqslant m} D^q\mu_q(\varphi)$$

as asserted in the theorem.

From (6.1) it easily follows that if T is of order $\leqslant m$ and if $D^\alpha\varphi_j \to 0$ for $|\alpha| \leqslant m$ uniformly with respect to $x \in \Omega$, where Ω is some neighborhood of N_0, then $T(\varphi_j) \to 0$. In general, Ω cannot be replaced by N_0. For some sets N_0, however, such as convex sets, the following is true: there exists an $m' = m'(m, N_0)$ such that if $D^\alpha\varphi_j \to 0$ uniformly on N_0, for $|\alpha| \leqslant m'$, then $T(\varphi_j) \to 0$. (The proof is omitted.) An example of a set N_0 not possessing this property is given in Problem 4.

In view of the above remarks, the following theorem is of interest.

Theorem 24. *If T is a distribution with a compact support N_0 and of order $\leqslant m$, then $T(\varphi) = 0$ for any $\varphi \in (D)$ for which $D^\alpha\varphi = 0$ on N_0 for $|\alpha| \leqslant m$.*

In other words, the values of φ and of its first m derivatives on N_0 determine $T(\varphi)$ uniquely.

Proof. Let V_d be the set of all points in R^n whose distance from N_0 is $\leqslant d$. Since all the directional derivatives of order m of φ (i.e., $\Delta_1 \cdots \Delta_m\varphi$

where the Δ_i are directional derivatives) vanish on N_0, their absolute values are smaller in V_d than an arbitrarily given $\eta > 0$, provided d is sufficiently small.

For any $x \in V_d$, $x \notin N_0$, let σ be a segment of length $\leqslant d$ connecting x to a point in N_0. Then, for any directional derivative of order $m - 1$, denote it for simplicity by D^{m-1},

$$D^{m-1}\varphi(x) = \int_\sigma \frac{\partial}{\partial\sigma} D^{m-1}\varphi(\xi_\sigma)\, d\sigma,$$

where ξ_σ varies on σ and $\partial/\partial\sigma$ is a derivative in the direction σ. We thus get the inequality $|D^{m-1}\varphi(x)| \leqslant \eta d$. Proceeding similarly step by step we get

$$|D^p\varphi(x)| \leqslant d^{m-|p|}\eta \qquad \text{for } x \in V_d, \tag{6.2}$$

where D^p here indicates any directional derivative of order $|p|$. The inequality (6.2) will be used for $D^p\varphi = D_1^{p_1} \cdots D_n^{p_n}\varphi$.

We are going to use $T(\varphi) = T(\alpha\varphi)$ for some special choice of $\alpha \in (D)$, $\alpha = 1$ in a neighborhood of N_0, namely $\alpha = \alpha_d$. To define α_d, let β_d be the characteristic function of $V_{(2/3)d}$ and let ρ_ϵ be the function defined in (2.1). Set

$$\alpha_d(x) = \int_{R^n} \beta_d(\xi)\rho_{d/4}(x - \xi)\, d\xi.$$

Clearly $\alpha_d \in (D)$, $0 \leqslant \alpha_d \leqslant 1$, $\alpha_d = 1$ on $V_{d/4}$ and $\alpha_d = 0$ outside V_d.

Since

$$|D^p\rho_\epsilon(x)| \leqslant A_p\epsilon^{-|p|} \qquad (A_p \text{ are constants}),$$

we get

$$|D^p\alpha_d(x)| \leqslant B_p d^{-|p|} \qquad (B_p \text{ are constants}). \tag{6.3}$$

Using Leibniz' rule and (6.2), (6.3) it follows, for $\psi_d = \alpha_d\varphi$, that

$$|D^p\psi_d(x)| \leqslant B\eta d^{m-|p|} \qquad (B \text{ a constant}) \tag{6.4}$$

for all $|p| \leqslant m$.

Recalling that $T(\varphi) = T(\alpha_d\varphi) = T(\psi_d)$ and observing that $D^p\psi_d \to 0$ uniformly in R^n as $d \to 0$, we conclude that $T(\varphi) = 0$.

We turn to distributions T with arbitrary support. One cannot expect T to be represented as a finite sum of derivatives of functions (see Problem 5). We shall obtain, however, representations of the form $\Sigma\, T_i$ where the series $\Sigma\, T_i$ is *locally finite*, i.e., every compact set in R^n intersects only a finite number of the supports of the T_i; the T_i will be derivatives of functions.

Theorem 25. *Let $\{\Omega_i\}$ be a countable covering, by open sets, of the support F of a distribution T. Then T can be decomposed into a locally finite* *series*

$$T = \sum_i T_i, \tag{6.5}$$

where the support of T_i is contained in $\Omega_i \cap F$.

Proof. Let Ω_0 be the complement of F. Then Ω_0 and $\{\Omega_i\}$ form a covering of R^n. Let α_0, $\{\alpha_i\}$ form a partition of unity subordinate to that covering. Define

$$T_0(\varphi) = T(\alpha_0\varphi), \qquad T_i(\varphi) = T(\alpha_i\varphi).$$

Since the support of $\alpha_0\varphi$ lies in the complement of F, $T(\alpha_0\varphi) = 0$; hence $T_0(\varphi) = 0$. T_i is a distribution whose support F_i is contained in $\Omega_i \cap F$. It follows that the series $\Sigma\, T_i$ is locally finite. Furthermore, it converges to T since

$$\sum_{i=1}^{\infty} T_i(\varphi) = \sum_{i=0}^{\infty} T(\alpha_i\varphi) = T\left(\sum_{i=0}^{\infty} \alpha_i\varphi\right) = T(\varphi).$$

Theorem 26. *Every distribution T can be written in the form*

$$T = \sum_j D^{\beta_j} G_j, \tag{6.6}$$

where each G_j is a continuous function with a compact support lying in any given neighborhood U of the support F of T, and the series (6.6) *is locally finite.*

Proof. Take a countable covering $\{\Omega_i\}$ of F by bounded open sets Ω_i and decompose T into a series (6.5) as in Theorem 25. Apply Theorem 22 to each T_i.

Theorem 27. *If T is a distribution of order $\leqslant m$ on an open set Ω then it can be decomposed into a finite sum of derivatives of orders $\leqslant m$ of measures.*

The converse is trivial.

Proof. Take a covering $\{\Omega_i\}$ of F as in the proof of the preceding theorem. Applying Theorem 23 to each T_i, we obtain

$$T = \sum_{|p| \leqslant m} D^p \mu_{p,i}.$$

Setting $\mu_p = \sum_i \mu_{p,i}$, we get the asserted representation

$$T = \sum_{|p| \leqslant m} D^p \mu_p. \tag{6.7}$$

We conclude this section with an extension of Theorem 24.

Theorem 28. *If all the derivatives of some* $\varphi \in (D)$ *vanish on the support of* T, *then* $T(\varphi) = 0$. *If* T *is of order* $\leqslant m$ *and if* φ *and its first* m *derivatives vanish on the support of* T *then* $T(\varphi) = 0$.

Both statements follow by Theorems 25, 24.

7. Distributions Having Support on Compact Sets or on Subspaces

Let $(\mathcal{E})$ be the space of C^∞ functions on R^n. We introduce a topology in $(\mathcal{E})$ in terms of a basis of neighborhoods $V(m, \epsilon, N)$ of 0. $\varphi \in V(m, \epsilon, N)$ if $|D^p\varphi(x)| < \epsilon$ for all $|p| \leqslant m$, $x \in N$. Here m is an arbitrary integer $\geqslant 0$, ϵ is an arbitrary number > 0, and N is an arbitrary compact set in R^n. The space $(\mathcal{E})$ is a complete countably normed space as follows by taking

$$\|\varphi\|_m = \sup_{|p| \leqslant m} \sup_{|x| \leqslant m} |D^p\varphi(x)| \qquad (m = 1, 2, \ldots).$$

$(\mathcal{E})$ is also easily seen to be a perfect space.

Let $(\mathcal{E}')$ be the conjugate of $(\mathcal{E})$. If $T \in (\mathcal{E}')$ then T is a distribution, call it T_0, since the topology of (D) is stronger than the topology of $(\mathcal{E})$. T_0 has a compact support, since otherwise there exists a sequence $\{\varphi_m\} \subset (D)$ such that the support of φ_m lies in $|x| \geqslant m$ and $T_0(\varphi_m) = 1$. This is impossible because $\varphi_m \to 0$ in $(\mathcal{E})$. Conversely, let T_0 be a distribution with a compact support. Then there exists at most one $T \in (\mathcal{E}')$ such that $T = T_0$ on (D). Indeed this follows by noting that (D) is a dense subspace of $(\mathcal{E})$. The existence of one such T is established by taking $T(\varphi) = T_0(\alpha\varphi)$ for any $\varphi \in (\mathcal{E})$ and for some $\alpha \in (D)$ such that $\alpha = 1$ in some neighborhood of the support of T_0 (compare with (6.1)).

We have thus proved the following:

Theorem 29. *The space of distributions with compact supports is the conjugate* $(\mathcal{E}')$ *of the space* $(\mathcal{E})$.

The topology of $(\mathcal{E}')$, as a conjugate space of $(\mathcal{E})$, is stronger than the topology induced upon it by (D'). Denoting by $(\mathcal{E}_N')$ the subspace of $(\mathcal{E}')$ consisting of the distributions whose support is contained in a compact set $N \subset R^n$, one can prove that $(\mathcal{E}')$ is the inductive limit of the $(\mathcal{E}_N')$. Also, the bounded sets of $(\mathcal{E}')$ are the sets $B' \cap (\mathcal{E}_N')$, where B' is an arbitrary bounded set of (D') and N is an arbitrary compact set in R^n. The proofs are left to the reader.

The structure of distributions with compact supports was given in Theorems 22, 23. If the support consists of one point, say the origin, we have:

Theorem 30. *If T is a distribution whose support consists of the origin then it can be represented, in a unique way, in the form*

$$T = \sum_{|p| \leqslant m} c_p D^p \delta, \tag{7.1}$$

where the c_p are complex numbers.

Proof. To prove uniqueness, apply both sides of (7.1) to x^q. Then,

$$T(x^q) = c_q\, D^q\delta(x^q) = (-1)^{|q|} c_q q!. \tag{7.2}$$

To prove the existence of a representation (7.1), we use the fact that T is of finite order $\leqslant m$. By Theorem 24, $T(\psi) = 0$ for $\psi \in (D)$ if $D^\alpha\psi(0) = 0$ for all $|\alpha| \leqslant m$. Applying this remark to

$$\psi(x) = \varphi(x) - \sum_{|p| \leqslant m} \frac{x^p}{p!} D^p\varphi(0),$$

for any $\varphi \in (D)$, we obtain

$$T(\varphi) = \sum_{|p| \leqslant m} \frac{D^p\varphi(0)}{p!} T(x^p) = \sum_{|p| \leqslant m} (-1)^{|p|} c_p D^p\varphi(0),$$

where the c_q are the constants defined in (7.2). The proof is thereby completed.

We shall extend Theorem 30 to distributions having supports on subspaces of R^n.

Let R^n be the (Cartesian) product $R^h \times R^k$ of two Euclidean spaces R^h and R^k. We take $x = (x_1, \ldots, x_h)$ to be a variable point in R^h and $y = (y_1, \ldots, y_k)$ to be a variable point in R^k. By $R^h \times 0$ we understand the subspace $\{(x, 0); x \in R^h\}$ of R^n. The spaces (D) corresponding to R^h, R^k, R^n will be denoted by $(D)_x$, $(D)_y$, and $(D)_{x,y}$ respectively. Similarly we define $(D')_x$, T_y, $T_{x,y}$, etc. It is clear that if $\varphi(x, y) \in (D)_{x,y}$ then $\varphi(x, 0) \in (D)_x$ and $\varphi(0, y) \in (D)_y$. To every $T_x \in (D')_x$ we define its *extension* $\bar{T}_{x,y} \in (D')_{x,y}$ by

$$\bar{T}_{x,y}(\varphi(x, y)) = T_x(\varphi(x, 0)) \tag{7.3}$$

for any $\varphi \in (D)_{x,y}$. $\bar{T}_{x,y}$ is a distribution on R^n.

Theorem 31. *If $T_{x,y}$ is a distribution on R^n whose support is contained in the subspace $R^h \times 0$, then it can be represented in a unique way as a locally finite sum*

$$T_{x,y} = \sum_q D_y^q \overline{(T_q)}_{x,y} \qquad ((T_q)_x \in (D')_x), \tag{7.4}$$

and the support of $(T_q)_x$ is contained in the support of $T_{x,y}$.

Proof. To prove uniqueness, apply both sides of (7.4) to

$$\varphi_q(x, y) = \psi(x)\frac{y^q}{q!} \qquad (\psi \in (D)_x),$$

noting that the support of φ_q intersects the support of each of the distributions in (7.4) in a compact set. We obtain

$$T_{x,y}(\varphi_q(x, y)) = (-1)^{|q|}(T_q)_x(\psi(x)), \tag{7.5}$$

which shows that $(T_q)_x$ is uniquely determined by $T_{x,y}$.

To prove the existence of the representation (7.4), we define linear functionals $(T_q)_x$ on$(D)_x$ by (7.5). It is easily verified that the $(T_q)_x$ are continuous functionals and that their supports are contained in the support of $T_{x,y}$. It remains to prove that the series in (7.4) is locally finite and that it represents $T_{x,y}$.

Let Ω be a bounded open set in R^n. $T_{x,y}$ is of finite order $\leqslant m$ on Ω. By Theorem 24, for any $\varphi(x, y) \in (D_\Omega)_{x,y}$, $T_{x,y}(\varphi^*) = 0$ for

$$\varphi^*(x, y) = \varphi(x, y) - \sum_{|q|\leqslant m} \frac{y^q}{q!} D_y^q\varphi(x, 0).$$

Hence, by (7.5),

$$T_{x,y}(\varphi) = \sum_{|q|\leqslant m} D_y^q\overline{(T_q)}_{x,y}(\varphi),$$

and the proof is thereby completed.

Remark on generalized functions. Let Φ be a complete countably normed fundamental space containing (D), i.e., containing all the C_c^∞ functions. By Theorem 1, Sec. 1, Chap. 2, it follows that if $T \in \Phi'$ then $T(\varphi_m) \to 0$ whenever $\varphi_m \to 0$ in $(D_N) \equiv K(N)$, N compact. Hence, by Theorem 5″, Sec. 3, T is a distribution. The structure theorems of Secs. 5 and 6 and Theorem 30 are therefore valid for T restricted to (D).

8. Tensor Product of Distributions

In this section x varies in R^m and y varies in R^n. By a tensor product of two functions $f(x)$, $g(y)$ we simply mean the product $f(x)g(y)$. We denote it also by $f(x) \otimes g(y)$. Using the notation of distributions, we have

$$f \otimes g[\varphi] = f(x)[g(y)[\varphi(x, y)]] = g(y)[f(x)[\varphi(x, y)]], \tag{8.1}$$

where $h[\psi] = h(\psi)$ denotes the application of h to ψ. If

$$\varphi(x, y) = \varphi_1(x)\varphi_2(y),$$

then

$$f \otimes g[\varphi] = f(\varphi_1)g(\varphi_2). \tag{8.2}$$

For measures μ_x, ν_y, $\mu_x \otimes \nu_y$ can be defined analogously.

We shall now prove that given any pair of distributions $S_x \in (D')_x$ and $T_y \in (D')_y$ there exists a unique distribution $S_x \otimes T_y$ which satisfies the analogues of (8.1) and (8.2).

Theorem 32. *Let $S_x \in (D')_x$, $T_y \in (D')_y$. Then there exists a unique distribution $W_{x,y} \in (D')_{xy}$ satisfying*

$$W_{x,y}[u(x)v(y)] = S(u)T(v) \tag{8.3}$$

for all $u \in (D)_x$, $v \in (D)_y$. We denote it by $S_x \otimes T_y$. For every $\varphi \in (D)_{x,y}$

$$S_x \otimes T_y[\varphi(x, y)] = S_x[T_y[\varphi(x, y)]] = T_y[S_x[\varphi(x, y)]]. \tag{8.4}$$

$S_x \otimes T_y$ is called the *tensor product* of S_x and T_y.

Proof. We first prove existence. We wish to define $W_{x,y}$ in the form

$$W_{x,y}[\varphi(x, y)] = T_y[I(y)], \qquad \text{where } I(y) = S_x[\varphi(x, y)], \tag{8.5}$$

but we first have to show that $I(y) \in (D)_y$. Let $\varphi \in (D)_{x,y}$. Noting that

$$\frac{\varphi(x, y + h) - \varphi(x, y)}{h_i} \to D_{y_i}\varphi(x, y)$$

in $(D)_x$ as $h = (0, \ldots, 0, h_i, 0, \ldots, 0) \to 0$, and using the continuity of S_x, one arrives at

$$D_y I(y) = S_x[D_y\varphi].$$

In the same way we obtain

$$D_y^q I(y) = S_x[D_y^q\varphi] \tag{8.6}$$

for any q. Hence $I(y) \in C^\infty$. Since the support of $I(y)$ is compact, $I \in (D)_y$ and the definition (8.5) is justified.

It is clear that $W_{x,y}$ is a linear functional. To prove that it is also continuous on $(D)_{x,y}$, it suffices to show (in view of Theorem 5′, Sec. 3) that $W_{x,y}$ is bounded on bounded sets $B_{x,y}$ of $(D)_{x,y}$. Since T_y is a distribution, it suffices (recalling (8.5)) to show that the set

$$B_y = \{I_\varphi;\ I_\varphi(y) = S_x[\varphi(x, y)],\ \varphi \in B_{x,y}\}$$

is a bounded set in $(D)_y$.

To prove the boundedness of B_y, we use the characterization of bounded sets given in Theorem 2, Sec. 1, from which we conclude, since $B_{x,y}$ is a bounded set, that the supports of the I_φ are contained in a compact set of R^n. Furthermore, for any q, the set

$$B_q = \{\psi_{qy};\ \psi_{qy}(x) = D_y^q\varphi(x, y),\ y \in R^n, \text{ and } \varphi \in B_{x,y}\}$$

is a bounded set in $(D)_x$ (y and φ being parameters), so that

$$S_x(B_q) < \infty.$$

By (8.6) it follows that

$$|D_y^q I_\varphi(y)| \leqslant M_q,$$

where M_q is a constant independent of $\varphi \in B_{x,y}$. We conclude that B_y is a bounded set in $(D)_y$. This completes the proof that $W_{x,y}$ is continuous.

To prove that $W_{x,y}$ satisfies (8.3) we substitute $\varphi(x, y) = u(x)v(y)$ into (8.5).

In a similar way we now define $W^*_{x,y} = S_x[T_y[\varphi(x, y)]]$ and show that it is a distribution in $(D')_{x,y}$ satisfying (8.3). Therefore, if we prove that there exists at most one distribution satisfying (8.3) then $W_{x,y} = W^*_{x,y}$ and the proof of the theorem is completed.

It suffices to show that the functions $u(x)v(y)$ (where $u \in (D)_x$, $v \in (D)_y$) span a linear space dense in $(D)_{x,y}$. Now, as is well known, a C^∞ function $\varphi(x, y)$ can be approximated, in the uniform sense, together with all its derivatives, on any given compact set, by a sequence of polynomials (the sequence depends on the compact set). Let N be the support of φ and let $\{P_i(x, y)\}$ be such a sequence which approximates φ in a set $|x| \leqslant a$, $|y| \leqslant a$ containing N in its interior. Let $\rho(x) \in (D)_x$, $\sigma(y) \in (D)_y$ be such that $\rho(x)\sigma(y) = 1$ in a neighborhood of N and $\rho(x)\sigma(x) = 0$ if $|x| \geqslant a$ or if $|y| \geqslant a$. Then $\rho(x)\sigma(y)P_i(x, y) \to \varphi(x, y)$ in $(D)_{x,y}$ and $\rho(x)\sigma(y)P_i(x, y)$ is a linear combination of functions $u(x)v(y)$ where $u \in (D)_x$, $v \in (D)_y$.

From the proof of Theorem 32 one easily infers that if $\{T_j\}$ is a sequence of distributions and $T_j \to 0$ in (D'), then $S_x \otimes (T_j)_y \to 0$ in $(D')_{x,y}$. It is also true (proof is omitted) that the mapping $(S, T) \to S_x \otimes T_y$ from $(D') \times (D')$ into $(D')_{x,y}$ is continuous.

Theorem 33. *The support of the tensor product $W_{x,y} = S_x \otimes T_y$ of two distributions $S_x \in (D')_x$, $T_y \in (D')_y$ is equal to the (Cartesian) product of the supports of S_x and T_y.*

Proof. Denote by $\hat{A}$, $\hat{B}$ the complements in R^n of the supports A, B of S and T, respectively. If the support of $\varphi \in (D)_{x,y}$ belongs to $\hat{A} \times R^n$ or to $R^m \times \hat{B}$ then $W_{x,y}(\varphi) = 0$ in virtue of (8.4). Hence the support of $W_{x,y}$ is contained in $A \times B$. On the other hand, if $z \in A \times B$, then $W_{x,y} \neq 0$ in any neighborhood Ω of z. Indeed, we can choose $u(x)$, $v(y)$ such that $u(x)v(y) \in (D_\Omega)_{x,y}$ and such that $S(u) \neq 0$, $T(v) \neq 0$, and then apply (8.3).

To verify that a certain equation involving tensor products is valid, one often applies both sides of the equation to "test functions" $u(x)v(y)$

and makes use of (8.3). In this way, for instance, one immediately verifies that

$$D_x^p D_y^q (S_x \otimes T_y) = D_x^p S_x \otimes D_y^q T_y, \tag{8.7}$$

and

$$\bar{T}_{x,y} = T_x \otimes \delta_y. \tag{8.8}$$

9. Product of Distribution by Functions. Applications to Differential Equations

Let $\alpha \in C^\infty$, $T \in (D')$. We define a distribution αT, or $\alpha \cdot T$, by

$$\alpha T(\varphi) = T(\alpha\varphi). \tag{9.1}$$

We call αT the *multiplicative product* of α and T (this fits in with the terminology of Chap. 2, Sec. 4). It is clear that αT is a distribution. If T is of order $\leqslant m$, then αT can be defined also for $\alpha \in C^m$ and αT will again be a distribution of order $\leqslant m$. In the future however we always take $\alpha \in C^\infty$.

Theorem 34. *The support of αT is contained in the intersection of the supports of α and T. The order of αT is less than or equal to the order of T.*

The proof is obvious.

Theorem 35. *If T is of order $\leqslant m$ and if α and all its derivatives of orders $\leqslant m$ vanish on the support of T, then $\alpha T = 0$. If α and all its derivatives vanish on the support of T, T being an arbitrary distribution, then $\alpha T = 0$.*

The proof follows from Theorem 28.

The following formulas are easily verified:

$$D_i(\alpha T) = D_i\alpha \cdot T + \alpha D_i T; \tag{9.2}$$

$$[\alpha(x) \otimes \beta(y)](S_x \otimes T_y) = \alpha(x)S_x \otimes \beta(y)T_y; \tag{9.3}$$

$$\alpha(\beta T)\cdot\varphi = \beta T\cdot\alpha\varphi = T\cdot\beta\alpha\varphi = (\alpha\beta)T\cdot\varphi; \tag{9.4}$$

$$\alpha\delta = \alpha(0)\delta, \qquad \alpha D_i\delta = \alpha(0)D_i\delta - (D_i\alpha(0))\delta. \tag{9.5}$$

If $\{\alpha_i\}$ forms a partition of unity for R^n then we can write

$$T = \sum_i (\alpha_i T). \tag{9.6}$$

Consider the linear differential operator of order $\leqslant m$

$$LT \equiv \sum_{|p| \leqslant m} a_p(x)\, D^p T, \tag{9.7}$$

where $p = (p_1, \ldots, p_n)$ and $a_p(x) \in C^\infty$. L is a linear continuous transformation from (D') into (D'). The *adjoint* of L, denote it by L^*, is defined by the requirement

$$LT \cdot \psi = T \cdot L^*\psi \qquad \text{for all } T \in (D'),\ \psi \in (D), \tag{9.8}$$

or, equivalently,

$$L^*T \equiv \sum_{|p| \leqslant m} (-1)^{|p|} D^p(a_p(x)T). \tag{9.9}$$

In order to consider a system of differential operators, we introduce the vector notation $T = (T_1, \ldots, T_\nu)$, where each component is a distribution. The topology and the other concepts and results derived for distributions can be extended, with minor modifications, to vector-distributions. Let $A_p(x) = (a_{ij}^p(x))$ be a $\nu \times \nu$ matrix whose elements are C^∞ functions.

The system of linear differential operators

$$\sum_{j=1}^{\nu} L_{ij}T_j \equiv \sum_{j=1}^{\nu} \sum_{|p| \leqslant m} a_{ij}^p(x)\, D^pT_j \tag{9.10}$$

can be written in the abbreviated form

$$LT = \sum_{|p| \leqslant m} A_p(x)\, D^pT. \tag{9.11}$$

The adjoint system can be defined either by (9.8) (with $f \cdot g = \sum_{i=1}^{\nu} f_i \cdot g_i$) or, equivalently, by

$$L^*T = \sum_{|p| \leqslant m} (-1)^{|p|} D^p[A_p^*(x)T], \tag{9.12}$$

where A_p^* is the transpose of the matrix A_p.

We shall be interested in systems of linear partial differential equations

$$\sum_{|p| \leqslant m} A_p(x)\, D^pT = B, \tag{9.13}$$

where $B = (B_1, \ldots, B_\nu)$ is a vector-distribution.

Consider the case $n = 1$, $m = 1$ and assume that the system is written in the (so-called) normal form

$$\frac{dT_i}{dx} + \sum_{j=1}^{\nu} A_{ij}(x)T_j = B_i \qquad (i = 1, \ldots, \nu). \tag{9.14}$$

Theorem 36. *If $B = 0$ then the only solutions of the system* (9.14) *are C^∞ functions. If $B \neq 0$ then there exist solutions of* (9.14). *The difference between any two solutions is a solution of the homogeneous system.*

Proof. Let C be a fundamental solution of the homogeneous system, i.e.,

$$dC/dx + AC = 0 \qquad (\det C(0) \neq 0),$$

where $A = (A_{ij})$. As is well known, $\det C(x)$ is then $\neq 0$ for all x. We try to find a solution T of (9.14) in the form $T = CS$ where S is a vector-distribution to be determined. T satisfies $dT/dx + AT = B$ if and only if

$$dS/dx = C^{-1}B. \tag{9.15}$$

Hence, if $B = 0$ then $dS/dx = 0$ and, by Theorem 12, Sec. 4, $S = k$ where k is a constant vector. $T = Ck$ is therefore a C^∞ function, thereby proving the first part of the theorem.

The existence of solutions S of (9.15) follows from Theorem 12, Sec. 4. In fact the most general solution is $S_0 + k$, where S_0 is one particular solution. Hence there exist solutions of (9.14), and the general solution is $T_0 + Ck$ where T_0 is any particular solution.

As another application consider the case $\nu = 1$, $m > 1$ and suppose that the differential system is of the form

$$Lu \equiv \frac{\partial^m u}{\partial x_n^m} + \sum_{|p| \leqslant m,\, p_n < m} a_p(x)\, D^p u = b(x) \qquad \text{for } x_n > 0, \tag{9.16}$$

and the boundary conditions

$$\partial^k u/\partial x_n^k = f^{(k)} \qquad \text{on } x_n = 0 \qquad (k = 0, 1, \ldots, m-1) \tag{9.17}$$

The problem of finding a solution of (9.16) and (9.17) is called a *Cauchy problem*. Suppose that $a_p \in C^\infty$ in R^n, that the $f^{(k)}$ are sufficiently smooth functions, and that b is a locally integrable function in the region $x_n \geqslant 0$. Suppose, further, that $u(x)$ is a smooth solution of the problem.

Let T be the function which is equal to $u(x)$ if $x_n \geqslant 0$ and to 0 if $x_n < 0$. Extend b in a similar way. T is a distribution, and upon replacing the classical derivatives of u which appear in (9.16) by distribution derivatives we obtain the distribution equation

$$LT = b + H, \tag{9.18}$$

where

$$H = \sum_{j=1}^{m} f^{(j-1)}_{x_1 \cdots x_{n-1}} \otimes \left(\frac{d}{dx_n}\right)^{m-j} \delta_{x_n} + \sum_{|p| \leqslant m} a_p \left[\sum_{j=1}^{p_n} \frac{\partial^{|p|-p_n}}{\partial x_1^{p_1} \cdots \partial x_{n-1}^{p_{n-1}}} f^{(j-1)}_{x_1 \cdots x_{n-1}} \otimes \left(\frac{d}{dx_n}\right)^{p_n-j} \delta_{x_n} \right]. \tag{9.19}$$

One can further verify that if $H = 0$ then $f^{(0)} = \cdots = f^{(m-1)} = 0$, thus proving that H determines the $f^{(k)}$ uniquely. We can now divide the Cauchy problem into three parts:

(a) find a distribution T which solves (9.18) and whose support lies in $x_n \geqslant 0$;
(b) prove that T is a sufficiently smooth function if $x_n > 0$;
(c) prove that T is a sufficiently smooth function up to the boundary $x_n = 0$.

It is then easily seen (by applying (9.18) to "test functions") that T is a solution of (9.16) in the classical sense. Denote by $g^{(k)}$ the classical derivatives $\partial^k T/\partial x_n^k$ on $x_n = 0$. Then, T satisfies (9.18) with H_0 which can be expressed in terms of the $g^{(k)}$ in the same way that H is expressed in terms of the $f^{(k)}$. Since $H_0 = H$, $g^{(k)} = f^{(k)}$ and T is the desired (classical) solution of the Cauchy problem.

Generally speaking, in most applications (b) is a harder problem than (a) and (c) is a harder problem than (b).

10. Convolutions of Distributions

If f and g are locally integrable functions, we define their convolution $f * g$ by

$$(f * g)(x) = \int_{R^n} f(x - t)g(t)\,dt, \tag{10.1}$$

provided the integral exists. As is well known, if $f \in L^p$ (in R^n) and $g \in L^q$, and if $1 \leqslant p \leqslant \infty$, $1 \leqslant q \leqslant \infty$, $(1/p) + (1/q) \geqslant 1$ then $f * g$ exists almost everywhere and belongs to L^r where

$$1/r = (1/p) + (1/q) - 1$$

and

$$\|f * g\|_r \leqslant \|f\|_p \|g\|_q \qquad \text{(Young's inequality).} \tag{10.2}$$

If $r = \infty$ then $f * g$ exists everywhere and is a continuous function.

Applying both sides of (10.1) to a function $\varphi \in (D)$ we get

$$(f * g)\cdot\varphi = \int dx \int f(x - t)g(t)\varphi(x)\,dt = \int g(t)\,dt \int f(x - t)\varphi(x)\,dx$$
$$= \int g(t)\,dt \int f(x)\varphi(x + t)\,dx,$$

provided the change of order of integration is justified. Hence,

$$(f * g)\cdot\varphi = g(t)[f(\tau_t\varphi)] = (g(t) \otimes f(x))\cdot\varphi(x + t), \tag{10.3}$$

where $g(t)[\alpha(t)]$ and $g(y)\cdot\alpha(y)$ mean the application of the distribution g to α.

To avoid difficulties of convergence, we shall first define and study $S * T$ in case at least one of the distributions has a compact support. Suppose that the support A of S is a compact set. By (10.3) we are motivated to define

$$S * T\cdot\varphi = S_\xi \otimes T_\eta\cdot\varphi(\xi + \eta) \qquad (\varphi \in (D)). \tag{10.4}$$

To show that the right side is well defined, note first that the support of $S_\xi \otimes T_\eta$ is contained in $A \times R^n$. Denote by N the support of φ. Then the support of $\varphi(\xi + \eta)$ in $R^n \times R^n$ intersects the support of $S_\xi \otimes T_\eta$ in a set of points (ξ, η) for which $(\xi, \eta) \in A \times R^n$ and $\xi + \eta \in N$, i.e., $(\xi, \eta) \in A \times (N - A)$. Since $A \times (N - A)$ is a bounded set, the right side of (10.4) is well defined. Furthermore, if $\alpha \in (D)$ and $\alpha = 1$ in some neighborhood of A, then

$$S * T\cdot\varphi = S_\xi \otimes T_\eta[\alpha(\xi)\varphi(\xi + \eta)]. \tag{10.5}$$

By (8.4) we have $S * T = T * S$.

Theorem 37. *If S and T are distributions and if at least one of them has a compact support, then there exists a distribution $S * T$, called the convolution of S and T, which is defined by*

$$(S * T)_x\cdot\varphi(x) = S_\xi \otimes T_\eta\cdot\varphi(\xi + \eta) \qquad (\varphi \in (D)).$$

Proof. Indeed, if φ varies in a bounded set $B \subset (D)$, then $\alpha(\xi)\varphi(\xi + \eta)$ varies in a bounded set of $(D)_{R^n\times R^n}$. Hence, the numbers obtained on the right side of (10.5), when φ varies in B, form a bounded set. Since the right side of (10.5) is equal to the right side of (10.4), we have proved that $S * T$ is bounded on bounded sets of (D). By Theorem 5′, Sec. 3, $S * T$ is a continuous functional.

Theorem 38. *Let S and T be distributions and let A and B be their supports. Assume that A is compact. Then the support of $S * T$ is contained in the (vector) sum $A + B$.*

Proof. $A + B$ is a closed set. We have to show that if $\varphi \in (D_\Omega)$ where Ω is the complement of $A + B$ in R^n, then $(S * T)\cdot\varphi = 0$. Now the support of $\varphi(\xi + \eta)$ in $R^n \times R^n$ is contained in the set of points (ξ, η) for which $\xi + \eta \in \Omega$, whereas the support of $S_\xi \otimes T_\eta$ is $A \times B$ (by Theorem 33, Sec. 8) and is therefore contained in the set of points (ξ, η) for which $\xi + \eta \in A + B$. Since $\Omega \cap (A + B) = \emptyset$, the support of $\varphi(\xi + \eta)$ does not intersect the support of $S_\xi \otimes T_\eta$ and consequently, by (10.4), $(S * T)\cdot\varphi = 0$.

Theorem 39. *Let A be the support of S and assume that either A or the support of T is compact. Then $S * T$ on any open set Ω depends only on T on $\Omega - A$, i.e., if $T_0 = T$ on $\Omega - A$ then $S * T_0 = S * T$ on Ω.*

Proof. It suffices to prove that if $T = 0$ in $\Omega - A$ then $S * T = 0$ on Ω. Let $\varphi \in (D_\Omega)$. The support of $\varphi(\xi + \eta)$ does not intersect the support of $S_\xi \otimes T_\eta$, for otherwise there is some point (ξ, η), $\xi + \eta \in \Omega$ whereas $\xi \in A$, $\eta \notin \Omega - A$, which is impossible. Hence, $(S * T)\cdot\varphi = 0$.

Theorem 40. *Let S be a distribution with a compact support and let $\{T_j\}$ be a sequence of distributions, $T_j \to 0$ in (D'). Then $S * T_j \to 0$ in (D'). Similarly, if the supports of the T_j are uniformly bounded and S is a distribution with an arbitrary support, then $T_j \to 0$ in (D') implies $S * T_j \to 0$ in (D').*

Proof. We shall only prove the first part of the theorem, leaving the proof of the second part to the reader. Let A be the support of S and take $\alpha \in (D)$ such that $\alpha = 1$ in some neighborhood of A. We have to show that for any bounded set $B \subset (D)$,

$$(S * T_j)(B) \to 0 \qquad \text{as } j \to \infty.$$

If φ varies in B, $\alpha(\xi)\varphi(\xi + \eta)$ varies in a bounded set B_1 of $(D)_{\xi,\eta}$. As in the proof of Theorem 32, Sec. 8, one can now show that the set

$$B_2 = \{I_\varphi;\, I_\varphi(\eta) = S_\xi[\alpha(\xi)\varphi(\xi + \eta)],\ \varphi \in B\}$$

is a bounded set in $(D)_\eta$. Hence $T_j(B_2) \to 0$ as $j \to \infty$. Since

$$T_j(B_2) = (S * T_j)(B),$$

the proof is completed.

One easily verifies that if A, B, C are distributions and at least two of them have compact supports, then

(10.6)
$$(A * B) * C\cdot\varphi = A * (B * C)\cdot\varphi = (A_\xi \otimes B_\eta \otimes C_\zeta)\cdot\varphi(\xi + \eta + \zeta).$$

For any distribution T,

$$\delta * T = T, \tag{10.7}$$

$$\delta_{(h)} * T = \tau_{-h}T, \tag{10.8}$$

$$\frac{\partial\delta}{\partial x_k} * T = \frac{\partial T}{\partial x_k}. \tag{10.9}$$

Equation (10.7) is a special case of (10.8). As for (10.8),

$$(\delta_{(h)} * T)\cdot\varphi = T_\xi[(\delta_{(h)})_\eta\cdot\varphi(\xi + \eta)] = T_\xi\cdot\varphi(\xi + h) = T\cdot\tau_h\varphi = \tau_{-h}T\cdot\varphi.$$

Similarly, (10.9) follows from

$$\left(\frac{\partial\delta}{\partial x_k} * T\right)\cdot\varphi = T_\xi\left[\left(\frac{\partial\delta}{\partial x_k}\right)_\eta\cdot\varphi(\xi+\eta)\right] = T_\xi\left(-\frac{\partial\varphi(\xi)}{\partial\xi_k}\right) = \frac{\partial T}{\partial x_k}\cdot\varphi(x).$$

If S has a compact support, then the following formulas hold:

$$\tau_h(S * T) = (\tau_h S) * T = S * \tau_h T; \tag{10.10}$$

$$\frac{\partial}{\partial x_k}(S * T) = \frac{\partial S}{\partial x_k} * T = S * \frac{\partial T}{\partial x_k}. \tag{10.11}$$

It will be enough to prove the first equality in (10.11). Using (10.9), (10.6),

$$\frac{\partial}{\partial x_k}(S * T) = \frac{\partial\delta}{\partial x_k} * (S * T) = \left(\frac{\partial\delta}{\partial x_k} * S\right) * T = \frac{\partial S}{\partial x_k} * T.$$

11. Convolutions of Distributions with Smooth Functions

Let $\alpha \in (D)$, $T \in (D')$ and consider the function

$$\theta(x) = T_t\cdot\alpha(x - t).$$

It is a C^∞ function (in fact, $D^p\theta(x) = T_t\cdot D_x^p\alpha(x - t)$) and for any $\varphi \in (D)$

$$\begin{aligned}(T * \alpha)\cdot\varphi &= T_\xi\cdot[\alpha_\eta\cdot\varphi(\xi+\eta)] = T_\xi\cdot\int\alpha(\eta)\varphi(\xi+\eta)\,d\eta \\ &= T_\xi\cdot\int\alpha(x-\xi)\varphi(x)\,dx = T_\xi\otimes\varphi_x\cdot\alpha(x-\xi) \\ &= \varphi_x\cdot[T_\xi\cdot\alpha(x-\xi)] = \varphi_x\cdot\theta_x = \theta\cdot\varphi,\end{aligned}$$

i.e., $T * \alpha = \theta$. Hence:

Theorem 41. *Let $T \in (D')$. For any $\alpha \in (D)$, $T * \alpha$ is a C^∞ function and*

$$(T * \alpha)_x = T_t\cdot\alpha(x - t). \tag{11.1}$$

One can also verify in the same way that if $T \in (D'^m)$ and $\alpha \in (D^m)$, then $T * \alpha$ is a continuous function given by (11.1).

$T * \alpha$ is called the *regularization* of T by α.

Taking now a sequence $\{\alpha_j\} \subset (D)$ such that $\alpha_j \to \delta$ in (D') (for instance, $\alpha_j = \rho_{1/j}$ where ρ_ϵ is defined in (2.1)) and using Theorem 40, Sec. 10, we conclude that

$$T * \alpha_j \to T \quad \text{in } (D'). \tag{11.2}$$

Every C^∞ function γ is a limit in the sense of (D') of a sequence of functions $\gamma_j(x) = \gamma(x)h\left(\frac{x}{j}\right)$ of (D), where $h \in (D)$, $h(x) = 1$ if $|x| < 1$.

Combining this remark with (11.2) we arrive at the following theorem.

Theorem 42. *(D) is a dense subset of the topological space (D').*

More precisely: every element of (D') is a limit (in the (D')-topology) of a *sequence* of elements of (D).

Introducing the notation

$$\check{\varphi}(x) = \varphi(-x), \qquad \check{T}(\varphi) = T(\check{\varphi}), \qquad \text{Tr.}\, f(x) = f(0)$$

(Tr. stands for "trace"), we have

$$T(\varphi) = \text{Tr.}\,(T * \check{\varphi}) = \text{Tr.}\,(\check{T} * \varphi). \tag{11.3}$$

Hence, if $T * \varphi = 0$ for all $\varphi \in (D)$, then $T = 0$.

One can verify directly the following formulas:

$$(\alpha T)^\vee = \check{\alpha}\check{T}, \qquad (S * T)^\vee = \check{S} * \check{T}, \tag{11.4}$$

$$(S * T)\varphi = T\cdot(\check{S} * \varphi) = S\cdot(\check{T} * \varphi). \tag{11.5}$$

Let $L = \sum_{|\alpha|\le m} a_\alpha D^\alpha$ be a linear differential operator with constant coefficients. A *fundamental solution* of $Lu = 0$ is a distribution E satisfying

$$LE = \delta \quad \text{or} \quad L\delta * E = \delta. \tag{11.6}$$

For the iterated Laplacian $\Delta^k u = 0$ $(\Delta = \sum_{i=1}^{n} D_i^2)$,

$$E = r^{2k-n}(A_{kn} \log r + B_{kn}), \tag{11.7}$$

where $r = |x|$ and A_{kn}, B_{kn} are constants. If $2k - n < 0$ or if n is an odd number then $A_{kn} = 0$, otherwise $B_{kn} = 0$ (for $k = 1$, compare Problem 3). Observe that if $2k > n$ then E belongs to C^{2k-n-1}. The use of fundamental solutions is an important tool in studying the differentiability properties of distributions and of solutions of differential equations. It is also used in studying the behavior of solutions near singular points. We shall bring in here four examples.

Theorem 43. *If all the derivatives of a distribution T belong to some fixed (D'^m), then T is a C^∞ function.*

Proof. It suffices to prove the theorem for T with compact support since, in the general case, we can decompose T as a series of the form (9.6) and apply the result to each term. We now claim that if $S = \Delta^k T$ then

$$T = E * S, \tag{11.8}$$

where E is the fundamental solution (11.7) of $\Delta^k u = 0$. Indeed,

$$T = \delta * T = (\Delta^k \delta * E) * T = E * (\Delta^k \delta * T) = E * \Delta^k T = E * S.$$

If $2k - n - 1 \geqslant m$ then $E \in C^m$. Since $S = \Delta^k T$ belongs to (D'^m), by assumption, the distribution

$$T = E * S = S_t \cdot E(x - t) \tag{11.9}$$

is a continuous function. Here we have made use of (11.8) and of the remark following Theorem 41. Thus, T is continuous and, by the same argument, each of its distribution derivatives is continuous; in particular, $S = \Delta^k T$ is continuous. By (11.9) we have

$$T(x) = \int S(t) E(x - t) \, dt,$$

from which we conclude that $T \in C^{2k-n-1}$. Since this is true for arbitrary k, $T \in C^\infty$.

Our next example is the following:

Theorem 44. *Let $T \in (D')$ and assume that, for some integer $m \geqslant 0$, $T * \alpha$ belongs to C^∞ whenever $\alpha \in (D^m)$. Then T is a C^∞ function.*

Proof. Take $\gamma \in (D)$ such that $\gamma = 1$ in some neighborhood of 0 in R^n. The function γE belongs to C^{2k-n-1} and has a compact support, and

$$\Delta^k \delta * \gamma E = \delta + \zeta \qquad \text{for some } \zeta \in (D). \tag{11.10}$$

We easily obtain

$$T = \Delta^k(\gamma E * T) - \zeta * T, \tag{11.11}$$

for T with an arbitrary support. Taking sufficiently large k so that $\gamma E \in C^m$ we conclude that the right side of (11.11) is a C^∞ function.

Note that Theorem 44 can also be proved by using (11.8) and, on the other hand, Theorem 43 can be proved using (11.11). In other problems, however, one of the formulas may be more suitable than the other.

The next example is concerned with the differentiability of solutions of $\Delta^k T = f$.

Theorem 45. *Let $f \in C^\infty$ in an open set $\Omega \subset R^n$ and let T be a distribution which satisfies, on Ω, the equation*

$$\Delta^k T = f.$$

Then T is a C^∞ function in Ω.

Proof. It is sufficient to prove the theorem in the case where Ω is a bounded set, since if the theorem is proved for bounded sets, then by applying it to αT ($\alpha \in (D)$, $\alpha = 1$ in any bounded open set of Ω) the theorem follows for unbounded Ω.

Next, it suffices to prove that $T \in C^\infty$ in any open subset Ω_0 of Ω such that $\overline{\Omega_0} \subset \Omega$. Let W be a sufficiently small neighborhood of 0 in R^n such that $\Omega_0 - W \subset \Omega$. Let $\gamma \in (D_W)$, $\gamma = 1$ in some neighborhood of 0 in R^n. By Theorem 39, the convolution $\gamma E * \Delta^k T$ on Ω_0 depends only on $\Delta^k T$ on $\Omega_0 - W$. Hence $\Delta^k T$ may be replaced by f. Noting that

$$\Delta^k(\gamma E * T) = \gamma E * \Delta^k T = \gamma E * f$$

and applying (11.11), the assertion of the theorem follows.

The last example is the following:

Theorem 46. *Let $u(x)$ be a solution of $\Delta^k u = 0$ in a bounded domain $\Omega - \{0\}$ and assume that*

$$|u(x)| \leqslant \frac{C}{|x|^m} \qquad \text{for } x \in \Omega - \{0\}, \tag{11.12}$$

where $C > 0$, $m \geqslant 0$ are constants and m is an integer. Then

$$u(x) = g(x) + \sum_{|p| \leqslant h} a_p D^p E \qquad \text{in } \Omega - \{0\}, \tag{11.13}$$

where g is a solution of $\Delta^k v = 0$ in Ω, E is the fundamental solution (11.7) *of $\Delta^k v = 0$, the a_p are constants, and h is some integer $\geqslant 0$.*

Proof. Consider the distribution

$$T(\varphi) = \int_\Omega u(x) \left[\varphi(x) - \sum_{|p| \leqslant m} \frac{x^p}{p!} D^p \varphi(0) \right] dx.$$

$T = u$ on $\Omega - \{0\}$ and therefore $\Delta^k T = 0$ in $\Omega - \{0\}$. By Theorem 30, Sec. 7, we then have

$$\Delta^k T = \sum_{|p| \leqslant h} a_p D^p \delta \qquad \text{on } \Omega.$$

Since the distribution $S = \sum_{|p| \leqslant h} a_p D^p E$ satisfies $\Delta^k(T - S) = 0$ on Ω, using Theorem 45 we obtain

$$T = S + g,$$

where g is a C^∞ solution of $\Delta^k v = 0$ in Ω. Recalling that $T = u$ on $\Omega - \{0\}$, the assertion (11.13) follows.

12. The Spaces $K_r\{M_p\}$, (D_{L^r}) and the Structure of Their Generalized Functions

In Chap. 2, Sec. 2 we considered spaces $K\{M_p\}$ and in Chap. 2, Sec. 5 we found the structure of the generalized functions over $K\{M_p\}$. In this section we consider related spaces $K_r\{M_p\}$ with M_p being as in Chap. 2 and r being any positive number $1 \leqslant r < \infty$. For simplicity we assume that the $M_p(x)$ are continuous (finite-valued) functions on the entire space R^n (satisfying (1.1), Chap. 2). The space $K_r\{M_p\}$ consists of all the complex-valued C^∞ functions on R^n for which the norms

$$(12.1) \qquad \|\varphi\|_{p,r} \equiv \sup_{|\alpha| \leqslant p} \left[\int_{R^n} M_p(x) |D^\alpha \varphi(x)|^r \, dx \right]^{1/r} \qquad (p = 1, 2, \ldots)$$

are finite.

We shall need Hölder's inequality

$$(12.2) \qquad \int_G |f(x)g(x)| \, dx \leqslant \left[\int_G |f(x)|^r \, dx \right]^{1/r} \left[\int_G |g(x)|^s \, dx \right]^{1/s}$$

where $1/r + 1/s = 1$, G is any open or any closed set, and f, g are any measurable functions for which the right side of (12.2) exists.

Another inequality that will be needed is (5.10) of Chap. 2. We give here an independent proof. Let T be a distribution having a compact support, and let

$$S = \frac{\partial^n T}{\partial x_1 \cdots \partial x_n}.$$

If $Y(x)$ is Heaviside's function (i.e., $Y(x) = Y_1(x_1) \cdots Y_n(x_n)$ where $Y_i(x_i) = 0$ if $x_i < 0$, $Y_i(x_i) = 1$ if $x_i > 0$) then

$$(12.3) \quad T = \delta * T = \left(\frac{\partial^n \delta}{\partial x_1 \cdots \partial x_n} * Y \right) * T = Y * \left(\frac{\partial^n \delta}{\partial x_1 \cdots \partial x_n} * T \right) = Y * S.$$

Taking $\alpha \in (D)$, $\alpha = 1$ if $|x| < \frac{1}{2}$ and $\alpha = 0$ if $|x| > 1$ we obtain from (12.3), with $T_\xi = \alpha(x - \xi)\varphi(\xi)$ (x fixed),

$$\alpha(x - y)\varphi(y) = \int_{R^n} Y(\xi) \frac{\partial^n[\alpha(x - y + \xi)\varphi(y - \xi)]}{\partial \xi_1 \cdots \partial \xi_n} \, d\xi,$$

from which it follows, upon taking $y = x$, that

$$(12.4) \qquad |\varphi(x)| \leqslant A \sum_{|q| \leqslant n} \int_{|\xi - x| < 1} |D^q \varphi(\xi)| \, d\xi.$$

Denote by Φ_p the completion of $\Phi \equiv K_r\{M_p\}$ with respect to the norm $\|\cdot\|_{p,r}$ and let $\{\varphi_m\}$ be a Cauchy sequence in $\|\cdot\|_{p,r}$ whose elements belong to Φ. Then,

(12.5)
$$\int_{R^n} M_p(x)|D^\alpha\varphi_m(x) - D^\alpha\varphi_k(x)|^r\,dx = (\|\varphi_m - \varphi_k\|_{p,r})^r \leqslant \epsilon_m \qquad (|\alpha| \leqslant p)$$

for $k \geqslant m$, where $\epsilon_m \to 0$ if $m \to \infty$.

As is well known, $L^r(R^n)$ is a complete space when the measure on R^n is given in terms of any positive density-function. Hence, there exist measurable functions $\varphi_{0,\alpha}$ such that $\|\varphi_{0,\alpha}\|_{p,r} < \infty$ and

(12.6) $$\int_{R^n} M_p(x)|D^\alpha\varphi_m(x) - \varphi_{0,\alpha}(x)|^r\,dx \to 0 \qquad \text{as } m \to \infty.$$

Since $M_p(x) \geqslant 1$ it follows, in particular, that

(12.7) $$\int_{R^n} |D^\alpha\varphi_m(x) - \varphi_{0,\alpha}(x)|^r\,dx \to 0 \qquad \text{as } m \to \infty.$$

We claim the following:

Lemma 2. *If* $\varphi_{0,0} = 0$ *almost everywhere then* $\varphi_{0,\alpha} = 0$ *almost everywhere, for all* $|\alpha| \leqslant p$.

Indeed, using Hölder's inequality (12.2) we have, for any $\psi \in (D)$,

$$\int_{R^n} [D^\alpha\varphi_m(x) - \varphi_{0,\alpha}(x)]\psi(x)\,dx \to 0 \qquad \text{as } m \to \infty.$$

Integrating by parts we obtain

$$(-1)^{|\alpha|}\int_{R^n} \varphi_m(x)\,D^\alpha\psi(x)\,dx \to \int_{R^n} \varphi_{0,\alpha}(x)\psi(x)\,dx.$$

The left side, however, converges to $(-1)^{|\alpha|}\int \varphi_{0,0}(x)\,D^\alpha\psi(x)\,dx$, as follows by using (12.7) (with $\alpha = 0$) in calculating the difference. Hence

(12.8) $$\varphi_{0,\alpha} = D^\alpha\varphi_{0,0}$$

in the sense of distributions, and the lemma follows immediately.

By (12.8), the elements of Φ_p can be identified with functions all of whose distribution derivatives of orders $\leqslant p$ are also functions having the property that their rth powers are integrable on R^n with respect to the measure $M_p(x)\,dx$. If $p \geqslant n$ then, by applying (12.4) to $D^\alpha\varphi_m - D^\alpha\varphi_k$, we find that $\{D^\alpha\varphi_m(x)\}$ is a uniformly convergent sequence if $|\alpha| \leqslant p - n$. Consequently, $\varphi_{0,0}$ is of differentiability class C^{p-n} on R^n. This shows that if $\varphi \in \bigcap_{p=1}^{\infty} \Phi_p$ then $\varphi \in C^\infty$ on R^n. Since the $D^\alpha\varphi = [D^\alpha\varphi]$ further satisfy

$$\int_{R^n} M_p(x)|D^\alpha\varphi(x)|^r\,dx < \infty,$$

φ belongs to $\Phi \equiv K_r\{M_p\}$. We have thus proved that

$$\Phi = \bigcap_{p=1}^{\infty} \Phi_p.$$

Note next that if $\varphi \in \Phi$,

$$\|\varphi\|_{1,r} \leqslant \|\varphi\|_{2,r} \leqslant \cdots \leqslant \|\varphi\|_{p,r} \leqslant \cdots.$$

We finally claim that any two norms $\|\cdot\|_{p,r}$, $\|\cdot\|_{q,r}$ are in concordance. The proof is analogous to that of Lemma 2, Sec. 2, Chap. 2. Indeed, let $\{\varphi_m\} \subset \Phi$, $\|\varphi_m\|_{q,r} \to 0$, $\|\varphi_m - \varphi_k\|_{p,r} \leqslant \epsilon_m \to 0$ if $k \geqslant m \to \infty$. Denote by $\varphi_{0,\alpha}$ the limit functions, in Φ_p, of $\{D^\alpha\varphi_m\}$. Since $\|\varphi_m\|_{q,r} \to 0$ and $M_q \geqslant 1$, we have, for any compact set $N \subset R^n$,

$$\int_N |\varphi_m(x)|^r\,dx \to 0 \qquad \text{as } m \to \infty.$$

Comparing with (12.7) for $\alpha = 0$, we conclude that $\varphi_{0,0} = 0$ almost everywhere. By (12.8), $\varphi_{0,\alpha} = 0$ almost everywhere and (12.6) becomes

$$\|\varphi_m\|_{p,r} \to 0 \qquad \text{as } m \to \infty.$$

We have thus completed the proof of the following theorem.

Theorem 47. *$K_r\{M_p\}$ is a complete countably normed space.*

Using the inequality (12.4) it can further be shown that if the M_p satisfy the condition (P) of Chap. 2, Sec. 2, then $K_r\{M_p\}$ is a perfect space.

The results of Chap. 2, Sec. 5, can also be extended to the spaces $K_r\{M_p\}$. Thus, in extending Theorem 10, we replace the l.u.b. norm used in Γ^ν by the norm

$$\|\theta\| = \sum_{|\alpha|\leqslant p} \left[\int_{R^n} [M_p(x)]^{1-r}|\theta_\alpha(x)|^r\,dx\right]^{1/r}.$$

Instead of (5.2), Chap. 2, we now obtain

$$\tilde{L}(\theta) = \sum_{|\alpha|\leqslant p} \int_{R^n} \theta_\alpha(x)g_\alpha(x)\,dx, \tag{12.9}$$

where the $g_\alpha(x)$ are measurable functions and

$$\|\tilde{L}\|^s = \sum_{|\alpha|\leqslant p} \int_{R^n} M_p(x)|g_\alpha(x)|^s\,dx < \infty \qquad \left(\frac{1}{r} + \frac{1}{s} = 1\right). \tag{12.10}$$

Substituting $\theta_\alpha = \psi_\alpha = M_p\, D^\alpha\varphi$ in (12.9) we get

$$(f, \varphi) = \sum_{|\alpha| \leqslant p} \int_{R^n} M_p(x)\, D^\alpha\varphi(x) \cdot g_\alpha(x)\, dx. \tag{12.11}$$

Taking $f_\alpha = (-1)^{|\alpha|} g_\alpha(x)$ we obtain the following:

Theorem 48. *Any generalized function f over $K_r\{M_p\}$ has the form, for some $p \geqslant 1$,*

$$f = \sum_{|\alpha| \leqslant p} D^\alpha[M_p(x) f_\alpha(x)], \tag{12.12}$$

where the f_α are measurable functions such that

$$\sum_{|\alpha| \leqslant p} \int_{R^n} M_p(x) |f_\alpha(x)|^s\, dx < \infty \qquad \left(\frac{1}{r} + \frac{1}{s} = 1\right). \tag{12.13}$$

The norm of f on Φ_p is equal to the sth root of the last sum.

We also have: if f varies in a bounded set B' of the conjugate space of $K_r\{M_p\}$, then the above representation is valid with the same p, and the numbers (12.13), when f varies on (B'), form a bounded set. The proof follows by using Theorem 12, Sec. 5, Chap. 1 and the proof of Theorem 48.

The rest of the present section is devoted to the special case $M_p(x) \equiv 1$ for all $p \geqslant 1$. We use the notation

$$K_r\{1\} = (D_{L^r}) \qquad (1 \leqslant r < \infty).$$

(D_{L^p}) $(1 \leqslant p \leqslant \infty)$ consists of all the C^∞ functions all of whose derivatives belong to $L^p(R^n)$. The topology of (D_{L^p}) can be given in terms of neighborhoods $V(m, \epsilon)$ of 0. $\varphi \in V(m, \epsilon)$ if

$$\|D^\alpha\varphi\|_p < \epsilon \qquad \text{for } |\alpha| \leqslant m,$$

where $\|\psi\|_p$ is now used to denote the L^p norm of ψ, i.e.,

$$\|\psi\|_p = \left[\int_{R^n} |\psi(x)|^p\, dx\right]^{1/p}.$$

We set $(B) = (D_{L^\infty})$ and denote by $(\dot{B})$ the subspace of (B) consisting of all the functions which converge uniformly to 0, as $|x| \to \infty$, together with each of their derivatives.

Theorem 49. *(D) is dense in $(\dot{B})$ and in (D_{L^p}) for $1 \leqslant p < \infty$.*

Proof. Let

$$\alpha_j(x) = \int_{|\xi| < j + (1/2)} \rho_\epsilon(x - \xi)\, d\xi, \qquad \epsilon = \tfrac{1}{4}, \tag{12.14}$$

where ρ_ϵ is the function defined in (2.1). Then $\alpha_j \in (D)$, $\alpha_j = 1$ if

$|x| \leqslant j$, $\alpha_j = 0$ if $|x| \geqslant j + 1$ and $|D^q\alpha_j| \leqslant B_q$ on R^n, where the constants B_q are independent of j. It is clear that if $\varphi \in (D_{L^p})$ $(1 \leqslant p < \infty)$ or if $\varphi \in (\dot{B})$ then $\alpha_j\varphi \to \varphi$ in (D_{L^p}) or in $(\dot{B})$ respectively. Since $\alpha_j\varphi \in (D)$, the proof is completed.

Theorem 50. $(D) \subset (D_{L^p}) \subset (D_{L^q}) \subset (\dot{B})$ *if* $p \leqslant q < \infty$.

Proof. The statement $(D_{L^q}) \subset (\dot{B})$ follows by using (12.4) with φ replaced by any $D^\alpha\varphi$ and thereby concluding that $D^\alpha\varphi(x) \to 0$ if $|x| \to \infty$. Since for large $|x|$, $|D^\alpha\varphi(x)| < 1$ if $\varphi \in (D_{L^p})$, it follows that $|D^\alpha\varphi(x)|^q \leqslant |D^\alpha\varphi(x)|^p$ and hence $(D_{L^p}) \subset (D_{L^q})$.

The conjugate of $(D_{L^{p'}})$ is denoted by (D'_{L^p}) for $1 < p \leqslant \infty$ where $p' = p/(p-1)$. The conjugate of $(\dot{B})$ is denoted by (D'_{L^1}), and the space (D'_{L^∞}) is denoted also by (B'). Since (D) is dense in $(\dot{B})$ and in (D_{L^q}) for $1 \leqslant q < \infty$, the elements of (D'_{L^p}) $(1 \leqslant p \leqslant \infty)$ can be identified in a one-to-one way with those linear functionals on (D) which are continuous on (D) provided with the topology of $(D_{L^{p'}})$, if $1 < p \leqslant \infty$, and provided with the topology of $(\dot{B})$ if $p = 1$. Since the topology of (D) is stronger than the topologies of $(\dot{B})$ and (D_{L^q}), the elements of (D'_{L^p}) are distributions. In the same way one shows that the elements of (D'_{L^p}) can be identified with elements of (D'_{L^q}) if $p \leqslant q$, and also with elements of (B'). We thus have

$$(12.15) \qquad (D'_{L^p}) \subset (D'_{L^q}) \subset (B') \subset (D') \qquad \text{if } p \leqslant q < \infty.$$

A distribution T is called a *bounded distribution* if it belongs to (B'). If $T_j \to 0$ in the strong topology of (B'), then we say that $T_j \to 0$ *uniformly* on R^n.

It is clear that

$$(12.16) \qquad (D_{L^p}) \subset L^p \subset (D'_{L^p}) \qquad (1 < p \leqslant \infty),$$

and that if $T \in (D'_{L^p})$ then $D^\alpha T \in (D'_{L^p})$ for any α. Hence $\sum_{|\alpha| \leqslant m} D^\alpha f_\alpha$ belongs to (D'_{L^p}) whenever all the $f_\alpha \in L^p$. The converse is also true, namely,

Theorem 51. *If* $T \in (D'_{L^p})$ $(1 < p \leqslant \infty)$ *then* T *is equal to a finite linear combination of derivatives of functions in* L^p.

The theorem is a special case of Theorem 48.

In view of Theorem 51 and (10.2), we can define the convolution $S * T$ if $S \in (D'_{L^p})$, $T \in (D'_{L^q})$ provided $1/p + 1/q - 1 \geqslant 0$. The convolution belongs to (D'_{L^r}), where $1/r = 1/p + 1/q - 1$. It can be shown without difficulty that the mapping $(S, T) \to S * T$ from $(D'_{L^p}) \times (D'_{L^q})$ into (D'_{L^r}) is continuous. Details are left to the reader.

13. Convolution Equations

Consider the convolution equation

$$A * T = B, \tag{13.1}$$

where A and B are given distributions and A has a compact support. If

$$A = \sum_{|\alpha| \leqslant m} a_\alpha D^\alpha \delta \qquad \text{(the } a_\alpha \text{ constants)}$$

then (13.1) is a differential equation. If $A = \sum_{|\alpha| \leqslant m} a_\alpha \delta_{(h_\alpha)}$ then (13.1) is a difference equation. If $A = K(x) + \delta$, the left side of (13.1) for $T = f(x)$ becomes

$$f(x) + \int K(x - t) f(t)\, dt.$$

Thus, convolution equations include difference-integro-differential equations.

A *fundamental solution* of (13.1) is a distribution E satisfying

$$A * E = \delta. \tag{13.2}$$

If E exists then we say that A is *inversible*. For any B having a compact support there exists then a solution of (13.1), namely $T = E * B$. Indeed,

$$A * (E * B) = (A * E) * B = \delta * B = B.$$

Theorem 52. *If* $\{T_j\}$ *is a sequence of solutions of* (13.1), *then its limit in* (D'), *if it exists, is also a solution. The convolution of a solution* T *of the homogeneous equation (i.e.,* $B = 0$) *with any distribution having a compact support is also a solution of the homogeneous equation. Every solution of the homogeneous equation is a limit, in* (D'), *of* C^∞ *solutions.*

The proof follows by using Theorem 40 and (11.2).

Consider the convolution inequality

$$A * T \geqslant 0, \tag{13.3}$$

where A has a compact support and is inversible. By Theorem 8, Sec. 3, (13.3) is equivalent to

$$A * T = \mu, \qquad \mu \geqslant 0, \tag{13.4}$$

where μ is a measure.

Theorem 53. *In order that* $T \in (D')$ *will be a solution of* (13.4) *it is necessary and sufficient that, for any bounded open set* Ω, T *can be decomposed as follows:*

$$T = E * \nu + S \qquad \text{on } R^n, \tag{13.5}$$

where $\nu \geqslant 0$ is a measure on R^n whose support lies in Ω, and

$$A * S = 0 \qquad \text{on } \Omega. \tag{13.6}$$

ν and S are then uniquely determined on R^n for a given Ω.

Proof. To prove the uniqueness of the representation (13.5), for a given Ω, convolve A with both sides of (13.5):

$$A * T = A * E * \nu + A * S = \begin{cases} \nu + A * S & \text{on } R^n, \\ \nu & \text{on } \Omega, \end{cases}$$

by using (13.6). Hence ν is determined uniquely on Ω. Since its support lies in Ω, it is determined uniquely on R^n. S is then uniquely determined on R^n, by (13.5).

We next prove that if T is a solution of (13.3) then it has the representation (13.5). We already know that T satisfies (13.4) for some measure $\mu \geqslant 0$. Let $\nu = \chi\mu$, where χ is the characteristic function of Ω. Then ν is also a measure, $\nu \geqslant 0$, and its support is contained in Ω. Finally, the distribution $S = T - (E * \nu)$ satisfies

$$A * S = A * T - A * E * \nu = \begin{cases} \mu - \nu & \text{on } R^n, \\ 0 & \text{on } \Omega. \end{cases}$$

To complete the proof of the theorem we have to show that if T can be represented in the form (13.5) for any Ω, then $A * T \geqslant 0$. From (13.5), (13.6), and $\nu \geqslant 0$ it follows that $A * T \geqslant 0$ on every Ω, hence $A * T \geqslant 0$ on R^n.

T is called a *subharmonic distribution* if $\Delta T \leqslant 0$. A locally integrable function $f(x)$ is called a *weakly subharmonic* function if for any $R > 0$ the inequality

$$f \geqslant \mu_R * f \quad \text{or} \quad (\delta - \mu_R) * f \geqslant 0 \tag{13.7}$$

is satisfied for almost all $x \in R^n$, where μ_R is the measure

$$\mu_R(\varphi) = \frac{1}{S_n(R)} \int_{|x|=R} \varphi(x)\, dS \qquad \left(S_n(R) = \int_{|x|=R} dS\right). \tag{13.8}$$

With the aid of Theorem 53 we shall prove the following:

Theorem 54. *T is a subharmonic distribution if and only if T is a weakly subharmonic function.*

Proof. Let f be a weakly subharmonic function. Then, in the sense of (D'),

$$\frac{\delta - \mu_R}{R^2} * f \geqslant 0 \qquad \text{for } R > 0. \tag{13.9}$$

As $R \to 0$,

$$\frac{\delta - \mu_R}{R^2} \cdot \varphi = \frac{1}{R^2}\left[\varphi(0) - \frac{1}{S_n(R)}\int_{|x|=R} \varphi\, dS\right] \to -\frac{1}{2n}\Delta\varphi(0),$$

where the convergence is uniform with respect to φ, if φ varies in any bounded set of (D). Use has been made here of Taylor's formula

$$\varphi(x) = \varphi(0) + \sum_i x_i D_i\varphi(0) + \tfrac{1}{2}\sum_{i,j} x_i x_j D_i D_j \varphi(0) + 0(|x|^3).$$

We conclude that $(\delta - \mu_R)/R^2 \to -(1/2n)\,\Delta\delta$ in (D'), as $R \to 0$. Taking $R \to 0$ in (13.9) and using Theorem 40, Sec. 10, the inequality $\Delta f \leqslant 0$ follows.

Suppose conversely that $\Delta T \leqslant 0$, or $\Delta\delta * T \leqslant 0$. By Theorem 53 T is a sum of $E * \nu$ and S. Since E is a locally integrable function and ν is a measure having a compact support, $E * \nu = \int E(x - \xi)\, d\nu(\xi)$ as verified without difficulty, and $E * \nu$ is therefore a locally integrable function. As for S, it satisfies $\Delta S = 0$ and, by Theorem 45, it is then a C^∞ function. It follows that T is a locally integrable function.

Consider the function

$$L(x) = \begin{cases} \dfrac{1}{N}\left(\dfrac{1}{r^{n-2}} - \dfrac{1}{R^{n-2}}\right) & \text{if } r = |x| \leqslant R, \\ 0 & \text{if } r > R, \end{cases}$$

for $n > 2$, and

$$L(x) = \begin{cases} \dfrac{1}{2\pi}\log\dfrac{R}{r} & \text{if } r \leqslant R, \\ 0 & \text{if } r > R, \end{cases}$$

for $n = 2$. If the constant $N > 0$ is appropriately chosen then

$$\delta - \mu_R = -\Delta\delta * L$$

and therefore

$$(\delta - \mu_R) * T = -\Delta\delta * L * T = -L * \Delta T \geqslant 0,$$

since $\Delta T \leqslant 0$ and $L \geqslant 0$. This completes the proof that T is a weakly subharmonic function.

14. The Spaces (S) and (S')

A function $\varphi \in C^\infty$ is said to be *fast decreasing* if

$$\lim_{|x|\to\infty} |x|^m |D^\alpha \varphi(x)| = 0 \tag{14.1}$$

for all α, m. The linear space of the C^∞ fast-decreasing functions is denoted by (S). A topology is defined by giving a basis of neighborhoods

$V(m, k, \epsilon)$ at 0 in (S). $\varphi \in V(m, k, \epsilon)$ if and only if

$$(1 + r^2)^k |D^\alpha \varphi(x)| \leqslant \epsilon \tag{14.2}$$

for all $|\alpha| \leqslant m$, $x \in R^n$, where $r = |x|$.

An equivalent topology is obtained by defining neighborhoods of 0 in (S) by the inequalities

$$\left| \frac{\partial^{mn}\varphi}{\partial x_1^m \cdots \partial x_n^m} \right| \leqslant \frac{\eta}{(1 + r^2)^\nu}. \tag{14.3}$$

Indeed, every set $\{\varphi\}$ satisfying (14.3) contains some neighborhood $V(m, k, \epsilon)$. To see this, write $\psi \in (S)$ in the form

$$\begin{aligned} \psi(x) &= \int_{-\infty}^{x_1} \frac{\partial \psi}{\partial t_1}(t_1, x_2, \ldots, x_n)\,dt_1 \qquad \text{if } x_1 \leqslant 0; \\ \psi(x) &= -\int_{x_1}^{\infty} \frac{\partial \psi}{\partial t_1}(t_1, x_2, \ldots, x_n)\,dt_1 \qquad \text{if } x_1 > 0. \end{aligned} \tag{14.4}$$

Following this principle for the other first derivatives and for higher derivatives, one can estimate each $D^\alpha \varphi$ in terms of $D_1^m D_2^m \cdots D_n^m \varphi$ provided $|\alpha| \leqslant m$. If ν is sufficiently large and η is sufficiently small, then we obtain (14.2) as a consequence of (14.3).

The proof of Theorem 49, Sec. 13, can be used to show that (D) is a dense subspace of (S).

The conjugate of (S) is denoted by (S'). The elements of (S') are called *tempered distributions*. Tempered distributions can be identified with those linear functionals on (D) which are continuous on (D) provided with the topology of (S). Since the topology of (S) is weaker than the topology of (D), tempered distributions are distributions.

In Chap. 2 we proved that S (or in the present notation (S)) is a perfect space and we found the structure of the generalized functions over S, i.e., of the tempered distributions. In the present terminology, Theorems 12 and 13 of Chap. 2, Sec. 5, can be stated as follows:

Theorem 55. *Every tempered distribution T has the form*

$$T = D^\beta[(1 + |x|^2)^m F(x)] \tag{14.5}$$

for some β, m, F, where F is a continuous bounded function on R^n. If T varies in a bounded set B' of (S') then there exist representations of the form (14.5) *with β and m independent of $T \in B'$, and with continuous functions F which are uniformly bounded as T varies in B'.*

Theorem 56. *If $T_j \to 0$ in (S'), then for some β and m*

$$T_j = D^\beta[(1 + |x|^2)^m F_j(x)], \tag{14.6}$$

where the $F_j(x)$ *are continuous bounded functions and*

$$\lim_{j\to\infty} \underset{x\in R^n}{\text{l.u.b.}} |F_j(x)| = 0. \tag{14.7}$$

Proof. We have to modify the proofs of Theorems 11, 12 of Sec. 5, Chap. 2, by providing Γ^ν with the L^2 norm instead of the L^1 norm. Instead of using (5.10), Chap. 2, we now use the (weaker) inequality

$$|\varphi(x)| \leqslant A \sum_{|\beta|\leqslant q} \left[\int_{|\xi-x|\leqslant 1} |D^\beta\varphi(\xi)|^2\, d\xi\right]^{1/2},$$

and instead of using the norms (5.8) we use the norms

$$\sup_{|\alpha|\leqslant p} \left[\int [M_p(x)]^2 |D^\alpha\varphi(x)|^2\, dx\right]^{1/2}.$$

Finally, instead of applying the Hahn-Banach extension to $L(\{\psi\})$, we use an extension similar to that employed in the proof of Theorem 19, Sec. 5. Details are left to the reader.

A measure μ is called a *tempered measure* if for some $k \geqslant 0$,

$$\int_{R^n} \frac{|d\mu|}{(1+|x|^2)^k} < \infty. \tag{14.8}$$

This is equivalent to assuming that

$$\int_{|x|<r} |d\mu| = 0(r^h) \qquad (r\to\infty) \tag{14.9}$$

for some $h \geqslant 0$.

Theorem 57. *In order that a measure* μ *will belong to* (S') *it is sufficient, and if* $\mu \geqslant 0$ *it is also necessary, that* μ *be a tempered measure.*

Proof. The sufficiency is rather obvious. Assume now that $\mu \in (S')$, $\mu \geqslant 0$. There exist neighborhoods of 0 in (S) on which μ is bounded. Thus, for some integers $k \geqslant 0$, $m \geqslant 0$ and for some $\eta > 0$, if $\varphi \in (S)$ and

$$|D^\alpha\varphi(x)|(1+|x|^2)^k \leqslant \eta \qquad (|\alpha| \leqslant m) \tag{14.10}$$

then $|\mu(\varphi)| \leqslant 1$. Let $\{\alpha_j\}$ be the sequence defined by (12.14). Then the function $\varphi = \epsilon\alpha_j(1+|x|^2)^{-k}$ satisfies (14.10) if ϵ is a sufficiently small positive number, independent of j. Hence

$$\mu\left(\frac{\epsilon\alpha_j}{(1+|x|^2)^k}\right) = \epsilon\int_{R^n} \frac{\alpha_j\, d\mu}{(1+|x|^2)^k} \leqslant 1.$$

Taking $j\to\infty$ and using $d\mu \geqslant 0$, the inequality (14.8) follows.

15. Fourier Transforms of Distributions

For any $\varphi \in (S)$ we define its Fourier transform $\mathfrak{F}\varphi$, or $\tilde{\varphi}$, by

$$(\mathfrak{F}\varphi)(\xi) = \tilde{\varphi}(\xi) = \int_{R^n} e^{ix\cdot\xi}\varphi(x)\,dx, \tag{15.1}$$

where $x \cdot \xi = x_1\xi_1 + \cdots + x_n\xi_n$. If $P(D) = \Sigma\, a_\alpha\, D^\alpha$ is a polynomial in $D = (D_1, \ldots, D_n)$, then

$$P(D)\tilde{\varphi}(\xi) = \widetilde{P(ix)\varphi(x)}, \tag{15.2}$$

$$\widetilde{P(D)\varphi(x)} = P(-i\xi)\tilde{\varphi}(\xi). \tag{15.3}$$

From these formulas one deduces that $\tilde{\varphi} \in (S)$.

As is well known,

$$\varphi(x) = \frac{1}{(2\pi)^n}\int_{R^n} e^{-ix\cdot\xi}\tilde{\varphi}(\xi)\,d\xi, \tag{15.4}$$

so that the operator $\mathfrak{F}$ has an inverse $\mathfrak{F}^{-1}$ ($\mathfrak{F}\mathfrak{F}^{-1}\varphi = \mathfrak{F}^{-1}\mathfrak{F}\varphi = \varphi$) given by

$$(\mathfrak{F}^{-1}\psi)(x) = \frac{1}{(2\pi)^n}\int_{R^n} e^{-ix\cdot\xi}\psi(\xi)\,d\xi. \tag{15.5}$$

An easy calculation shows that if φ varies in a bounded set of (S), then $\mathfrak{F}\varphi$ and $\mathfrak{F}^{-1}\varphi$ will also vary in some bounded set of (S). Hence, $\mathfrak{F}$ is a topological mapping from (S) onto itself. Furthermore,

$$\|\mathfrak{F}\varphi\|_2 = (2\pi)^{n/2}\|\varphi\|_2 \qquad \text{(Plancherel's equality)}. \tag{15.6}$$

Using (15.6), one can extend $(2\pi)^{-n/2}\mathfrak{F}$ into an isometric mapping of $L^2(R^n)$ onto itself.

From (15.6) one deduces

$$(\mathfrak{F}f, \mathfrak{F}g) = (2\pi)^n(f, g) \tag{15.7}$$

if $f \in L^2$, $g \in L^2$, where $(f, g) = \int f(x)\overline{g(x)}\,dx$. We are thus motivated to define for any distribution T

$$\mathfrak{F}T \cdot \overline{\mathfrak{F}\varphi} = (2\pi)^n T \cdot \bar{\varphi} \tag{15.8}$$

or, equivalently (since $\mathfrak{F}\varphi = \psi$ implies $\bar{\varphi} = (2\pi)^{-n}\mathfrak{F}\bar{\psi}$),

$$\mathfrak{F}T \cdot \varphi = T \cdot \mathfrak{F}\varphi. \tag{15.9}$$

In this section we consider only the Fourier transforms of tempered distributions. The Fourier transform of any generalized function will be considered in the next chapter.

Defining the Fourier transform $\mathfrak{F}T$ of $T \in (S')$ by (15.8), or (15.9), and recalling that $\mathfrak{F}$ defines a topological mapping of (S) onto itself, we easily get the following:

Theorem 58. *$T \to \mathfrak{F}T$ is a topological mapping of (S') onto itself.*

We next have

$$P(D)\mathfrak{F}T = \mathfrak{F}(P(ix)T), \tag{15.10}$$

$$\mathfrak{F}(P(D)T) = P(-i\xi)\mathfrak{F}T. \tag{15.11}$$

Indeed, note first that the right sides of these equalities exist since a polynomial is a multiplier in (S) (see Chap. 2, Sec. 4). Now, to prove (15.10) we use (15.9), (15.3), and get, with $\varphi = \varphi(\xi)$,

$$\begin{aligned} P(D)\mathfrak{F}T\cdot\varphi &= \mathfrak{F}T\cdot P(-D)\varphi = T\cdot\mathfrak{F}(P(-D)\varphi) = T\cdot P(ix)\mathfrak{F}\varphi \\ &= P(ix)T\cdot\mathfrak{F}\varphi = \mathfrak{F}(P(ix)T)\cdot\varphi. \end{aligned}$$

As for (15.11), using (15.9), (15.2), we find that

$$\begin{aligned} \mathfrak{F}(P(D)T)\cdot\varphi &= P(D)T\cdot\mathfrak{F}\varphi = T\cdot P(-D)\mathfrak{F}\varphi = T\cdot\mathfrak{F}(P(-i\xi)\varphi) \\ &= \mathfrak{F}T\cdot P(-i\xi)\varphi = P(-i\xi)\mathfrak{F}T\cdot\varphi, \end{aligned}$$

and the proof is completed.

The following formulas are easily verified:

$$\mathfrak{F}\delta = 1, \qquad \mathfrak{F}1 = (2\pi)^n\delta; \tag{15.12}$$

$$\mathfrak{F}\delta_{(h)} = e^{ih\cdot\xi}, \qquad \mathfrak{F}(e^{ih\cdot x}) = (2\pi)^n\delta_{(h)}; \tag{15.13}$$

$$\mathfrak{F}(P(D)\delta) = P(-i\xi), \qquad \mathfrak{F}(P(ix)\delta) = (2\pi)^nP(D)\delta. \tag{15.14}$$

(15.14) follows from (15.12) and the general rules (15.10), (15.11). (15.12) is a special case of (15.13). As for the first equality in (15.13),

$$\mathfrak{F}\delta_{(h)}\cdot\varphi = \delta_{(h)}\cdot\mathfrak{F}\varphi = (\mathfrak{F}\varphi)(h) = \int_{R^n} e^{ix\cdot h}\varphi(x)\,dx = e^{ih\cdot x}\cdot\varphi(x).$$

The second equality in (15.13) follows similarly.

A useful way to verify equalities involving Fourier transforms of distributions is first to verify these equalities for functions in (S) (or in (D)) and then to extend the equalities to distributions in (S') by continuity, noting that (S) (or (D)) is dense in (S'). As an obvious example we mention the formula

$$\mathfrak{F}(T_x \otimes T_y) = \mathfrak{F}T_x \otimes \mathfrak{F}T_y. \tag{15.15}$$

(15.10), (15.11) also follow, in this way from (15.2) and (15.3), respectively.

In the classical theory of Fourier transforms, one of the most useful theorems is the convolution theorem stating that

$$\mathfrak{F}(f * g) = \mathfrak{F}f \cdot \mathfrak{F}g \tag{15.16}$$

under suitable growth conditions of f and g. With the aid of this theorem, convolution equations and, in particular, differential equations can be simplified. It is therefore desirable to extend (15.16) to wide classes of distributions. Note that if $\mathfrak{F}f$ (or $\mathfrak{F}g$) is to be a distribution then $\mathfrak{F}g$ (or $\mathfrak{F}f$) must be a smooth function. With the purpose of extending (15.16) to a large class of distributions, we shall now introduce a new class of functions and a new class of distributions.

A C^∞ function is said to be of *slow increase* if each of its derivatives is bounded by some power of $|x|$, as $|x| \to \infty$. The space of C^∞ slow-increasing functions is denoted by (O_M). We say that $\alpha_j \to 0$ in (O_M) if, for any p and $f(x) \in (S)$,

$$f(x) D^p \alpha_j(x) \to 0 \qquad \text{uniformly on } R^n,$$

as $j \to \infty$. (O_M) can be made into a locally convex topological space by giving a nondenumerable neighborhood basis at 0.

A distribution T is said to be of *fast decrease* if, for any $k \geqslant 0$, $(1 + |x|^2)^k T$ is a bounded distribution (i.e., it belongs to (B'); see Sec. 12). The space of all fast-decreasing distributions is denoted by (O'_C). We say that $T_j \to 0$ in (O'_C) if, for any $k \geqslant 0$,

$$(1 + |x|^2)^k T_j \to 0 \qquad \text{in } (B').$$

Theorem 59. *T belongs to (O'_C) if and only if, for any $k \geqslant 0$, T has a representation (on (D))*

$$T = \sum_{|\alpha| \leqslant p} D^\alpha f_\alpha, \tag{15.17}$$

where the f_α are measurable functions and the $(1 + |x|^2)^k f_\alpha(x)$ are essentially bounded functions in R^n. (p depends on k.)

Proof. If representations of the form (15.17) are valid for any $k \geqslant 0$, then for any $h \geqslant 0$, $(1 + |x|^2)^h T$ is a sum $\Sigma\, (1 + |x|^2)^h D^\alpha f_\alpha$ and each term is in (B') if $k \geqslant h$, since

$$\int_{R^n} f_\alpha(x)\, D^\alpha[(1 + |x|^2)^h \varphi(x)]\, dx$$

is a continuous linear functional on (D_{L^1}). Conversely, let $T \in (O'_C)$. Then, for any $k \geqslant 0$, $(1 + |x|^2)^k T \in (B')$ and by Theorem 51, Sec. 12,

$$(1 + |x|^2)^k T = \sum_{|\alpha| \leqslant q} D^\alpha g_\alpha,$$

where the g_α are essentially bounded functions. Writing

(15.18)
$$(1 + |x|^2)^{-k}\, D\, D^\beta g_\alpha = D[(1 + |x|^2)^{-k}\, D^\beta g_\alpha] - [D(1 + |x|^2)^{-k} \cdot D^\beta g_\alpha],$$

then writing $D^\beta g_\alpha = D\, D^{\beta-1} g_\alpha$ and proceeding to treat each of the brackets on the right side of (15.18) in the same way that we treated the left side, then proceeding again to lower the derivatives of g_α, etc., we finally obtain $\Sigma\, D^\gamma\, h_{\alpha\gamma}$, where $(1 + |x|^2)^k h_{\alpha\gamma}$ are essentially bounded functions. The proof is thereby completed.

It is obvious that if $\alpha \in (O_M)$ then α is a multiplier in (S). Hence if $T \in (S')$ also $\alpha T \in (S')$. We also have the following:

Theorem 60. *If $\alpha_j \to \alpha$ in (O_M) and $T \in (S')$ then $\alpha_j T \to \alpha T$ in (S'). If $\alpha \in (O_M)$ and $T_j \to T$ in (S') then $\alpha T_j \to \alpha T$ in (S').*

Proof. Since (S) is a perfect space (see Sec. 2, Chap. 2), (strong) convergence in (S') is equivalent to weak convergence (by Theorem 23, Sec. 6, Chap. 1). Since we may also assume in the first part of the theorem that $\alpha = 0$ and in the second part that $T = 0$, it thus suffices to prove that

(15.19) $$\text{if } \alpha_j \to 0 \text{ in } (O_M) \quad \text{then} \quad \alpha_j T \cdot \varphi \to 0;$$

(15.20) $$\text{if } T_j \to 0 \text{ in } (S') \quad \text{then} \quad \alpha T_j \cdot \varphi \to 0$$

for any $\varphi \in (S)$. (15.19) follows by noting that $\alpha_j \varphi \to 0$ in (S), as $j \to \infty$, and (15.20) follows by noting that $\alpha\varphi \in (S)$ and hence $T_j \cdot \alpha\varphi \to 0$ as $j \to \infty$.

Theorem 61. *If $S \in (O'_C)$, $T \in (S')$, then the expression*

$$S * T \cdot \varphi \equiv S_\xi \cdot [T_\eta \cdot \varphi(\xi + \eta)]$$

is well defined for $\varphi \in (S)$, and $S * T \in (S')$. If $S_j \to S$ in (O'_C) then $S_j * T \to S * T$ in (S'). If $T_j \to T$ in (S'), then $S * T_j \to S * T$ in (S').

Proof. By Theorem 55, Sec. 14,

$$T_\eta = D^\beta[(1 + |\eta|^2)^m F(\eta)],$$

where $F(\eta)$ is a continuous bounded function. Introducing the function $I(\xi) = T_\eta \cdot \varphi(\xi + \eta)$, where φ belongs to any bounded set B in (S), we have

$$D^q I(\xi) = (-1)^{|\beta|} \int_{R^n} (1 + |\eta|^2)^m F(\eta)\, D^{\beta+q} \varphi(\xi + \eta)\, d\eta.$$

Using the inequality

$$1 + |\eta|^2 \leqslant C_1(1 + |\xi|^2)(1 + |\xi + \eta|^2)$$

and noting that

$$\int_{R^n} |D^{\beta+q}\varphi(x)|(1 + |x|^2)^m \, dx \leqslant C'_q \qquad \text{for all } \varphi \in B,$$

we obtain

$$|D_\xi^q I(\xi)| \leqslant B_q(1 + |\xi|^2)^m, \tag{15.21}$$

where the C'_q, B_q are constants independent of the particular φ in B.

Writing $S_\xi \cdot [T_\eta \cdot \varphi(\xi + \eta)]$ in the form

$$(1 + |\xi|^2)^h S \left(\frac{I(\xi)}{(1 + |\xi|^2)^h} \right) \tag{15.22}$$

and noting, by (15.21), that $\psi(\xi) \equiv I(\xi)(1 + |\xi|^2)^{-h}$ belongs to (D_{L^1}) if $h > m + n/2$, whereas $(1 + |\xi|^2)^h S$ belongs to (B'), the conjugate of (D_{L^1}), for any h, we conclude that (15.22) is defined and, in fact, the numbers (15.22) form a bounded set when $\varphi \in B$. Hence, by the notation of the theorem, $S * T \cdot \varphi$ is well defined and $S * T$ is a continuous linear functional on (S).

To prove the second part of the theorem, we may assume that $S = 0$. Instead of (15.22) we now have

$$(1 + |\xi|^2)^h S_j \cdot \psi.$$

Since, by definition of convergence in (O'_C), $(1 + |\xi|^2)^h S_j \to 0$ in (B'), we conclude that $S_j * T \cdot \varphi \to 0$ uniformly with respect to φ in bounded sets of (S) (the uniformity with respect to φ is actually not needed since, (S) being a perfect space, weak convergence in (S') implies strong convergence). Hence $S_j * T \to 0$ in (S'). The last part of the theorem can be established along the same lines, using Theorem 56, Sec. 14. We leave the details to the reader.

We set $T * S \cdot \varphi \equiv T_\eta \cdot [S_\xi \cdot \varphi(\xi + \eta)]$. Then Theorem 61 remains true, i.e., $T * S \in (S')$ if $T \in (S')$, $S \in (O'_C)$, and $T * S$ is continuous in T and S in the (sequential) sense of Theorem 61. (Use is being made of Theorem 59.) Since $S * T = T * S$ when either S or T has a compact support, we conclude, by continuity, that

$$S * T = T * S \qquad (T \in (S'),\ S \in (O'_C)).$$

We can now state an extension of the convolution theorem.

Theorem 62. *If $\alpha \in (O_M)$, $U \in (S')$, then $\mathfrak{F}\alpha \in (O'_C)$, $\mathfrak{F}U \in (S')$, and*

$$\mathfrak{F}(\alpha U) = \mathfrak{F}\alpha * \mathfrak{F}U. \tag{15.23}$$

If $T \in (O'_C)$, $U \in (S')$, then $\mathfrak{F}T \in (O_M)$, $\mathfrak{F}U \in (S')$, and

$$\mathfrak{F}(T * U) = \mathfrak{F}T \cdot \mathfrak{F}U. \tag{15.24}$$

Proof. Using the rules (15.10), (15.11) one easily finds that $\mathfrak{F}$ maps (O_M) into (O'_C) and (O'_C) into (O_M) and $\mathfrak{F}$ is, in fact, a continuous operator. From Theorem 42, Sec. 11, it follows that (S) is dense in (O'_C). Recalling that $\mathfrak{F}$ maps (S) onto itself it further follows that (S) is dense in (O_M). In virtue of the continuity of $S * T$ in $S \in (O'_C)$, $T \in (S')$ and the continuity of αT in $\alpha \in (O_M)$, $T \in (S')$ as established by Theorems 61, 60, we conclude that it is enough to prove (15.23), (15.24) for α, U, T in (S). But then we have

$$\mathfrak{F}(T * U) = (T * U)_x \cdot e^{ix \cdot y} = (T_\xi \otimes U_\eta) \cdot e^{i(\xi+\eta)\cdot y}$$
$$= [T_\xi \cdot e^{i\xi\cdot y}][U_\eta \cdot e^{i\eta\cdot y}] = \mathfrak{F}T \cdot \mathfrak{F}U,$$

which proves (15.24). The proof of (15.23) is similar.

We conclude this section with an application to differential equations. Let

$$P(\xi) = \sum_{|\alpha| \leqslant m} a_\alpha \xi^\alpha$$

be a polynomial which satisfies the condition

$$P(\xi) \neq 0 \qquad \text{if } \xi \neq 0,\ \xi \text{ real.} \tag{15.25}$$

Theorem 63. *If u is a solution of $P\left(i \dfrac{\partial}{\partial x}\right) u = 0$ in R^n and $|u(x)| \leqslant A(1 + |x|)^h$ for some positive constants A, h, then u is a polynomial in x of degree $\leqslant [h]$ (the integer part of h).*

Proof. u belongs to (S') and therefore also $T \equiv \mathfrak{F}u$ belongs to (S'), and $P(\xi)T = 0$. For any $\psi \in (S)$ we then have

$$T \cdot P(\xi)\psi = P(\xi)T \cdot \psi = 0.$$

We now observe that for any open set $\Omega \subset R^n$, $0 \notin \Omega$, if $\varphi \in (D_\Omega)$ then

$$T \cdot \varphi = T \cdot P(\xi) \frac{\varphi(\xi)}{P(\xi)} = T \cdot P(\xi)\psi = 0.$$

Hence the support of T consists at most of the origin. By Theorem 30, Sec. 7,

$$T = \sum_{|p| \leqslant m} c_p\, D^p \delta,$$

from which it follows that $u = \mathfrak{F}^{-1}T$ is a polynomial. Its degree is clearly $\leqslant [h]$.

PROBLEMS

1. Prove that (D) does not have a denumerable basis of neighborhoods at 0.
2. Verify formula (4.6).

3. Let $|x| = r$, $\Delta = \sum_{i=1}^{n} D_i^2$. Prove that

$$\Delta \log \frac{1}{r} = -2\pi\delta \qquad \text{if } n = 2,$$

$$\Delta \frac{1}{r^{n-2}} = -(n-2)\omega_n \delta \qquad \text{if } n > 2,$$

where $\omega_n = \dfrac{2\pi^{(n/2)}}{\Gamma(n/2)}$ is the area of the unit sphere in R^n.

4. Show that $T(\varphi) = \lim_{m\to\infty} \left[\left(\sum_{j=1}^{m} \varphi\left(\frac{1}{j}\right) \right) - m\varphi(0) - (\log m)\varphi'(0) \right]$ is a distribution on R^1 whose support is the set $N = \{0, 1, \frac{1}{2}, \ldots, 1/n, \ldots\}$. Next verify that for any sequence $\{\varphi_k\} \subset (D)$ such that $0 \leqslant \varphi_k(x) \leqslant 1/\sqrt{k}$ and

$$\varphi_k(x) = \begin{cases} \dfrac{1}{\sqrt{k}} & \text{if } x \geqslant \dfrac{1}{k}, \\ 0 & \text{if } x \leqslant \dfrac{1}{k+1}, \end{cases}$$

$\varphi_k \to 0$ uniformly and $D^\alpha \varphi_k = 0$ on N for any α ($|\alpha| > 0$), but $T(\varphi_k) \to \infty$.

5. Prove that $T = \sum_{i=1}^{\infty} \dfrac{d^i}{dx^i} \delta_{(\alpha_i)}$, where $\alpha_i \to \infty$ as $i \to \infty$, is a distribution on R^1 which cannot be represented as a finite sum $\sum_{j=1}^{N} \dfrac{d^j}{dx^j} f_j$, where the f_j are continuous functions.

6. Let $\{F_i\}$ be a countable covering of R^n by closed sets F_i. Prove that every distribution T can be represented in the form $T = \sum T_i$ where the support of T_i lies in F_i. [*Hint:* Use Theorem 26 and decompose $G_j = \sum G_{jk}$ where the support of G_{jk} lies in F_k.]

7. Verify: $D_y^q T_{x,y} = T_x \otimes D_y^q \delta_y$.

8. Let $Y_{x_i}(x_i)$ be Heaviside's function ($= 0$ if $x_i < 0$, $= 1$ if $x_i > 0$, $= \frac{1}{2}$ if $x_i = 0$) and let

$$Y_{(k)}(x) = Y_{x_1} \otimes \cdots \otimes Y_{x_k} \otimes \delta_{x_{k+1}} \otimes \cdots \otimes \delta_{x_n}.$$

Prove:

$$\frac{\partial}{\partial x_k} Y_{(k)} = Y_{(k-1)}, \qquad \frac{\partial^k}{\partial x_1 \cdots \partial x_k} Y_{(k)} = \delta.$$

9. Verify the equality (which generalizes (9.5))

$$\alpha D^p \delta = \sum_{q \leqslant p} (-1)^{|p-q|} \binom{p}{q} D^{p-q}\alpha(0) \cdot D^q \delta.$$

Applying it to x^p, one obtains

$$x^p D^p \delta = (-1)^{|p|} p! \delta.$$

10. Prove that if u satisfies (9.16), (9.17) and

$$T = \begin{cases} u & \text{if } x_n \geqslant 0, \\ 0 & \text{if } x_n < 0, \end{cases}$$

then T satisfies (9.18) with H defined by (9.19).

11. Prove that if $H = 0$ then $f^{(k)} = 0$ for $k = 0, 1, \ldots, m - 1$.

12. Prove that if A and C have compact supports, then

$$(A_x \otimes C_y) * (B_x \otimes D_y) = (A_x * B_x) \otimes (C_y * D_y).$$

13. Show that $\left(1 * \dfrac{d\delta}{dx}\right) * Y \neq 1 * \left(\dfrac{d\delta}{dx} * Y\right)$ where Y is the Heaviside function ($n = 1$).

14. Let $e(x) = \exp\,(\sum_{i=1}^{n} a_i x_i)$. Prove:

$$e(x) \cdot (S * T) = [e(x)S] * [e(x)T];$$

$$e(x) * T = (T \cdot \check{e})e(x).$$

15. Show that if α is a polynomial of degree $\leqslant m$ and T is a distribution having a compact support, then $T * \alpha$ is also a polynomial of degree $\leqslant m$.

16. Show that (D_{L^p}) is not a perfect space.

17. A continuous function $f(x)$ on R^n is of *positive type* ($f \gg 0$) if for any set of points $x^{(1)}, \ldots, x^{(k)}$ in R^n and for any set of complex numbers $z_1, \ldots, z_k$, $\sum_{i,j} f(x_i - x_j) z_i \bar{z}_k \geqslant 0$. A distribution T is of *positive type* ($T \gg 0$) if $T \cdot (\varphi * \hat{\varphi}) \geqslant 0$ for any $\varphi \in (D)$, where $\hat{\varphi}(x) = \overline{\varphi(-x)}$. Prove:

(a) If $f \gg 0$ then $f(0) \geqslant 0$, $|f(x)| \leqslant f(0)$ and $f \cdot (\varphi * \hat{\varphi}) \geqslant 0$.
(b) If $f \cdot (\varphi * \hat{\varphi}) \geqslant 0$ for any $\varphi \in (D)$ and if f is continuous, then $f \gg 0$.
(c) If $T = \mathfrak{F}\mu$ where $\mu \geqslant 0$ is a tempered measure, then $T \gg 0$. (Incidentally, the converse is also true.)
(d) If S and T have compact supports and are of positive type, then $S * T \gg 0$.

18. Prove that $(\mathfrak{F}T)^{\vee} = \mathfrak{F}\check{T}$.

CHAPTER 4

CONVOLUTIONS AND FOURIER TRANSFORMS OF GENERALIZED FUNCTIONS

1. Fourier Transforms of Fundamental Functions

Throughout this chapter we shall assume Φ is a fundamental space satisfying the following properties:

(Φ_1) Φ admits differentiation;
(Φ_2) x_j ($j = 1, \ldots, n$) (and hence all polynomials in x) are multipliers in Φ;
(Φ_3) the elements of Φ are fast-decreasing C^∞ functions, i.e., $\Phi \subset S$;
(Φ_4) if $\varphi(x) \in \Phi$ then also $\bar{\varphi}(x)$ and $\varphi(-x)$ belong to Φ.

We recall the definition of the Fourier transform of a function $\varphi \in \Phi$:

$$\psi(\sigma) \equiv (\mathfrak{F}\varphi)(\sigma) \equiv \tilde{\varphi}(\sigma) = \int_{R^n} e^{ix\cdot\sigma}\varphi(x)\, dx.$$

In Chap. 3, Sec. 15, we mentioned some basic properties of the operator $\mathfrak{F}$ (see (15.2) through (15.7)). In particular,

$$P(D)\tilde{\varphi}(\xi) = \widetilde{P(ix)\varphi(x)}, \tag{1.1}$$

$$\widetilde{P(D)\varphi(x)} = P(-i\xi)\tilde{\varphi}(\xi). \tag{1.2}$$

In Chap. 2, Sec. 1, we introduced the spaces $K(a)$, $Z(a)$. The inductive limit of the $K(a)$ is denoted by K. In view of Theorem 6, Sec. 3, Chap. 3, K is precisely the topological space (D). Note also that K is, in fact, the strict inductive limit of $K(a^{(m)})$ where $\{a^{(m)}\}$ is any sequence of vectors with components $a_j^m \to \infty$ as $m \to \infty$ ($j = 1, \ldots, n$). Hence K is a fundamental space. We shall denote the union space of the $Z(a)$ by Z. Z is also a fundamental space.

Let $\varphi \in K$. Then $\psi(\sigma) = \tilde{\varphi}(\sigma)$ is the restriction to the real axes of the function

$$\psi(s) = \int_{R^n} e^{ix\cdot s}\varphi(x)\, dx = \int_{R^n} e^{ix\cdot\sigma - x\cdot\tau}\varphi(x)\, dx,$$

where $s = \sigma + i\tau$. If $\varphi \in K(a)$ then, using the notation of Chap. 2,

Sec. 1, we obtain

$$|s^k\psi(s)| = \left| \int_{G_a} e^{ix\cdot s} D^k\varphi(x)\, dx \right| \leqslant C'_k e^{a|\tau|},$$

where

$$G_a = \{x; |x_1| \leqslant a_1, \ldots, |x_n| \leqslant a_n\} \tag{1.3}$$

and the C'_k are constants. It follows that

$$(1 + |s|)^m |\psi(s)| \leqslant C_m e^{a|\tau|} \qquad (0 \leqslant m < \infty). \tag{1.4}$$

Comparing (1.4) with (1.11), Chap. 2, we conclude that $\mathfrak{F}$ maps $K(a)$ into $Z(a)$. In fact, it maps bounded sets in $K(a)$ into bounded sets in $Z(a)$. Hence $\mathfrak{F}$ is a continuous linear operator from $K(a)$ into $Z(a)$, and consequently, $\mathfrak{F}$ is also a continuous linear operator from K into Z.

To show that $\mathfrak{F}$ maps $K(a)$ onto $Z(a)$, we first define for any $\psi \in Z(a)$

$$\hat{\varphi}(x) = \frac{1}{(2\pi)^n} \int_{R^n} e^{-ix\cdot\sigma}\, \psi(\sigma)\, d\sigma = \frac{1}{(2\pi)^n} \int_{R^n} e^{-ix\cdot(\sigma + i\tau)}\, \psi(\sigma + i\tau)\, d\sigma;$$

the change of the domain of integration is permissible because of the inequality (1.11), Chap. 2 (used with $k > n$). Taking $\tau_j = -\delta_j \operatorname{sgn} x_j$ $(j = 1, \ldots, n)$ we obtain

$$|D^k\hat{\varphi}(x)| \leqslant B_k |s^k| e^{-\delta|x|} e^{\delta\cdot a}.$$

If $x_j > a_j$ for some j $(j = 1, \ldots, n)$ then $\hat{\varphi}(x) \to 0$ as $\delta_j \to \infty$. We conclude that $\hat{\varphi} \in K(a)$. Since $\hat{\varphi}$ is the function whose Fourier transform coincides with ψ, it follows that $\mathfrak{F}$ maps $K(a)$ in a one-to-one way *onto* $Z(a)$. It is also easily seen that $\mathfrak{F}^{-1}$ maps bounded sets in $Z(a)$ into bounded sets in $K(a)$. Hence $\mathfrak{F}^{-1}$ also maps Z continuously onto K. We have proved the following:

Theorem 1. $\mathfrak{F}[K(a)] = Z(a)$, $\mathfrak{F}^{-1}[Z(a)] = K(a)$, $\mathfrak{F}[K] = Z$, $\mathfrak{F}^{-1}[Z] = K$ *and* $\mathfrak{F}$, $\mathfrak{F}^{-1}$ *are continuous linear operators.*

Now let Φ be any fundamental space (satisfying the properties (Φ_1) through (Φ_4)). Denote by $\Psi \equiv \mathfrak{F}[\Phi] \equiv \tilde{\Phi}$ the set of all functions

$$\psi \equiv \mathfrak{F}\varphi \equiv \tilde{\varphi},$$

where $\varphi \in \Phi$. Ψ is a linear space and $\mathfrak{F}$ is a one-to-one mapping of Φ onto Ψ. If Φ is a topological space, we introduce a topology on Ψ in such a way that $\mathfrak{F}$ becomes a topological mapping from Φ onto Ψ. Thus, the neighborhoods of 0 in Ψ are the sets $\mathfrak{F}(V)$ where V is an arbitrary neighborhood of 0 in Φ. If Φ is a union space, we define convergence in Ψ as follows: $\mathfrak{F}\varphi_m \to 0$ in Ψ if and only if $\varphi_m \to 0$ in Φ.

$\psi(\sigma) = \mathfrak{F}[\varphi(x)]$ implies that

$$\mathfrak{F}^{-1}[\varphi(x)] = (2\pi)^{-n}\psi(-\sigma), \qquad \mathfrak{F}[\bar{\psi}(\sigma)] = (2\pi)^n\bar{\varphi}(x). \tag{1.5}$$

Using the first equality and a part of the assumption (Φ_4), it follows that

$$\mathfrak{F}^{-1}[\Phi] = \Psi, \qquad \mathfrak{F}[\Psi] = \Phi. \tag{1.6}$$

In particular,

$$\mathfrak{F}^{-1}[K(a)] = Z(a), \qquad \mathfrak{F}^{-1}[K] = Z.$$

It also follows from (1.5) that Ψ satisfies the property (Φ_4). In view of the rules (1.1), (1.2), Ψ also satisfies the properties (Φ_1) and (Φ_2). Finally, since $\mathfrak{F}[S] = S$, Ψ satisfies (Φ_3).

2. Fourier Transforms of Generalized Functions

For any $f \in \Phi'$ we define

$$\mathfrak{F}f \cdot \psi = f \cdot \mathfrak{F}\psi. \tag{2.1}$$

Since $\mathfrak{F}$ is a topological mapping from Φ onto Ψ, $\mathfrak{F}f$ is clearly an element of Ψ', the conjugate space of Ψ. In fact, $\mathfrak{F}$ is seen to be a continuous linear operator, in both the weak and the strong topologies, of Φ' into Ψ' (the proof is left to the reader). $\mathfrak{F}$ is clearly a one-to-one mapping. $\mathfrak{F}^{-1}g$ $(g \in \Psi')$ is also a one-to-one continuous mapping from Ψ' into Φ' $(\mathfrak{F}^{-1}g \cdot \varphi = g \cdot \mathfrak{F}^{-1}\varphi)$. Hence, $\mathfrak{F}$ establishes a homeomorphism from Φ' onto Ψ' (and, in view of (1.6), also from Ψ' onto Φ').

The definition (2.1) is consistent with the definition of Fourier transforms of tempered distributions (i.e., the case $\Phi \equiv S$) given in Chap. 3, Sec. 15. We also have (compare Chap. 3)

$$\mathfrak{F}f \cdot \overline{\mathfrak{F}\varphi} = (2\pi)^n f \cdot \bar{\varphi}; \tag{2.2}$$

$$P(D)\mathfrak{F}(f) = \mathfrak{F}[P(ix)f]; \tag{2.3}$$

$$\mathfrak{F}[P(D)f] = P(-i\xi)\mathfrak{F}(f). \tag{2.4}$$

Consider a distribution f having its support in some n-rectangle G_a defined in (1.3). By Theorem 22, Sec. 6, Chap. 3,

$$f = \sum_{q \leqslant p} (-1)^{|q|} D^q g_q, \tag{2.5}$$

where the g_q are continuous functions which vanish outside $G_{a+\epsilon}$ for any given $\epsilon = (\epsilon_1, \ldots, \epsilon_n) > 0$. Using the rule (2.4) we have

$$\tilde{f} = \sum_{q \leqslant p} (i\sigma)^q h_q(\sigma),$$

where the $h_q(\sigma)$ can be extended into entire functions $h_q(s)$ satisfying

$$|h_q(s)| \leqslant C_q e^{(a+\epsilon)|\tau|}.$$

We have thus proved the following:

Theorem 2. *The Fourier transform of any distribution whose support lies in an n-rectangle G_a is a functional, of Z', of function type $h(\sigma)$, where $h(\sigma)$ can be extended into an entire function $h(s)$ of exponential type $\leqslant a + \epsilon$ and of slow increase on $s = \sigma$; more precisely,*

$$|h(s)| \leqslant A(1 + |\sigma|)^{\lambda} e^{(a+\epsilon)|\tau|} \qquad (A > 0, \lambda \geqslant 0). \tag{2.6}$$

The converse is also true; see Theorem 5, Sec. 3, Chap. 6.

The tool of Fourier transform can be used in order to find the structure of generalized functions over $\tilde{\Phi}$, provided one knows the structure of generalized functions over Φ. We give here one example. If $f \in K'(a)$, then, by the proof of Theorem 17, Sec. 5, Chap. 3,

$$f = (-1)^{|m|} D^m g,$$

where $g(x)$ is a continuous function on R^n whose support lies in G_a. Taking $\varphi \in K(a) \equiv (D_{G_a})$ and using (2.1) we get

$$\begin{aligned} \tilde{f} \cdot \psi &= (-1)^{|m|} D^m g \cdot \mathfrak{F}\psi = g \cdot D^m(\mathfrak{F}\psi) = g \cdot \mathfrak{F}[(i\sigma)^m \psi] \\ &= \int_{G_a} g(x) \left[\int_{R^n} e^{ix\cdot\sigma} (i\sigma)^m \psi(\sigma)\, d\sigma \right] dx = \int_{R^n} F(\sigma)\psi(\sigma)\, d\sigma, \end{aligned}$$

where

$$F(\sigma) = (i\sigma)^m \int_{G_a} e^{ix\cdot\sigma} g(x)\, dx.$$

We have thus proved the following:

Theorem 3. *Every generalized function k in $Z'(a)$ is of function type $F(\sigma)$, i.e.,*

$$k \cdot \psi = \int_{R^n} F(\sigma)\psi(\sigma)\, d\sigma, \tag{2.7}$$

where $F(s)$ is an entire function satisfying

$$|F(s)| \leqslant A(1 + |\sigma|)^{\lambda} e^{a|\tau|} \qquad (A > 0, \lambda \geqslant 0).$$

The converse is trivial.

3. Convolutions of Generalized Functions

Definition. We say that Φ admits a *continuous translation* if, for any $h \in R^n$, the operator

$$\tau_h \varphi(x) = \varphi(x + h) \tag{3.1}$$

maps Φ into Φ and is bounded uniformly with respect to h where $|h| \leqslant 1$. Obviously, it is then bounded uniformly with respect to h in any bounded set of R^n. The operator τ_h is called a *translation* by h.

Definition. We say that Φ admits a *differentiable* translation if, for any $h = (0, \ldots, 0, h_j, 0, \ldots, 0)$ $(1 \leqslant j \leqslant n)$,

$$\frac{\varphi(x+h) - \varphi(x)}{h_j} \rightarrow \frac{\partial \varphi}{\partial x_j} \quad \text{in } \Phi \tag{3.2}$$

if $h \rightarrow 0$. It is obvious that Φ then admits a continuous translation.

Theorem 4. *If Φ is a perfect space which admits a continuous translation, or if Φ is a countable union space of such spaces $\Phi^{(m)}$, then $h \rightarrow \tau_h\varphi$ is a continuous function from R^n into Φ, for any fixed $\varphi \in \Phi$.*

Proof. It suffices to prove the continuity of $h \rightarrow \tau_h\varphi$ for $h = 0$. The set $\{\tau_h\varphi;\ |h| \leqslant 1\}$ is a bounded set in Φ and (by definition) it belongs to some perfect space $\Phi^{(k)}$ (or to Φ, if Φ itself is a perfect space). Hence this set is relatively compact, and all we have to show is that there cannot be two different limit elements as $h \rightarrow 0$. Since if $\tau_{h^{(m)}}\varphi \rightarrow \varphi_0$, for some sequence $h^{(m)} \rightarrow 0$, then, for each point x,

$$\tau_{h^{(m)}}\varphi(x) = \varphi(x + h^{(m)}) \rightarrow \varphi_0(x) \qquad \text{as } h^{(m)} \rightarrow 0,$$

the uniqueness of the limit element φ_0 follows.

Theorem 5. *If Φ is a perfect space which admits a continuous translation then Φ admits also a differentiable translation.*

Proof. If we prove that the set

$$\left\{\frac{\varphi(x+h) - \varphi(x)}{h_j};\ |h| \leqslant 1\right\} \tag{3.3}$$

is bounded in Φ, then the proof is completed by observing that there cannot be two different limit elements as $h \rightarrow 0$ since, for any fixed x, any limit element φ_1 at the point x must coincide with $\partial\varphi(x)/\partial x_j$. To prove the boundedness in Φ of the functions in (3.3), write these functions in the form

$$\varphi_h(x) \equiv \frac{1}{h_j}\int_0^{h_j} \frac{\partial\varphi(x+t)}{\partial x_j}\,dt_j \qquad (t = (0, \ldots, 0, t_j, 0, \ldots, 0)). \tag{3.4}$$

For every t_j, $|t_j| \leqslant 1$, the function

$$\tau_t \frac{\partial\varphi}{\partial x_j} = \frac{\partial\varphi(x+t)}{\partial x_j}$$

is an element of Φ and (by Theorem 4) it varies continuously in the topology of Φ when t varies in $|t| \leqslant 1$. Hence the abstract integral (see

(6.1), Chap. 1)

$$\int_0^{h_j} \tau_t \frac{\partial \varphi}{\partial x_j}\, dt_j$$

exists. From its definition and from property 2, Sec. 1, Chap. 2, of Φ it follows that its value for any fixed $x \in R^n$ is the (standard) integral

$$\int_0^{h_j} \tau_t \frac{\partial \varphi(x)}{\partial x_j}\, dt_j, \quad \text{i.e.,} \quad h_j \varphi_h(x).$$

We can thus say that φ_h is the function

$$\frac{1}{h_j} \int_0^{h_j} \tau_t \frac{\partial \varphi}{\partial x_j}\, dt_j,$$

where integration is taken in the abstract sense. Since, for $|t| \leqslant 1$, $\rho_{t_j} \equiv \tau_t \dfrac{\partial \varphi}{\partial x_j}$ belongs to a bounded set $B \subset \Phi$, which we may assume to be convex, the partial sums

$$\frac{1}{h_j} \Sigma\, \rho_{t_{j,m}}\, \Delta t_{j,m} \in \frac{1}{h_j} (\Sigma\, \Delta t_{j,m}) B = B$$

for any $|h| \leqslant 1$. Hence, the limit φ_h belongs to the bounded set $\bar{B}$, and the proof is completed.

The space $K(a)$ does not admit translations. K, S and $Z(a)$ admit a differentiable translation. Every $K\{M_p\}$ space admits a continuous translation if $R_M = \phi$ and the property (N_2) of Sec. 5, Chap. 2, is satisfied.

Definition. Let Φ admit a continuous translation and consider, for any $f \in \Phi'$, the function

$$(f * \varphi)(x) \equiv f \cdot \tau_x \varphi \equiv (f(\xi), \varphi(x + \xi)).$$

This is evidently a continuous function of x. We say that f_0 ($f_0 \in \Phi'$) *convolves* with Φ if for any $\varphi \in \Phi$

$$f_0 * \varphi \equiv f_0 \cdot \tau_x \varphi \equiv (f_0(\xi), \varphi(x + \xi))$$

belongs to Φ and the mapping $\varphi \to f_0 * \varphi$ is a continuous mapping from Φ into Φ, i.e., if $\varphi_m \to 0$ in Φ then $f_0 * \varphi_m \to 0$ in Φ.

We then define $f_0 * f$, for any $f \in \Phi'$, by

$$f_0 * f \cdot \varphi = f \cdot f_0 * \varphi, \tag{3.5}$$

or $(f_0 * f, \varphi) = (f, f_0 * \varphi)$, and call $f_0 * f$ the *convolution* of f_0 with f. $f_0 * f$ is clearly an element of Φ'. The present definition of convolution is consistent with the definition given in Chap. 3, Sec. 10. The present definition is, of course, more general and, generally, if $f_0 * f$ exists, the convolution $f * f_0$ need not exist since f may not convolve with Φ. We may think of $f_0 * f$ $(f_0 * \varphi)$ as the mapping of f (φ) by an operator f_0*. When taking this point of view we call f_0 a *convolutor*.

δ convolves with any space Φ since

$$\delta \cdot \tau_x \varphi = \varphi(x).$$

This shows also that $\delta * \varphi = \varphi$ and, therefore,

$$\delta * f = f \qquad (f \in \Phi'). \tag{3.6}$$

Theorem 6. *Let Φ admit a differentiable translation and assume that f_0 convolves with Φ. Then, for any polynomial $P(D)$ in $D = (D_1, \ldots, D_n)$, $P(D)f_0$ also convolves with Φ and*

$$P(D)(f_0 * f) = P(D)f_0 * f = f_0 * P(D)f. \tag{3.7}$$

Proof. If we prove the theorem for $P(D) = D_j$, then, by successively applying this result, we obtain the proof for any polynomial $P(D)$. The assertion that $\partial f_0/\partial x_j$ convolves with Φ follows from the equalities

$$\begin{aligned} \frac{\partial f}{\partial x_j} * \varphi &= \left(\frac{\partial f(\xi)}{\partial \xi_j}, \varphi(x + \xi)\right) = -\left(f(\xi), \frac{\partial \varphi(x + \xi)}{\partial \xi_j}\right) \\ &= -f * \frac{\partial \varphi}{\partial x_j}, \end{aligned} \tag{3.8}$$

which hold for any $f \in \Phi'$, when we take $f = f_0$.

Since f_0* is a continuous operator from Φ into Φ,

$$f_0 * \frac{\varphi(x + h) - \varphi(x)}{h_j} \to f_0 * \frac{\partial \varphi}{\partial x_j}$$

in the sense of Φ, as $h = (0, \ldots, 0, h_j, 0, \ldots, 0) \to 0$. But the left side is $[\lambda(x + h) - \lambda(x)]/h_j$, where $\lambda(x) = (f_0 * \varphi)(x)$. Hence, $\lambda(x)$ is a differentiable function and

$$\frac{\partial}{\partial x_j} (f_0 * \varphi) = f_0 * \frac{\partial \varphi}{\partial x_j}. \tag{3.9}$$

The proof of (3.7) now follows with the aid of (3.8), (3.9):

$$\begin{aligned} \left(\frac{\partial}{\partial x_j} (f_0 * f) \ \varphi\right) &= -\left(f_0 * f, \frac{\partial \varphi}{\partial x_j}\right) = -\left(f, f_0 * \frac{\partial \varphi}{\partial x_j}\right) \\ &= \left(f, \frac{\partial f_0}{\partial x_j} * \varphi\right) = \left(\frac{\partial f_0}{\partial x_j} * f, \varphi\right); \end{aligned}$$

$$\begin{aligned} \left(\frac{\partial}{\partial x_j} (f_0 * f), \varphi\right) &= -\left(f_0 * f, \frac{\partial \varphi}{\partial x_j}\right) = -\left(f, f_0 * \frac{\partial \varphi}{\partial x_j}\right) \\ &= -\left(f, \frac{\partial}{\partial x_j} (f_0 * \varphi)\right) = \left(\frac{\partial f}{\partial x_j}, f_0 * \varphi\right) \\ &= \left(f_0 * \frac{\partial f}{\partial x_j}, \varphi\right). \end{aligned}$$

Definition. Let $f \in \Phi'$. We say that f is *supported* by a closed set G if $f \cdot \varphi = 0$ for any $\varphi \in \Phi$ which vanishes in some open neighborhood of G. For a distribution T, the smallest closed set which supports T exists and is called the *support* of T (see Chap. 3, Sec. 5).

A functional $f \in \Phi'$ is called a *finite* functional if it is supported by a compact set.

Theorem 7. *Assume that Φ is a perfect space which contains C_c^∞ as a dense subset, and which admits a continuous translation. Then any finite functional in Φ' convolves with Φ.*

Proof. By the remark at the end of Sec. 7, Chap. 3,

$$f = \sum_{q \leqslant p} D^q f_q \qquad \text{on } C_c^\infty, \tag{3.10}$$

where the f_q are continuous functions having compact supports. By definition of the functional f_q,

$$f_q \cdot \varphi = \int_R f_q(\xi)\varphi(\xi)\, d\xi,$$

where R is a compact set containing the support of f. Hence

$$f_q * \varphi = \int_R f_q(\xi)\varphi(x + \xi)\, d\xi. \tag{3.11}$$

We shall show that $\varphi \to f_q * \varphi$ is a continuous operator from Φ into Φ. (Note that even though $f_q \cdot \varphi$, $f_q * \varphi$ are defined for any $\varphi \in \Phi$ (since $\Phi \subset S$), we do not know yet that f_q is a continuous functional on Φ.)

As in the proof of Theorem 5, the integral in (3.11) may be considered as the value at x of the abstract integral

$$I(\varphi) \equiv \int_R f_q(\xi)\tau_\xi\varphi\, d\xi. \tag{3.12}$$

The existence of the integral (3.12) implies of course that $f_q * \varphi$ belongs to Φ. Now let φ vary in a bounded set $B \subset \Phi$. Recalling the definition of the integral (3.12) (see (6.1), Chap. 1) we find that $I(\varphi)$ varies in the set

$$B_0 \equiv \{\mu B_1;\ |\mu| \leqslant M|R|\},$$

where M = l.u.b. $|f_q|$, $|R|$ is the measure of R, and B_1 is a bounded convex set containing all the elements $\tau_\xi\varphi$ for $\varphi \in B$, $\xi \in R$. Since B_0 is a bounded set, $f_q *$ maps bounded sets in Φ into bounded sets in Φ. Taking $x = 0$ and using property 2, Sec. 1, Chap. 2, of Φ it follows that f_q is a continuous functional on Φ.

Having proved that each f_q convolves with Φ, it follows, by Theorem 6, that $D^q f_q$ convolves with Φ, and the proof is completed by noting that (3.10) holds on Φ, since both sides are continuous on Φ.

Note that if f_0 is a C_c^∞ function, then $f_0 * f$ is a C^∞ function. Indeed each $f \in \Phi'$ is a distribution (by the remark at the end of Sec. 7, Chap. 3) and the result therefore follows from Theorem 41, Sec. 11, Chap. 3.

We next consider the question of the continuity of $f_0 * f$ in f_0.

Theorem 8. *Assume that Φ is a perfect space which contains C_c^∞ as a dense subset, and which admits a continuous translation. Let f and f_j be generalized functions over Φ which are supported by a compact set R. If $f_j \to f$ in Φ', then, for any $g \in \Phi'$.*

$$f_j * g \to f * g \qquad \text{in } \Phi'.$$

Proof. Without loss of generality we may assume that $f = 0$. By the remark at the end of Sec. 7, Chap. 3, f and the f_j belong to $K'(a)$ for some $a < \infty$. Applying Theorem 19, Sec. 5, Chap. 3, and the proof of Theorem 22, Sec. 6, Chap. 3, we have the representation

$$f_j = \sum_{q \leqslant p} D^q g_{jq},$$

where p is independent of j, the supports of the g_{jq} are contained in some compact set R^0, and the g_{jq} are continuous functions satisfying

$$\text{(3.13)} \qquad \operatorname*{l.u.b.}_{x} |g_{jq}(x)| \to 0 \qquad \text{as } j \to \infty.$$

The above representation holds on C_c^∞, but, by the proof of Theorem 7, it remains true on Φ.

From the proof of Theorem 7 and from (3.7) we have

$$\text{(3.14)} \qquad (-1)^{|q|} D^q g_{jq} * \varphi = g_{jq} * D^q \varphi = \int_{R^0} g_{jq}(\xi)\, D^q\varphi(x + \xi)\, d\xi,$$

where the integration is taken in the abstract sense. Let $\varphi \in \Phi$. Then $\tau_\xi D^q \varphi$ varies in a bounded set, say,

$$\|\tau_\xi D^q \varphi\|_m \leqslant C_m \qquad (1 \leqslant m < \infty).$$

Applying (6.3) of Chap. 1 to the integral of (3.14) we obtain

$$\|D^q g_{jq} * \varphi\|_m \leqslant C_m \operatorname*{l.u.b.}_{\xi} |g_{jq}(\xi)| \leqslant C_m \epsilon_j \to 0$$

as $j \to \infty$. Hence $D^q g_{jq} * \varphi \to 0$ and therefore also $f_j * \varphi \to 0$ in Φ, as $j \to \infty$. It follows that, for any $g \in \Phi'$,

$$(f_j * g, \varphi) = (g, f_j * \varphi) \to 0 \qquad \text{as } j \to \infty,$$

i.e., $f_j * g \to 0$ weakly and, since Φ is a perfect space, also strongly.

4. The Convolution Theorems

We shall prove in this section an extension of the classical convolution theorem.

Theorem 9. *Assume that Φ admits a continuous translation and set $\Psi = \mathfrak{F}\Phi$. Let $g \in \Psi'$ be of function type $g(\sigma)$ and assume further that $g(\sigma)$ is a multiplier in Ψ. Then $f = \mathfrak{F}^{-1}(g)$ convolves with Φ, and, for any $f_1 \in \Phi'$,*

$$\mathfrak{F}(f * f_1) = \mathfrak{F}(f) \cdot \mathfrak{F}(f_1). \tag{4.1}$$

Proof. We first show that f convolves with Φ. For any $\varphi \in \Phi$, let $\mathfrak{F}\varphi = \psi$. Then

$$\mathfrak{F}(\tau_h\varphi) = e^{-ih\cdot\sigma}\psi(\sigma). \tag{4.2}$$

Since $\mathfrak{F}\varphi_1 = \psi_1$ implies $(2\pi)^n\bar{\varphi}_1 = \mathfrak{F}\bar{\psi}_1$, we obtain, taking $\varphi_1 = \tau_h\varphi$ and using (4.2),

$$(2\pi)^n\overline{\tau_h\varphi} = \mathfrak{F}(e^{ih\cdot\sigma}\bar{\psi}(\sigma)). \tag{4.3}$$

Using (2.1) and (4.3) we obtain

$$f * \bar{\varphi} = f \cdot \overline{\tau_x\varphi} = \frac{1}{(2\pi)^n} f \cdot \mathfrak{F}(e^{ix\cdot\sigma}\bar{\psi}(\sigma)) = \frac{1}{(2\pi)^n} g \cdot e^{ix\cdot\sigma}\bar{\psi}$$

$$= \frac{1}{(2\pi)^n} \int_{R^n} g(\sigma)e^{ix\cdot\sigma}\bar{\psi}(\sigma)\, d\sigma,$$

i.e.,

$$(f * \bar{\varphi})(x) = \frac{1}{(2\pi)^n} \mathfrak{F}(g\bar{\psi})(x) = \mathfrak{F}^{-1}(g\bar{\psi})(-x). \tag{4.4}$$

Since g is a multiplier in Ψ, $g\bar{\psi}$ belongs to Ψ, and the right side of (4.4) therefore belongs to Φ (recall that both Φ and Ψ satisfy the property (Φ_4) of Sec. 1). The continuity of $f * \bar{\varphi}$ in $\bar{\varphi}$ also follows from (4.4). Hence f convolves with Φ.

Using (4.4), the proof of (4.1) is easily obtained as follows:

$$(\mathfrak{F}(f * f_1), \bar{\psi}) = (f * f_1, \mathfrak{F}\bar{\psi}) = (2\pi)^n(f_1, f * \bar{\varphi}) = (f_1, \mathfrak{F}(g\bar{\psi}))$$

$$= (\mathfrak{F}f_1, g\bar{\psi}) = (g \cdot \mathfrak{F}f_1, \bar{\psi}).$$

In Chapter 7 we shall need a more general concept of convolution than the one introduced in Sec. 3. Let $f_0 \in \Phi'$ and assume that $f_0 * \varphi$ belongs to some space Φ_1 for any $\varphi \in \Phi$, and that $\varphi \to f_0 * \varphi$ is a continuous mapping from Φ into Φ_1. We then say that f_0 is a convolutor from Φ into Φ_1. We also say that f_0 convolves Φ into Φ_1. We then further define $f_0 * f$, for any $f \in \Phi_1'$, by

$$(f_0 * f, \varphi) = (f, f_0 * \varphi) \qquad (\varphi \in \Phi).$$

Clearly, $f_0 * f$ belongs to Φ'.

We can now state an extension of Theorem 9, which can be proved by slightly modifying the proof of Theorem 9.

Theorem 9′. *Assume that Φ, Φ_1 admit continuous translations and set $\Psi = \mathfrak{F}\Phi$, $\Psi_1 = \mathfrak{F}\Phi_1$. Let $g \in \Psi'$ be of function type $g(\sigma)$ and assume further that $\Psi \to g\Psi$ is a continuous mapping from Ψ into Ψ_1. Then $f = \mathfrak{F}^{-1}(g)$ is a convolutor from Φ into Φ_1 and for any $f_1 \in \Phi_1'$,*

$$\mathfrak{F}(f * f_1) = \mathfrak{F}(f) \cdot \mathfrak{F}(f_1),$$

where, by definition, $(g \cdot \mathfrak{F}(f_1), \Psi) = (\mathfrak{F}(f_1), g\Psi)$ for $\Psi \in \Psi$.

Theorem 9 is analogous to the first part of Theorem 62, Sec. 15, Chap. 3. An analogue of the second part is the following:

Theorem 10. *Let Φ be a perfect space which contains C_c^∞ as a dense linear subspace and which admits a continuous translation. Then, for any finite functional $f \in \Phi'$, $\mathfrak{F}f \equiv g$ is a multiplier in $\Psi = \mathfrak{F}[\Phi]$ and*

$$\mathfrak{F}(f * f_1) = g \cdot g_1 \tag{4.5}$$

for any $f_1 \in \Phi'$, where $g_1 = \mathfrak{F}f_1$.

Proof. By Theorem 7, f convolves with Φ. Let $\varphi \in C_c^\infty$. Then

$$\mathfrak{F}(f * \varphi) = \mathfrak{F}[(f(\xi), \varphi(x + \xi))] = \int e^{ix\cdot\sigma}(f(\xi), \tau_x\varphi(\xi))\, dx$$

$$= \left(f(\xi), \int e^{ix\cdot\sigma}\tau_x\varphi\, dx\right)$$

where the last integral is taken in the abstract sense of Sec. 6, Chap. 1 (note that the integration is on a bounded set of R^n). The abstract integral is a function in Φ and its value at any point ξ is the standard integral when the element φ is replaced by $\varphi(\xi)$ (this follows from the definition (6.1) and property 2, Sec. 1, Chap. 2, of Φ). Writing

$$\tau_x\varphi(\xi) = \tau_\xi\varphi(x)$$

and using (4.2) we obtain

$$\mathfrak{F}(f * \varphi) = (f(\xi), e^{-i\sigma\cdot\xi}\psi(\sigma)) = \psi(\sigma)(f(\xi), e^{-i\sigma\cdot\xi}). \tag{4.6}$$

We now use the following formula:

$$(f(\xi), e^{i\sigma\cdot\xi}) = \mathfrak{F}f. \tag{4.7}$$

The proof follows by using the representation (3.10) for f and noting that (4.7) holds for functions (by the definition of $\mathfrak{F}$), and finally using (3.7) and (2.4). Using (4.7), (4.6) becomes

$$\mathfrak{F}(f * \varphi) = g(-\sigma)\psi(\sigma). \tag{4.8}$$

Since the left side of (4.8) varies continuously in Ψ when φ varies in Φ, the right side can be extended as a continuous operator from Ψ into Ψ. Hence g is a multiplier in Ψ and therefore also in Ψ'.

It remains to prove (4.5). Using (4.8) with φ replaced by $\bar{\varphi}$ (and $\psi = \mathfrak{F}\varphi$), i.e.,

$$\mathfrak{F}(f * \bar{\varphi}) = g(-\sigma)\bar{\psi}(-\sigma), \tag{4.9}$$

and using the obvious identity $\mathfrak{F}\varphi_0(\sigma) = (2\pi)^n(\mathfrak{F}^{-1}\varphi_0)(-\sigma)$, we get

$$(\mathfrak{F}(f * f_1), \bar{\psi}) = (f * f_1, \mathfrak{F}\bar{\psi}) = (2\pi)^n(f_1, f * \bar{\varphi}) = (2\pi)^n(\mathfrak{F}^{-1}g_1, f * \bar{\varphi})$$
$$= (2\pi)^n(g_1, \mathfrak{F}^{-1}(f * \bar{\varphi})) = (g_1(\sigma), g(\sigma)\bar{\psi}(\sigma)) = (g \cdot g_1, \bar{\psi}),$$

which completes the proof.

PROBLEMS

1. Prove that the functional of function type $e^{ib\cdot\sigma}$ over $Z(a)$ is the zero functional if $b > a$.
2. Show that if $f_j \to f$ in the weak (strong) sense of Φ', and if f_0 convolves with Φ, then
$$f_0 * f_j \to f_0 * f$$
in the weak (strong) sense of Φ'.
3. Prove that
$$\mathfrak{F}(\tau_h f) = e^{ih\cdot\sigma}\mathfrak{F}f,$$
where $\tau_h f$ is defined by $(\tau_h f, \varphi) = (f, \tau_{-h}\varphi)$ for all $\varphi \in \Phi$.
4. Let f_0 convolve with Φ. Verify that
$$\tau_h(f_0 * f) = \tau_h f_0 * f = f_0 * \tau_h f.$$

CHAPTER 5

W SPACES

In this chapter we introduce a very important class of fundamental spaces, namely, W spaces. These spaces will be used extensively in the following chapters. Some theorems on entire functions, which will be needed throughout this and the following chapters, are given in Sec. 1.

1. Theorems on Complex Analytic Functions

The notation of Chap. 2, Sec. 1, will be freely used in this and in subsequent chapters.

An entire function $f(z)$ $(z = (z_1, \ldots, z_n))$ is said to be of order $\leqslant p$ $(p = (p_1, \ldots, p_n))$ if for any $\epsilon > 0$

$$|f(z)| \leqslant C \exp [a|z|^{p+\epsilon}]$$

for some constants C, a. $f(z)$ is said to be of order $\leqslant p$ and type $\leqslant b$ if for any $\epsilon > 0$

$$|f(z)| \leqslant C \exp [(b + \epsilon)|z|^p]$$

for some constant C. If $p = 1$, we say that f is of exponential type $\leqslant b$.

Theorem 1 (*Phragmén-Lindelöf*). *Let $f(z)$ be a complex analytic function of z when each z_j varies in an angular domain of opening $\varphi_j < \pi/p_j$, say*

$$G_j = \left\{z_j; |\arg z_j| \leqslant \frac{\varphi_j}{2}\right\},$$

and let Γ_j be the boundary of G_j. If

$$(1.1) \qquad |f(z)| \leqslant C \exp [b|z|^p] \qquad \text{when } z_j \in G_j,\ j = 1, \ldots, n,$$

$$(1.2) \qquad |f(z)| \leqslant C_1 \qquad \text{when } z_j \in \Gamma_j,\ j = 1, \ldots, n,$$

where $b = (b_1, \ldots, b_n) > 0$, $p = (p_1, \ldots, p_n) > 0$, then

$$(1.3) \qquad |f(z)| \leqslant C_1 \qquad \text{when } z_j \in G_j,\ j = 1, \ldots, n,$$

Proof. We first give the proof in the case $n = 1$. Let q be any number satisfying

$$\varphi < \frac{\pi}{q} < \frac{\pi}{p}$$

and consider the function

$$f_\epsilon(z) = \exp\,[-\epsilon z^q] f(z)$$

for any fixed $\epsilon > 0$. $f_\epsilon(z)$ is an analytic function in G, and on Γ

$$|f_\epsilon(z)| \leqslant C_1 \exp\,[-\epsilon r^q \cos q\varphi/2] \leqslant C_1,$$

where $r = |z|$, since $q\varphi/2 < \pi/2$. For $z = re^{i\theta} \in G$,

$$|f_\epsilon(z)| \leqslant C \exp\,[br^p - \epsilon r^q \cos q\theta] \to 0 \qquad \text{as } r \to \infty$$

uniformly in θ, since $q > p$. Applying the maximum principle for analytic functions we conclude that $|f_\epsilon(z)| \leqslant C_1$ in G. Taking $\epsilon \to 0$ the inequality (1.3) follows.

We proceed to prove the theorem for any n in n steps. In the first step we fix $z_2, \ldots, z_n$ on $\Gamma_2, \ldots, \Gamma_n$ respectively, and apply the result for $n = 1$ to $f(z)$ as an analytic function in $z_1 \in G_1$. We then get

$$|f(z_1, z_2, \ldots, z_n)| \leqslant C_1 \tag{1.4}$$

for $z_1 \in G_1$, $z_2 \in \Gamma_2, \ldots, z_n \in \Gamma_n$. Next we consider $z_1, z_3, \ldots, z_n$ fixed on $G_1, \Gamma_3, \ldots, \Gamma_n$ respectively, and apply the result for $n = 1$ to $f(z)$ as a function of z_2, making use of (1.4). Proceeding in this way, step by step, we finally obtain (1.3).

We can apply Theorem 1 to

$$\frac{f(z)}{\prod_{j=1}^{n} (1 + z_j)^{h_j}} \qquad \text{and} \qquad \frac{f(z)}{(1 + z_1 + \cdots + z_n)^h}$$

where the h_j, h are $\geqslant 0$, provided $p > \frac{1}{2}$ or $p > 1$ respectively, and obtain the following results.

Theorem 1′. *Let* $p > \frac{1}{2}$. *If* (1.2) *is replaced by*

$$|f(z)| \leqslant C_1 \prod_{k=1}^{n} (1 + |z_k|)^{h_k} \qquad (z_j \in \Gamma_j,\ j = 1, \ldots, n), \tag{1.5}$$

then the assertion of the theorem is valid with (1.3) *replaced by*

$$|f(z)| \leqslant C_2 \prod_{k=1}^{n} (1 + |z_k|)^{h_k}. \tag{1.6}$$

Theorem 1″. *Let $p > 1$. If (1.2) is replaced by*

$$|f(z)| \leqslant C_1'(1 + |z|)^h \qquad (z_j \in \Gamma_j,\, j = 1, \ldots, n), \tag{1.5'}$$

then the assertion of the theorem is valid with (1.3) *replaced by*

$$|f(z)| \leqslant C_2'(1 + |z|)^h. \tag{1.6'}$$

Theorem 1 will be used in proving the next three theorems, which consider the question of how the growth behavior of $f(z)$ on $z = x$ influences the growth behavior of $f(z)$ "near" $z = x$.

Theorem 2. *Let $f(z)$ be an entire function satisfying*

$$|f(z)| \leqslant C_1 \exp\,[b|z|^p] \qquad \text{for } z \in C^n, \tag{1.7}$$

$$|f(x)| \leqslant C_2 \exp\,[a|x|^h] \qquad \text{for } x \in R^n, \tag{1.8}$$

where $b > 0$, $p > 0$, $a_j \neq 0$, and $0 \leqslant h_j \leqslant p_j$ $(j = 1, \ldots, n)$. Then there exists a region H_μ defined by

$$|y_j| \leqslant K_j(1 + |x_j|)^{\mu_j} \qquad (\mu_j = 1 - (p_j - h_j), \qquad j = 1, \ldots, n) \tag{1.9}$$

such that if $z = x + iy \in H_\mu$ then

$$|f(x + iy)| \leqslant C'\, C_3 \exp\,[a'|x|^h] \qquad (C_3 = \max\,(e^{|a|}C_1, C_2)), \tag{1.10}$$

where a' $(> a)$ can be taken arbitrarily close to a, K_j depends only on a_j', a_j, b_j, p_j, h_j, and C' depends only on a', a, b, p, h.

Proof. We first give the proof for $n = 1$. Consider the function

$$f_1(z) = f(z)\,\{\exp\,[-az^h]\}\,\{\exp\,[icz^p]\}, \tag{1.11}$$

where $c = b + |a|$. $f_1(z)$ is analytic if $0 \leqslant \arg z \leqslant \pi/2p$, and

$$|f_1(x)| \leqslant C_2 \qquad (x \geqslant 0),$$

$$|f_1(re^{i\pi/2p})| \leqslant C_1\,\{\exp\,[br^p + |a|r^h]\}\,\{\exp\,[-cr^p]\}$$

$$\leqslant \begin{cases} C_1 & \text{if } r \geqslant 1, \\ e^{|a|}C_1 & \text{if } r < 1. \end{cases}$$

Furthermore, if $|z| \geqslant 1$,

$$|f_1(z)| \leqslant C_1 \exp\,[d|z|^p] \qquad \text{if } 0 \leqslant \arg z \leqslant \pi/2p,$$

where $d = 2b + 2|a|$. Applying Theorem 1 we get the inequality $|f_1(z)| \leqslant C_3$ for $0 \leqslant \arg z \leqslant \pi/2p$. Hence

$$|f(re^{i\theta})| \leqslant C_3 \exp\,[ar^h \cos h\theta + cr^p \sin p\theta].$$

To prove (1.10) it thus suffices to show that for any $a' > a$

$$ar^h \cos h\theta + cr^p \sin p\theta \leqslant a'r^h \cos^h \theta.$$

This inequality can also be written in the form

$$r^{p-h} \leqslant \frac{a' \cos^h \theta - a \cos h\theta}{c \sin p\theta}. \tag{1.12}$$

If $h = p$ then (1.12) is satisfied for all $0 \leqslant \theta \leqslant \min(\theta_0, \pi/2p)$, where θ_0 is the smallest positive solution of the equation

$$a' \cos^h \theta_0 - a \cos h\theta_0 = c \sin p\theta_0.$$

If $h < p$, then the denominator in (1.12) must tend to zero uniformly with respect to r, as $r \to \infty$; the same is then true also of θ. Hence, for large r,

$$\sin p\theta = p\frac{y}{x}(1 + o(1)), \quad \cos^h \theta = 1 + o(1), \quad \cos h\theta = 1 + o(1),$$

$$r = x(1 + o(1)),$$

where $o(1) \to 0$ as $r \to \infty$, uniformly with respect to θ. Inequality (1.12) now takes the form

$$x^{p-h} \leqslant \frac{a' - a}{c}\frac{x}{y}(1 + o(1)).$$

Hence, if $0 \leqslant \theta \leqslant \pi/2p$ and

$$0 \leqslant y \leqslant \frac{a' - a}{c}(1 + o(1))x^\mu, \quad 0 \leqslant x < \infty, \quad r \geqslant r_0,$$

then (1.10) is satisfied.

If $x \geqslant 1$, $0 \leqslant y \leqslant K(1 + x)^\mu$ then the preceding calculations are valid (without any restriction on r) provided K is sufficiently small. If, however, $0 \leqslant x \leqslant 1$, $0 \leqslant y \leqslant K(1 + x)^\mu$, then from the inequality obtained above for $|f(re^{i\theta})|$ we directly infer that (1.10) holds provided C' is appropriately chosen.

A similar argument applies to the cases $y \leqslant 0$, $x \geqslant 0$; $y \geqslant 0$, $x \leqslant 0$, and $y \leqslant 0$, $x \leqslant 0$. The proof of (1.10) for $n = 1$ is thus completed.

The proof for any n can be given by applying the previous proof step by step as in the proof of Theorem 1. It can also be given more directly in the following way: for $y_1 \geqslant 0, \ldots, y_n \geqslant 0, x_1 \geqslant 0, \ldots, x_n \geqslant 0$ we take

$$f_1(z) = f(z)\,\{\exp[-a_1 z_1^{h_1} - \cdots - a_n z_n^{h_n}]\}\,\{\exp[ic_1 z_1^{p_1} + \cdots ic_n z_n^{p_n}]\},$$

where $c_j = b_j + |a_j|$, and proceed by the argument for $n = 1$. Then, instead of (1.12) we get n inequalities which, if satisfied, imply (1.10).

For $\epsilon_j y_j \geqslant 0$ ($\epsilon_j = \pm 1, j = 1, \ldots, n$), $x_1 \geqslant 0, \ldots, x_n \geqslant 0$, we replace c_j by $\epsilon_j c_j$. To obtain the inequality (1.10) for $\delta_j x_j \geqslant 0$ ($\delta_j = \pm 1, j = 1, \ldots, n$) we apply the result for $x_1 \geqslant 0, \ldots, x_n \geqslant 0$ to $f_0(z) \equiv f(\delta_1 z_1, \ldots, \delta_n z_n)$.

Theorem 2′. *Assume that* $p > \frac{1}{2}$. *Let* $f(z)$ *be an entire function satisfying* (1.7) *and assume further that, for some* $h_j > 0$,

$$(1.13) \qquad |f(x)| \leqslant C_2 \prod_{j=1}^{n} (1 + |x_j|)^{h_j} \qquad \text{for } x \in R^n.$$

Then, for any region H_μ *defined by* (1.9) *with* $\mu_j = 1 - p_j$, *if* $z = x + iy \in H_\mu$ *then*

$$(1.14) \qquad |f(x + iy)| \leqslant C_3 \prod_{j=1}^{n} K_j'(1 + |x_j|)^{h_j} \qquad (C_3 = \max(C_1, C_2)),$$

where K_j' *depends only on* b_j, p_j, h_j, K_j.

Proof. It will be enough to give the proof for $n = 1$. Instead of (1.11) we introduce the function

$$f_1(z) = f(z)(1 + z)^{-h} \exp[ibz^p],$$

noting that $z = -1$ does not lie in $0 \leqslant \arg z \leqslant \pi/2p$. We then obtain

$$(1.15) \qquad |f(re^{i\theta})| \leqslant A'C_3(1 + |z|)^h \exp[br^p \sin p\theta]$$

where A' is a constant depending on p, h. If $0 \leqslant y \leqslant A_1 x^{1-p}$ then $y/x \to 0$ as $x \to \infty$. Consequently

$$r^p = x^p(1 + o(1)), \qquad \sin p\theta = p\frac{y}{x}(1 + o(1)),$$

and therefore

$$br^p \sin p\theta \leqslant bpA_1(1 + o(1)).$$

(1.15) then yields the desired inequality

$$|f(re^{i\theta})| \leqslant A_1'C_3(1 + |z|)^h.$$

The proof, for $n = 1$, can now be easily completed by an argument used in the proof of Theorem 2.

Later on we shall need to apply Theorem 2′, for $p > 1$, in a slightly different form; namely, (1.13) is replaced by

$$(1.13') \qquad |f(x)| \leqslant C_2(1 + |x|)^h \qquad (h > 0)$$

and (1.14) is replaced by

$$(1.14') \qquad |f(x + iy)| \leqslant C_3K'(1 + |x|)^h.$$

We can prove this version for any n, in one step, by considering the function

$$f_1(z) = \frac{f(\delta_1 z_1, \ldots, \delta_n z_n)}{(1 + z_1 + \cdots + z_n)^h} \exp [i\epsilon_1 c_1 z_1^{p_1} + \cdots + i\epsilon_n c_n z_n^{p_n}]$$

(where $\epsilon_j = \pm 1$, $\delta_j = \pm 1$) in $\epsilon_j y_j \geqslant 0$, $x_j \geqslant 0$ ($j = 1, \ldots, n$). (Compare the end of the proof of Theorem 2.) We thus have the following:

Theorem 2″. *If in the assumptions of Theorem 2′ we replace* (1.13) *by* (1.13′) *and if* $p > 1$ *then the inequality* (1.14′) *holds.*

If $\mu > 0$ in Theorems 2, 2′, and if $n = 1$, then further information can be given on the growth behavior of $f(z)$ in the whole complex space.

Theorem 3. *Let $f(z)$ be an entire function of one complex variable z satisfying* (1.7) *and suppose that in some region H_μ defined by*

$$|y| \leqslant K(1 + |x|)^\mu \qquad (0 < \mu \leqslant 1), \tag{1.16}$$

$f(z)$ satisfies

$$|f(z)| \leqslant C_2 \exp [a|x|^h], \tag{1.17}$$

where $h \leqslant p$ and $a \neq 0$. Then

$$|f(x + iy)| \leqslant C_3 \exp [a|x|^h + b'|y|^{p/\mu}], \tag{1.18}$$

where $C_3 = \max (C_1, C_2)$ and b' depends only on a, b, h, p, μ, K.

Proof. The function

$$M(y) = \sup_{x \in R^1} |f(x + iy)| \exp [-a|x|^h] \tag{1.19}$$

is finite valued. Indeed, each line $y = \bar{y}$ intersects H_μ in a set $H_\mu(\bar{y})$ whose complement on $y = \bar{y}$ is a bounded interval $B(\bar{y})$, whereas on $H_\mu(\bar{y})$ (1.17) is satisfied. In view of (1.17), either $M(\bar{y}) \leqslant C_2$ or else there exists a point $\bar{x} \in B(\bar{y})$ such that the supremum on the right side of (1.19), for $y = \bar{y}$, is attained at $\bar{x}$. In the first case we obtain (1.18) for $y = \bar{y}$, $x \in R^1$ with any $b' > 0$. In the second case

$$|\bar{x}| \leqslant K_1^{1/\mu} |\bar{y}|^{1/\mu},$$

and therefore, assuming that $|\bar{y}| \geqslant 1$,

$$|\bar{z}| = |\bar{x} + i\bar{y}| = (\bar{x}^2 + \bar{y}^2)^{1/2} \leqslant K_1' |\bar{y}|^{1/\mu}.$$

Using (1.7) we obtain

$$\begin{aligned} M(\bar{y}) = |f(\bar{x} + i\bar{y})| \exp [-a|\bar{x}|^h] &\leqslant C_1 \exp [b|\bar{z}|^p + |a| K_1' |\bar{y}|^{h/\mu}] \\ &\leqslant C_1 \exp [b'|\bar{y}|^{p/\mu}], \end{aligned}$$

from which (1.18) follows. The assumption $|\bar{y}| \geqslant 1$ is satisfied if, in the definition of H_μ, $K \geqslant 1$. If $K < 1$ we apply the previous argument to the function $f_0(z) \equiv f(Kz)$. We then also obtain (1.18) (with a different b').

Theorem 3′. *Let $f(z)$ be an entire function of one complex variable z satisfying* (1.7) *and suppose that in some region H_μ, defined by* (1.16),

$$|f(z)| \leqslant C_2(1 + |x|^h). \tag{1.20}$$

Then

$$|f(z)| \leqslant C_3(1 + |x|^h) \exp [b'|y|^{p/\mu}] \qquad (C_3 = \max (C_1, C_2)), \tag{1.21}$$

where b' depends only on b, p, h, μ, K.

The proof is obtained by modifying the proof of Theorem 3, replacing in the definition of $M(y)$

$$\exp [a|x|^h] \quad \text{by} \quad (1 + |x|^h).$$

Theorems 3, 3′ *are not true if $n > 1$.* A counterexample is given by the function

$$f(z) = \exp [-(z_1^2 + 1)(z_2^2 + 1)].$$

From

$$|f(z)| = e^{a(z)} \quad \text{where} \quad a(z) = -(x_1^2 - y_1^2 + 1)(x_2^2 - y_2^2 + 1) + 4x_1x_2y_1y_2$$

one sees that if

$$y_1^2 < K(x_1^2 + 1), \qquad y_2^2 < K(x_2^2 + 1),$$

and K is sufficiently small, then $a(z) < 0$. Hence,

$$p = 4, \qquad h = 0, \qquad \mu = 1.$$

We shall now show that no inequality of the form

$$|f(z)| \leqslant A \exp [\alpha(x_1^4 + x_2^4)^\epsilon + \beta(y_1^4 + y_2^4)] \qquad (A > 0, \alpha > 0, \beta > 0)$$

can be satisfied if $\epsilon < 1$. Indeed, if such an inequality is satisfied, then, for some constant A_1,

$$-(x_1^2 - y_1^2 + 1)(x_2^2 - y_2^2 + 1) + 4x_1x_2y_1y_2 < \alpha(x_1^4 + x_2^4)^\epsilon + \beta(y_1^4 + y_2^4) + A_1,$$

which is clearly false for $y_1 = 0$, $x_2 = 0$, $|x_1| = |y_2|^\delta$, $y_2 \to \infty$, where δ is any number satisfying the inequalities

$$\delta > 1, \qquad \frac{1}{2\delta} + \frac{1}{2} > \epsilon.$$

In the next four theorems we estimate the derivatives of $f(z)$ on $z = x$ in terms of quantities appearing in growth assumptions on f.

Theorem 4. *If an entire function $f(z)$ satisfies*

(1.22) $$|f(x + iy)| \leqslant C \exp [a|x|^h + b|y|^\gamma] \qquad (h \leqslant \gamma),$$

then, for any q,

(1.23) $$|D^q f(x)| \leqslant CC_0 B^q q^{q(1-(1/\gamma))} \exp [a'|x|^h] \qquad \left(\frac{1}{\gamma} = \left(\frac{1}{\gamma_1}, \ldots, \frac{1}{\gamma_n}\right)\right),$$

where a' ($> a$) can be taken arbitrarily close to a, and B, C_0 depend on a', a, b, h, γ. If $a = 0$ then $a' = 0$ and B is any vector $> (1/e)(be\gamma)^{1/\gamma}$.

Recall that by $q^{\beta q}$ we understand the expression $q_1^{\beta_1 q_1} \cdots q_n^{\beta_n q_n}$.

Proof. Using the n-dimensional Cauchy formula

(1.24) $$D^q f(x) = \frac{q!}{(2\pi i)^n} \int_\Gamma \frac{f(\zeta)d\zeta}{(\zeta - x)^{n+q}},$$

where Γ is the product of n circles $|\zeta_j - x_j| = R_j$ ($j = 1, \ldots, n$), we get

$$|D^q f(x)| \leqslant C \frac{q!}{R^q} \{\exp [bR^\gamma]\} \{\exp [a|x'|^h]\}$$

for some x' satisfying $|x'_j - x_j| \leqslant R_j$. Noting that

(1.25) $$R = \left(\frac{q}{b\gamma}\right)^{1/\gamma} \quad \text{minimizes} \quad \frac{\exp [bR^\gamma]}{R^q},$$

and that, with this choice of R,

$$R^h \leqslant A_1 q \qquad (A_1 \text{ a constant})$$

(since $h \leqslant \gamma$), we obtain the inequality (1.23) for any $a' > a$. The assertion for the case $a = 0$ follows easily.

Theorem 5. *Let $f(z)$ be an entire function satisfying, for each integer $k \geqslant 0$,*

(1.26) $$(1 + |x|^k)|f(x + iy)| \leqslant C_k \exp [b|y|^\gamma] \qquad (\gamma > 1).$$

Then, for any q,

(1.27) $$(1 + |x|^k)|D^q f(x)| \leqslant C'_k B^q q^{q(1-(1/\gamma))}$$

for any $B > (1/e)(be\gamma)^{1/\gamma}$, where the C'_k depend only on b, γ, B, k and C_j, $0 \leqslant j \leqslant k$.

Proof. Let $\{B_k\}$ be a monotone-increasing sequence such that

$$B_0 > \frac{1}{e}(be\gamma)^{1/\gamma}, \qquad B_k < B \qquad (k = 0, 1, 2, \ldots).$$

We shall prove by induction on k that

$$(1.28) \qquad (1 + |x|^k)|D^q f(x)| \leqslant C'_k B_k^q q^{q(1-(1/\gamma))} \qquad (0 \leqslant |q| < \infty).$$

Assuming then that (1.28) holds with k replaced by k', for all $k' < k$, we shall prove (1.28) (the case $k = 0$ follows from Theorem 4).

If we prove that

$$(1.29) \qquad |x_1^k D^q f(x)| \leqslant C''_k B_k^q q^{q(1-(1/\gamma))} \qquad (0 \leqslant |q| < \infty),$$

then, since similar inequalities can then be proved with x_j instead of x_1, and since (1.29) with $k = 0$ is already known, (1.28) follows. To prove (1.29) we use the identity

$$(1.30) \qquad x_1^k D^q f = D^q(x_1^k f) - \sum_{0<\alpha_1 \leqslant k} \binom{q}{\alpha} D^{\alpha_1} x_1^k \cdot D^{q-\alpha} f$$

$$(\alpha = (\alpha_1, 0, \ldots, 0)).$$

By applying Theorem 4 to $x_1^k f$ we find that

$$(1.31) \qquad |D^q(x_1^k f(x))| \leqslant A_k B_{k-1}^q q^{q(1-(1/\gamma))}.$$

Noting that the number of nonzero terms in the sum of (1.30) is at most k, we can bound this sum by

$$A'_k B_{k-1}^q \operatorname*{l.u.b.}_{0<|\alpha|\leqslant k} q^\alpha (q - \alpha)^{(q-\alpha)(1-(1/\gamma))}.$$

The expression inside the l.u.b. is not larger than

$$q^\alpha q^{q(1-(1/\gamma))} \leqslant A'_{k\delta}(1 + \delta)^q q^{q(1-(1/\gamma))}$$

for any $\delta > 0$, where $A'_{k\delta}$ depends only on k and δ. Taking

$$1 + \delta = B_k/B_{k-1}$$

we obtain the bound

$$A''_k B_k^q q^{q(1-(1/\gamma))}$$

for the sum which appears in (1.30). Combining this estimate with (1.31) and taking $C''_k = A_k + A''_k$, the proof of (1.29) is completed.

We shall now prove an analogue of Theorem 4 for $\mu \leqslant 0$.

Theorem 6. *Let $f(z)$ be an analytic function in a region H_μ defined by*

$$(1.32) \qquad |y_j| \leqslant K_j(1 + |x_j|)^{\mu_j} \qquad (\mu_j \leqslant 0, j = 1, \ldots, n),$$

which satisfies the inequality

$$(1.33) \qquad |f(x + iy)| \leqslant C \exp [a|x|^h] \qquad (a_j \neq 0, j = 1, \ldots, n).$$

Then, for any q,

$$|D^q f(x)| \leqslant CC'B^q q^{q(1-(\mu/h))} \exp [a'|x|^h], \tag{1.34}$$

where a' ($> a$) can be taken arbitrarily close to a, and C', B depend only on K, μ, h, a, a'.

Proof. We first give the proof for the case $n = 1$. We shall denote by C_j and B_i constants which depend on K, μ, h, a, a' but are independent of C. If $|\zeta - x| = R$ and $R = \delta(1 + |x|)^\mu$, where δ is sufficiently small (depending on K), then ζ is contained in H_μ. Using Cauchy's formula (1.24) for $n = 1$ we then obtain

$$|D^q f(x)| \leqslant CC_1 \frac{q!}{R^q} \exp [a''|x|^h],$$

where a'' is arbitrarily close to a. If $|x| \geqslant 1$, then $R \geqslant \delta_1 |x|^\mu$, and we get

$$|D^q f(x)| \leqslant q! CC_2 B_1^q |x|^{-\mu q} \exp [a''|x|^h] \leqslant q! CC_3 B_2^q q^{-\mu q/h} \exp [a'|x|^h] \tag{1.35}$$

for any $a' > a''$, where the last inequality is obtained by maximizing

$$|x|^{-\mu q} \exp [-\epsilon |x|^h],$$

where $\epsilon = a' - a''$ (compare (1.25)). From (1.35) we obtain (1.34) provided $B > B_2 e$.

If $|x| < 1$, then by Cauchy's inequalities

$$|D^q f(x)| \leqslant C_4 C \frac{q!}{\rho^q}$$

for any $\rho > 0$ such that the disc of radius ρ and center x is contained in H_μ. Thus in this case too (1.34) is satisfied.

To prove the theorem for any n, we apply the previous proof to $f(z)$ when $z_2, \ldots, z_n$ are fixed numbers satisfying (1.32). We obtain

$$|D_1^{q_1} f(x_1, z_2, \ldots, z_n)| \leqslant \hat{C} C' B^{q_1} q_1^{q_1(1-(\mu_1/h_1))} \exp [a_1'|x_1|^{h_1}],$$

where
$$\hat{C} = C \exp [a_2|x_2|^{h_2} + \cdots + a_n|x_n|^{h_n}].$$

We next fix $x_1, z_3, \ldots, z_n$ and apply the proof for $n = 1$ to

$$D_1^{q_1} f(x_1, z_2, z_3, \ldots, z_n)$$

as a function of z_2. Proceeding in this way step by step, we finally obtain (1.34) (with a different C').

Theorem 6′. *Let $f(z)$ be an analytic function in the domain* (1.32) *and let it satisfy the inequality*

$$|f(x + iy)| \leqslant C \prod_{j=1}^{n} (1 + |x_j|^{h_j}). \tag{1.36}$$

Then, for any q,

$$(1.37)\qquad |D^q f(x)| \leqslant C'CB^q q^q \prod_{j=1}^{n} (1 + |x_j|^{h_j - \mu_j q_j}),$$

where B and C′ depend only on K, μ, h.

Proof. The proof is similar to that of Theorem 6. Thus, if $n = 1$ and $|x| \geqslant 1$ we get

$$|D^q f(x)| \leq CC_1 B_1^q q^q \frac{(1 + |x|^h)}{R^q} \leqslant CC_2 B^q q!(1 + |x|^{h-\mu q})$$

if we take $R = \delta(1 + |x|)^\mu$. Further details are left to the reader.

We conclude this section with an existence theorem concerning entire functions having a prescribed growth behavior. This theorem will be used in Sec. 5 in proving that some very important W spaces are nontrivial spaces.

Theorem 7. *Let* $1 < p < 2$. *There exists an entire function* $f(z) \not\equiv 0$ *which satisfies the inequality*

$$(1.38)\qquad |f(x + iy)| \leqslant C \exp[-a|x|^p + b|y|^p]$$

for some constants $C > 0$, $a > 0$, $b > 0$.

From Theorem 15, in Sec. 4 below, it follows that the Fourier transform $\tilde{f}$ of $f(x)$ can be extended into an entire function $\tilde{f}(z)$ which satisfies the inequality

$$|\tilde{f}(x + iy)| \leqslant C' \exp[-a'|x|^{p'} + b'|y|^{p'}] \qquad (p' = p/(p - 1))$$

for some constants $C' > 0$, $a' > 0$, $b' > 0$. Hence, *Theorem 7 holds also for* $2 < p < \infty$.

The function

$$f(z) \equiv \exp[-z^2] \equiv \{\exp[-z_1^2]\} \cdots \{\exp[-z_n^2]\}$$

is of order $p = 2$ and

$$|f(x)| \leqslant \exp[-|x|^2].$$

Applying Theorems 2, 3 we conclude that f satisfies (1.38) for $p = 2$. Thus, *Theorem 7 holds also for* $p = 2$. The theorem is false if $p < 1$ ($p = 1$), as follows by Theorem 1 (Theorems 2, 1).

Proof. If $f_0(z_1)$ satisfies (1.38) for $n = 1$, then $f(z) = f_0(z_1) \cdots f_0(z_n)$ satisfies (1.38). Hence it suffices to prove the theorem in the case $n = 1$.

We shall prove that the entire function

$$\varphi(z) = \prod_{m=1}^{\infty}\left(1 - \frac{z}{m^{\rho}}\right) \qquad \left(\rho = \frac{2}{p} > 1\right), \tag{1.39}$$

(the infinite product is convergent since $\rho > 1$) satisfies

$$|\varphi(z)| \leqslant C \exp\,[\beta|z|^{1/\rho}] \qquad (\beta > 0), \tag{1.40}$$

$$|\varphi(x)| \leqslant C \exp\,[\alpha|x|^{1/\rho}] \qquad (\alpha < 0, x > 0). \tag{1.41}$$

It would then follow that the function $f_0(z) = \varphi(z^2)$ satisfies

$$|f_0(z)| \leqslant C \exp\,[\beta|z|^{p}], \qquad |f_0(x)| \leqslant C \exp\,[\alpha|x|^{p}]$$

for all complex z and real x. Applying Theorems 2, 3 we obtain the desired inequality.

If $z = re^{i\theta}$ and $\epsilon \leqslant \theta \leqslant \pi$, for some $\epsilon > 0$, then each factor in $\varphi(z)$ is $\neq 0$ and, since $\rho > 1$, the infinite product of $\varphi(z)$ is convergent to a nonzero value. We therefore obtain a single-valued function $\log \varphi(z)$ by fixing one branch of the logarithm function. Take this branch to be the principal branch $\operatorname{Log} \varphi(z)$ ($\operatorname{Log} \varphi(-r)$ is real). We have

$$\operatorname{Log} \varphi(z) = \log|\varphi(z)| + i \arg \varphi(z). \tag{1.42}$$

Denoting by $n(t)$ a monotone-increasing step function whose jumps occur at $t = n^{\rho}$ $(n = 1, 2, \dots)$ and are equal to 1 in magnitude, and normalizing $n(t)$ by $n(0) = 0$, we have

$$\begin{aligned}\operatorname{Log} \varphi(re^{i\theta}) &= \sum_{n=1}^{\infty} \operatorname{Log}\left(1 - \frac{re^{i\theta}}{n^{\rho}}\right) = \int_{1-0}^{\infty} \operatorname{Log}\left(1 - \frac{re^{i\theta}}{t}\right) dn(t) \\ &= \left[n(t) \operatorname{Log}\left(1 - \frac{re^{i\theta}}{t}\right)\right]_{1-0}^{\infty} - re^{i\theta} \int_{1-0}^{\infty} \frac{n(t)\,dt}{t(t - re^{i\theta})}.\end{aligned} \tag{1.43}$$

Since

$$n(t) = t^{1/\rho} - \sigma(t) \qquad (0 \leqslant \sigma(t) < 1), \tag{1.44}$$

the expression in brackets on the right side of (1.43) is equal to zero.

Substituting $t = ur$ into the second integral in (1.43) and using (1.44) we get

$$\begin{aligned}\operatorname{Log} \varphi(re^{i\theta}) &= -r^{1/\rho}e^{i\theta} \int_{1/r}^{\infty} \frac{u^{(1/\rho)-1}}{u - e^{i\theta}}\,du + e^{i\theta} \int_{1/r}^{\infty} \frac{\sigma(ur)}{u(u - e^{i\theta})}\,du \\ &\equiv I_1 + I_2.\end{aligned} \tag{1.45}$$

It is obvious that

$$|I_2| \leqslant C_1 \log r \qquad (r \geqslant 1). \tag{1.46}$$

By the method of residues one finds that

$$(1.47)\qquad I \equiv \int_0^\infty \frac{u^{(1/\rho)-1}}{u - e^{i\theta}}\,du = -e^{-i\theta}\frac{\pi}{\sin(\pi/\rho)}\,e^{i(\theta-\pi)/\rho}.$$

Substituting this into I_1 and using (1.46), we obtain from (1.45) and (1.42),

$$(1.48)\qquad \begin{aligned}\log|\varphi(re^{i\theta})| &= \operatorname{Re}\{-r^{1/\rho}e^{i\theta}I\} + 0(\log r)\\ &= r^{1/\rho}\frac{\pi}{\sin(\pi/\rho)}\cos\frac{\pi-\theta}{\rho} + 0(\log r).\end{aligned}$$

Observing that

$$(1.49)\qquad |\varphi(r)| \equiv |\varphi(|z|)| \leqslant |\varphi(z)| \leqslant \varphi(-|z|) \equiv \varphi(-r),$$

we obtain (1.40) by taking $\theta = \pi$ in (1.48), and (1.41) by taking in (1.48) any fixed θ such that $\cos(\pi - \theta)/\rho < 0$ (it is possible to choose such a θ since $\rho < 2$) and using the first inequality of (1.49).

2. Definition of *W* Spaces

In this section we introduce three types of spaces: $W_{M,a}$ (and their union space W_M), $W^{\Omega,b}$ (and their union space W^Ω) and $W^{\Omega,b}_{M,a}$ (and their union space W^Ω_M). All these spaces will be referred to as W spaces.

Let $\mu_j(\xi_j)$ $(j = 1, \ldots, n)$ be a continuous monotone-increasing function for $0 \leqslant \xi_j < \infty$ $(j = 1, \ldots, n)$ and let $\mu_j(0) = 0$. The function

$$M_j(x_j) = \int_0^{x_j} \mu_j(\xi_j)\,d\xi_j \qquad (x_j \geqslant 0)$$

is then a convex function and

$$(2.1)\qquad M_j(x_j) + M_j(x_j') \leqslant M_j(x_j + x_j').$$

We define

$$M_j(-x_j) = M_j(x_j) \qquad \text{if } x_j > 0$$

and set

$$\mu(\xi) = (\mu_1(\xi_1), \ldots, \mu_n(\xi_n)), \qquad M(x) = (M_1(x_1), \ldots, (M_n(x_n)).$$

Note that if $\mu(\infty) = \infty$ (i.e., $\mu_j(\infty) = \infty$ for $j = 1, \ldots, n$) then $M_j(x_j)/|x_j| \to \infty$ as $|x_j| \to \infty$ $(j = 1, \ldots, n)$.

For any $a > 0$, we denote by $W_{M,a}$ the space of all C^∞ complex-valued functions $\varphi(x)$ on R^n which satisfy the inequalities

$$(2.2)\qquad |D^q\varphi(x)| \leqslant C_{qa'}\exp[-M(a'x)] \qquad (0 \leqslant |q| < \infty)$$

for any $a' < a$, where the $C_{qa'}$ are constants (depending on φ) and where

$$\exp[-M(a'x)] = \exp[-M_1(a_1'x_1) - \cdots - M_n(a_n'x_n)].$$

$W_{M,a}$ is a linear space and we introduce in it a topology in terms of the sequence of norms

$$\|\varphi\|_p = \sup_{|\alpha| \leqslant p} \sup_{x \in R^n} M_p(x)|D^\alpha \varphi(x)|, \tag{2.3}$$

where

$$M_p(x) = \exp\left\{M\left[a\left(1 - \frac{1}{p}\right)x\right]\right\} \qquad (p = 1, 2, \ldots). \tag{2.4}$$

Recalling the results of Chap. 2, Sec. 2, and observing that the M_p satisfy the condition (P), we conclude that $W_{M,a}$ is a perfect space. We also have the following (see Lemma 3, Chap. 2, Sec. 2):

Lemma 1. *Let* $\{\varphi_m\} \subset W_{M,a}$. *If* $\varphi_m(x) \to 0$ *properly and if* $\|\varphi_m\|_p \leqslant C_p$ $(p = 1, 2, \ldots)$ *then* $\varphi_m \to 0$ *in* $W_{M,a}$.

We denote by W_M the union space of the spaces $W_{M,a}$ when $a > 0$ and a varies in R^n. W_M is evidently also the countable union space of any sequence $\{W_{M,a^m}\}$ where each component of $\{a^m\}$ decreases monotonically to 0, as $m \to \infty$. Thus, W_M is a fundamental space. By definition, a sequence $\{\varphi_m\}$ is convergent to 0 in W_M if and only if $\{\varphi_m\} \subset W_{M,a}$ for some $a > 0$ and $\varphi_m \to 0$ in $W_{M,a}$. But then $\varphi_m(x) \to 0$ properly and $\{\varphi_m\}$ is a bounded set, i.e.,

$$|D^\alpha \varphi_m(x)| \leqslant C_\alpha \exp\left[M(a'x)\right] \qquad (0 \leqslant |\alpha| < \infty) \tag{2.5}$$

for some constants C_α, a'. From Lemma 1 it follows that, conversely, if $\varphi_m(x) \to 0$ properly and (2.5) is satisfied for some constants C_α, a', then $\varphi_m \to 0$ in some $W_{M,a}$, hence also in W_M.

If $M(\gamma x) \leqslant M'(\gamma' x)$ (i.e., if $M_j(\gamma_j x_j) \leqslant M'_j(\gamma'_j x_j)$ $(j = 1, \ldots, n)$) then $W_{M'} \subset W_M$ and the topology of $W_{M'}$ is stronger than the topology induced upon it by W_M. If

$$M(\gamma x) \leqslant M'(\gamma' x) \leqslant M(\gamma'' x),$$

then we say that $M(x)$ and $M'(x)$ are *equivalent*. The sets W_M, $W_{M'}$ then coincide and the topologies are equivalent.

We next introduce the spaces $W^{\Omega,b}$. Let $\omega_j(\eta_j)$ be a continuous monotone-increasing function in $0 \leqslant \eta_j < \infty$, such that $\omega_j(0) = 0$, and set

$$\Omega_j(y_j) = \int_0^{y_j} \omega_j(\eta_j)\, d\eta_j \qquad (y_j \geqslant 0).$$

Then Ω_j is a convex function and

$$\Omega_j(y_j) + \Omega_j(y'_j) \leqslant \Omega_j(y_j + y'_j). \tag{2.6}$$

We define

$$\Omega_j(-y_j) = \Omega_j(y_j) \qquad \text{if } y_j > 0,$$

and set

$$\omega(\eta) = (\omega_1(\eta_1), \ldots, \omega_n(\eta_n)), \qquad \Omega(y) = (\Omega_1(y_1), \ldots, \Omega_n(y_n)).$$

Note that if $\omega(\infty) = \infty$ (i.e., $\omega_j(\infty) = \infty$ for $j = 1, \ldots, n$) then $\Omega_j(y_j)/|y_j| \to \infty$ as $|y_j| \to \infty$ $(j = 1, \ldots, n)$.

For any $b > 0$ we denote by $W^{\Omega,b}$ the set of all complex-valued functions $\varphi(x)$ on R^n which can be extended into entire functions $\varphi(z)$, on C^n, satisfying

$$(2.7) \qquad (1 + |z|^k)|\varphi(z)| \leqslant C_{kb'} \exp[\Omega(b'y)] \qquad (0 \leqslant k < \infty)$$

for any $b' > b$, where the $C_{kb'}$ are constants depending on φ and where

$$\exp[\Omega(b'y)] = \exp[\Omega_1(b_1'y_1) + \cdots + \Omega_n(b_n'y_n)].$$

$W^{\Omega,b}$ is a linear space, and we introduce in it a topology in terms of the sequence of norms

$$(2.8) \qquad \|\varphi\|_p = \sup_{z \in C^n} M_p(z)|\varphi(z)|,$$

where

$$(2.9) \quad M_p(z) = (1 + |z|)^p \exp\left\{-\Omega\left[b\left(1 + \frac{1}{p}\right)y\right]\right\} \qquad (p = 1, 2, \ldots).$$

$W^{\Omega,b}$ is a $Z\{M_p\}$ space satisfying the condition (P_0), and from the results of Chap. 2, Sec. 3 we conclude that $W^{\Omega,b}$ is a perfect space. By Lemma 4, Sec. 3, Chap. 2 we also have the following: $\varphi_m \to 0$ in $W^{\Omega,b}$ if and only if $\varphi_m(x) \to 0$ for any $x \in R^n$ and

$$(2.10) \qquad (1 + |z|^k)|\varphi_m(z)| \leqslant C_{kb'} \exp[\Omega(b'y)] \qquad (0 \leqslant k < \infty)$$

for any $b' > b$.

We denote by W^Ω the union space of the spaces $W^{\Omega,b}$ $(0 < b \in R^n)$. W^Ω is also the countable union space of any sequence $\{W^{\Omega,b^m}\}$ where each component of $\{b^m\}$ increases monotonically to ∞ as $m \to \infty$. W^Ω is therefore a fundamental space. A sequence $\{\varphi_m\}$ is convergent to zero in W^Ω if and only if $\varphi_m(x) \to 0$ for any $x \in R^n$ and (2.10) is satisfied for some b', $C_{kb'}$ independent of m. The concept of equivalence for functions $\Omega(y)$, $\Omega'(y)$ is defined analogously to the definition for $M(x)$, $M'(x)$.

We finally introduce the spaces of the third type, namely the spaces $W_{M,a}^{\Omega,b}$. M and Ω are the functions previously defined and $a > 0$, $b > 0$. We denote by $W_{M,a}^{\Omega,b}$ the space of all the functions in $W^{\Omega,b}$ which satisfy the inequality

$$(2.11) \qquad |\varphi(x + iy)| \leqslant C_{a'b'} \exp[-M(a'x) + \Omega(b'y)]$$

for any $a' < a$, $b' > b$. A topology is defined in terms of the norms

$$\|\varphi\|_p = \sup_{z \in C^n} M_p(z)|\varphi(z)| \qquad (p = 1, 2, \ldots), \tag{2.12}$$

where

$$M_p(z) = \exp\left\{M\left[a\left(1 - \frac{1}{p}\right)x\right] - \Omega\left[b\left(1 + \frac{1}{p}\right)y\right]\right\}. \tag{2.13}$$

$W_{M,a}^{\Omega,b}$ is then a $Z\{M_p\}$ space and, since the condition (P_0) is satisfied, it is a perfect space. We also have the following: a sequence $\{\varphi_m\}$ is convergent to 0 in $W_{M,a}^{\Omega,b}$ if and only if $\varphi_m(x) \to 0$ for any $x \in R^n$ and

$$|\varphi(x + iy)| \leqslant C_p \exp\left\{-M\left[a\left(1 - \frac{1}{p}\right)x\right] + \Omega\left[b\left(1 + \frac{1}{p}\right)y\right]\right\} \qquad (p = 1, 2, \ldots). \tag{2.14}$$

We denote by W_M^Ω the union space of the spaces $W_{M,a}^{\Omega,b}$ $(0 < a \in R^n$, $0 < b \in R^n)$. W_M^Ω is also the countable union space of any sequence $\{W_{M,a^m}^{\Omega,b^m}\}$ where each component of $\{b^m\}$ tends monotonically to ∞ and each component of $\{a^m\}$ tends monotonically to 0. It is clear that $\varphi_j \to 0$ in W_M^Ω if and only if $\varphi_j \to 0$ in some $W_{M,a}^{\Omega,b}$, or, if and only if $\varphi_j(x) \to 0$ for any $x \in R^n$ and (2.14) is satisfied for all $\varphi = \varphi_j$, where a, b, and the C_p are constants independent of j.

We recall the following property shared by all the fundamental spaces (see Chap. 2, Sec. 1): a linear functional f is continuous on any W space if and only if $f \cdot \varphi_m \to 0$ whenever $\varphi_m \to 0$ in that W space.

3. Operators in W Spaces

Theorem 8. *Differentiation and multiplication by polynomials are bounded operators in all the W spaces.*

Proof. We first give the proof for the space $W_{M,a}^{\Omega,b}$. Let $\varphi \in W_{M,a}^{\Omega,b}$. Using Cauchy's formula we obtain

$$|D_1\varphi(z)| \leqslant \frac{1}{R_1} \max_{|\zeta_1 - z_1| \leqslant R_1} |\varphi(\zeta_1, z_2, \ldots, z_n)|,$$

for any $R_1 > 0$. Hence, if $y_1 \geqslant 0$, $x_1 \geqslant 0$,

$$|D_1\varphi(z)| \leqslant C_{a'b'R} \exp\{-M[a'(x - R)] + \Omega[b'(y + R)]\}$$

for any $a' < a$, $b' > b$, where $R = (R_1, 0, \ldots, 0)$. Since

$$M_j[a_j'(x_j - R_j)] \geqslant C_{R_j}' + M_j[(a_j' - R_j)x_j], \tag{3.1}$$

$$\Omega_j[b_j'(y_j + R_j)] \leqslant C_{R_j}'' + \Omega_j[(b_j' + R_j)y_j], \tag{3.2}$$

for all positive x_j, y_j, we obtain, upon taking R_1 arbitrarily small,

$$|D_1\varphi(z)| \leqslant C_{a''b''} \exp[-M(a''x) + \Omega(b''y)]$$

for any $a'' < a'$, $b'' > b'$. A similar inequality can be proved in the same way if (x_1, y_1) belongs to any of the other three quadrants. Since a', b' can be taken arbitrarily close to a and b respectively, the same is true of a'', b''. This completes the proof that $D_1\varphi \in W_{M,a}^{\Omega,b}$. The proof that $D_j\varphi \in W_{M,a}^{\Omega,b}$ $(j = 2, \ldots, n)$ is similar, and the proof that any $D^\alpha\varphi$ belongs to $W_{M,a}^{\Omega,b}$ follows by successively applying the result for $|\alpha| = 1$.

From the inequalities appearing in the above proof it follows that if φ varies in a bounded set B of $W_{M,a}^{\Omega,b}$ (i.e., $\|\varphi\|_p \leqslant C_p$ for $p = 1, 2, \ldots$ and $\varphi \in B$) then the set $\{D_j\varphi;\ \varphi \in B\}$ is also a bounded set in $W_{M,a}^{\Omega,b}$. Hence the operators D_j, and consequently also the D^α for any α, are bounded operators.

We next prove that multiplication by any polynomial in z is a bounded operator in $W_{M,a}^{\Omega,b}$. It suffices to consider the polynomials z_j $(j = 1, \ldots, n)$.

Let $\varphi \in W_{M,a}^{\Omega,b}$. Then,

$$|\varphi(z)| \leqslant C_{a'b'} \exp[-M(a'x) + \Omega(b'y)]$$

for any $a' < a$, $b' > b$. Since

$$|x_j| \leqslant C_\epsilon \exp[M(\epsilon x)], \qquad |y_j| \leqslant C_\epsilon \exp[\Omega(\epsilon y)]$$

for any $\epsilon > 0$ and for some constant C_ϵ,

$$\begin{aligned}|z_j\varphi(z)| &\leqslant C_{a'b'\epsilon} \exp[M(\epsilon x) - M(a'x) + \Omega(\epsilon y) + \Omega(b'y)] \\ &\leqslant C_{a'b'\epsilon} \exp\{-M[(a' - \epsilon)x] + \Omega[(b' + \epsilon)y]\},\end{aligned}$$

where use has been made of (2.1) and (2.6). Since $\epsilon > 0$ can be taken arbitrarily small and a', b' can be taken arbitrarily close to a and b respectively, we get

$$|z_j\varphi(z)| \leqslant C'_{a''b''} \exp[-M(a''x) + \Omega(b''y)]$$

for any $a'' < a$, $b'' > b$. Hence $z_j\varphi(z) \in W_{M,a}^{\Omega,b}$. From the inequalities obtained in the proof it also follows that if φ varies in a bounded set of $W_{M,a}^{\Omega,b}$, then the same is true of $z_j\varphi$. This completes the proof.

We next prove that $\varphi \to D_j\varphi$ is a bounded operator in $W^{\Omega,b}$. Using Cauchy's formula we obtain

$$(1 + |z|^k)|D_j\varphi(z)| \leqslant \frac{1}{R_j}\left[\max_{|\zeta_j - z_j| = R_j} \left(\frac{1 + |z|^k}{1 + |\zeta|^k}\right)\right]\left[\max_{|\zeta_j - z_j| = R_j} (1 + |\zeta|^k)|\varphi(\zeta)|\right],$$

where $\zeta = (z_1, \ldots, z_{j-1}, \zeta_j, z_{j+1}, \ldots, z_n)$. The second factor on the right side is bounded by some constant C_k. The last factor is bounded by

$$C_{kb'} \exp \{\Omega[b'(y + R)]\} \qquad (R = (0, \ldots, 0, R_j, 0, \ldots, 0)) \tag{3.3}$$

if $y_j \geqslant 0$. Using (3.2) and taking R_j arbitrarily small we find that the function in (3.3) is bounded by

$$C'_{kb''} \exp [\Omega(b''y)]$$

for any $b'' > b'$. A similar inequality is obtained in case $y_j < 0$. It follows that $D_j\varphi \in W^{\Omega,b}$. Furthermore, from the inequalities derived one sees that D_j maps bounded sets of $W^{\Omega,b}$ into bounded sets of $W^{\Omega,b}$.

The proof that multiplication by a polynomial is a bounded operator in $W^{\Omega,b}$ is obvious.

The proof that D^α is a bounded operator in $W_{M,a}$ is trivial. Thus, in order to complete the proof of the theorem it remains to show that multiplication by each x_j is a bounded operator in $W_{M,a}$. This follows from the identity

$$D^\alpha(x_j\varphi(x)) = x_jD^\alpha\varphi(x) + \alpha_jD^{\alpha'}\varphi(x)$$

(where $\alpha' = (\alpha_1, \ldots, \alpha_{j-1}, \alpha_j - 1, \alpha_{j+1}, \ldots, \alpha_n)$), and from the inequality

$$|x_j| \leqslant C_\epsilon \exp [M(\epsilon x)] \qquad (\epsilon > 0).$$

Having proved Theorem 8 we see that the properties (Φ_1) through (Φ_4) of Chap. 4, Sec. 1 are satisfied for any W space. It is also easily verified that W spaces admit a differentiable translation. Hence *all the results of Chap. 4 hold for W spaces.*

Definition. α is a *multiplier* from a test space Φ into a test space Φ_1 if $\varphi \to \alpha\varphi$ is a continuous mapping from Φ into Φ_1. We next define $\varphi \cdot f$ (or αf) for $f \in \Phi'_1$ by $(\alpha \cdot f, \varphi) = (f, \alpha\varphi)$ $(\varphi \in \Phi)$. $f \to \alpha f$ is a continuous mapping from Φ'_1 into Φ'. We say that α is a *multiplier* from Φ'_1 into Φ'.

Theorem 9. *Let $f(z)$ be an entire function satisfying*

$$|f(z)| \leqslant C\,(1 + |x|^h) \exp [\Omega(b_0y)] \qquad (b_0 > 0, h > 0). \tag{3.4}$$

Then f is a multiplier from $W^{\Omega,b}$ into $W^{\Omega,b+b_0}$.

Proof. Let $\varphi \in W^{\Omega,b}$. Then

$$(1 + |z|^k)|\varphi(z)| \leqslant C_{kb'} \exp [\Omega(b'y)] \qquad (0 \leqslant k < \infty)$$

for any $b' > b$. Using (2.6) it follows that for any $k \geqslant h$

$$(1 + |z|^{k-h})|f(z)\varphi(z)| \leqslant C'C_{kb'} \exp [\Omega((b' + b_0)y)],$$

thus proving that $f\varphi \in W^{\Omega,b_0+b}$. From the proof it follows that if φ varies in bounded sets of $W^{\Omega,b}$ then $f\varphi$ varies in bounded sets of $W^{\Omega,b+b_0}$. This completes the proof of the theorem.

Theorem 10. *Let $f(z)$ be an entire function satisfying*

$$(3.5) \qquad |f(z)| \leqslant C \exp\,[M(a_0x) + \Omega(b_0y)] \qquad (a_0 > 0,\, b_0 > 0).$$

Then f is a multiplier from $W_{M,a}^{\Omega,b}$ into $W_{M,a-a_0}^{\Omega,b+b_0}$, provided $a > a_0$.

The proof is similar to that of Theorem 9, and is left to the reader.

Corollary 1. *If $f(z)$ is an entire function satisfying, for any $\epsilon > 0$,*

$$|f(z)| \leqslant C_\epsilon \exp\,[\Omega(\epsilon y)]$$

(or

$$|f(z)| \leqslant C_\epsilon \exp\,[M(\epsilon x) + \Omega(\epsilon y)]),$$

then f is a multiplier in $W^{\Omega,b}$ (or in $W_{M,a}^{\Omega,b}$).

Corollary 2. *For any $\sigma \in R^n$, $f(z) = e^{i\sigma\cdot z}$ is a multiplier in $W^{\Omega,b}$ and in $W_{M,a}^{\Omega,b}$, provided $\omega(\infty) = \infty$.*

Indeed,

$$|f(z)| \leqslant e^{|\sigma\cdot y|} \leqslant C_\epsilon \exp\,[\Omega(\epsilon y)]$$

for any $\epsilon > 0$, since $\Omega_j(y_j)/|y_j| \to \infty$ as $|y_j| \to \infty$ $(j = 1, \ldots, n)$.

The reader will easily verify that, *for any $\sigma \in R^n$, $f(x) = e^{i\sigma\cdot x}$ is a multiplier in $W_{M,a}$ provided $\mu(\infty) = \infty$.*

4. Fourier Transforms of W Spaces

In this and in the following section we shall assume that $\mu(\infty) = \infty$ and $\omega(\infty) = \infty$ so that, for $j = 1, \ldots, n$,

$$(4.1) \qquad \frac{M_j(x_j)}{|x_j|} \to \infty \quad \text{as } |x_j| \to \infty, \qquad \frac{\Omega_j(y_j)}{|y_j|} \to \infty \quad \text{as } |y_j| \to \infty.$$

If μ and ω are inverse functions of each other, i.e., if

$$\mu_j[\omega_j(\eta_j)] = \eta_j \quad \text{or} \quad \omega_j[\mu_j(\xi_j)] = \xi_j \qquad (j = 1, \ldots, n),$$

then we say that $M(x)$ is the *conjugate* of $\Omega(y)$ and that $\Omega(y)$ is the conjugate of $M(x)$. In that case, if $\mu(\infty) = \infty$ then $\omega(\infty) = \infty$ and vice versa. We also have the inequalities

$$(4.2) \qquad x_jy_j \leqslant M_j(x_j) + \Omega_j(y_j) \qquad (x_j \geqslant 0,\, y_j \geqslant 0),$$

where equality holds if and only if $y_j = \mu_j(x_j)$. Indeed, the right side represents the sum of two areas. The first area is bounded by the curve $\eta_j = \mu(\xi_j)$, by the ξ_j-axis and by $\xi_j = x_j$. The second area is bounded by the same curve $\eta_j = \mu(\xi_j)$ (since ω_j is the inverse function of μ_j), by the η_j-axis and by $\eta_j = y_j$. Hence the sum of the two areas is $\leqslant x_jy_j$ and equality occurs if and only if $y_j = \mu_j(x_j)$.

Theorem 11. *If $M(x)$ and $M^0(x)$ are the conjugate functions of $\Omega(y)$ and $\Omega^0(y)$ respectively, and if $M(x) \geqslant M^0(x)$ for all $x \geqslant x^0$, then $\Omega(y) \leqslant \Omega^0(y)$ for all y sufficiently large, say $y \geqslant y_0$.*

Proof. We take $y_j^0 = \mu_j(x_j^0)$. If x_j varies in the interval $x_j^0 < x_j < \infty$, then $y_j = \mu_j(x_j)$ varies in $y_j^0 < y_j < \infty$, and

$$x_j y_j = M_j(x_j) + \Omega_j(y_j).$$

We further have, by (4.2),

$$x_j y_j \leqslant M_j^0(x) + \Omega_j^0(y_j).$$

From these inequalities it follows that

$$0 \leqslant M_j(x_j) - M_j^0(x_j) \leqslant \Omega_j^0(y_j) - \Omega_j(y_j),$$

and the proof is completed.

We shall now find the Fourier transforms of W spaces.

Theorem 12. *Let $\Omega(y)$ be the conjugate of $M(x)$. Then*

$$\mathfrak{F}[W_{M,a}] \subset W^{\Omega,1/a}. \tag{4.3}$$

Proof. Let

$$\psi(\sigma) = \int_{R^n} e^{i\sigma\cdot x}\varphi(x)\,dx$$

be the Fourier transform of $\varphi \in W_{M,a}$. Since, for any $a > 0$, $|\varphi(x)| \to 0$ faster than $e^{-a|x|}$, as $|x| \to \infty$, the integral

$$\psi(\sigma + i\tau) \equiv \int_{R^n} e^{i(\sigma+i\tau)\cdot x}\varphi(x)\,dx$$

exists for any $\tau \in R^n$, and $\psi(s)$ is an entire function which coincides for $s = \sigma$ with the previous function $\psi(\sigma)$. We further have

$$(is)^k\psi(s) = \int_{R^n} e^{ix\cdot s}D^k\varphi(x)\,dx,$$

since each of the derivatives $D^k\varphi(x)$ tends to 0 faster than any $e^{-a|x|}$, as $|x| \to \infty$.

Since $\varphi \in W_{M,a}$ we obtain

$$|s^k|\,|\psi(s)| \leqslant C_{ka'}\int_{R^n}\{\exp\,[-M(a'x) + |\tau x|]\}\,dx$$

for any $a' < a$. By (4.2) and (2.1),

$$\begin{aligned}-M_j(a_j'x_j) + |\tau_j x_j'| &\leqslant -M_j(a_j'x_j) + M_j(\gamma_j x_j) + \Omega_j(\tau_j/\gamma_j)\\ &\leqslant -M_j((a_j' - \gamma_j)x_j) + \Omega(\tau_j/\gamma_j).\end{aligned}$$

Hence, taking $\gamma_j < a_j'$ and arbitrarily close to a_j' we obtain

$$|s^k|\,|\psi(s)| \leqslant C_{k\gamma}' \exp\,[\Omega(\tau/\gamma)], \tag{4.4}$$

for any $\gamma < a$. Using (4.4) with $k = 0$ and with $s^k = s_1^{|k|}, \ldots, s^k = s_n^{|k|}$, we get (with $|k| = j$)

$$(1 + |s|^j)|\psi(s)| \leqslant C_{j\gamma}'' \exp\,[\Omega(\tau/\gamma)] \qquad (0 \leqslant j < \infty)$$

for any $\gamma < a$, thereby proving that $\psi \in W^{\Omega,1/a}$.

From the inequalities derived in the previous proof we infer:

Corollary. *$\mathfrak{F}$ is a linear bounded operator from $W_{M,a}$ into $W^{\Omega,1/a}$.*

Theorem 13. *Let $M(x)$ be the conjugate of $\Omega(y)$. Then*

$$\mathfrak{F}[W^{\Omega,b}] \subset W_{M,1/b}. \tag{4.5}$$

Proof. Let

$$\psi(\sigma) = \int_{R^n} e^{i\sigma\cdot x}\varphi(x)\,dx$$

be the Fourier transform of $\varphi \in W^{\Omega,b}$. Since $\varphi(z)$ is an entire function and since, for any $m > 0$, $|x|^m|\varphi(x + iy)| \to 0$ as $|x| \to \infty$ uniformly with respect to y in bounded sets, we may change the domain of integration and thus get

$$\psi(\sigma) = \int_{R^n} e^{i(x+iy)\cdot\sigma}\varphi(x + iy)\,dx.$$

Furthermore,

$$D^q\psi(\sigma) = \int_{R^n} (iz)^q\, e^{i\sigma\cdot z}\varphi(z)\,dx,$$

since the integral is uniformly convergent with respect to σ in bounded sets. Taking $y_j = |y_j| \operatorname{sgn} \sigma_j$ we get

$$\begin{aligned} |D^q\psi(\sigma)| &\leqslant C_q\,\{\exp\,[-|\sigma y|]\} \int_{R^n} (1 + |z|^{|q|+n+1})|\varphi(z)|\,\frac{dx}{(1 + |x|)^{n+1}} \\ &\leqslant C_{qb'} \exp\,[-|\sigma y| + \Omega(b'y)] \end{aligned} \tag{4.6}$$

for any $b' > b$. By the statement containing (4.2),

$$\sigma_j y_j = M_j\left(\frac{\sigma_j}{b_j'}\right) + \Omega_j(b_j'y_j) \tag{4.7}$$

provided

$$b_j'|y_j| = \mu_j\left(\frac{|\sigma_j|}{b_j'}\right).$$

Substituting (4.7) into (4.6) we get

$$|D^q\psi(\sigma)| \leqslant C_{qb'} \exp\,[-M(\sigma/b')].$$

Hence, $\psi \in W_{M,1/b}$.

From the proof one easily infers that $\mathfrak{F}$ is a linear bounded mapping from $W^{\Omega,b}$ into $W_{M,1/b}$. Combining this result with the corollary to Theorem 12 and using (1.6) of Chap. 4, we obtain the following theorem.

Theorem 14. *Let $M(x)$ be the conjugate of $\Omega(y)$. Then, for any $a > 0$, $b > 0$,*

$$\mathfrak{F}[W_{M,a}] = W^{\Omega,1/a}, \qquad \mathfrak{F}[W^{\Omega,b}] = W_{M,1/b}, \tag{4.8}$$

and $\mathfrak{F}$ is a linear topological mapping.

Corollary.

$$\mathfrak{F}[W_M] = W^{\Omega}, \qquad \mathfrak{F}[W^{\Omega}] = W_M, \tag{4.9}$$

and $\mathfrak{F}$ is a linear topological mapping.

We turn to spaces $W^{\Omega,b}_{M,a}$.

Theorem 15. *Let $\Omega^0(y)$, $M^0(x)$ be the conjugates of $M(x)$ and $\Omega(y)$ respectively. Then*

$$\mathfrak{F}[W^{\Omega,b}_{M,a}] = W^{\Omega^0,1/a}_{M^0,1/b} \tag{4.10}$$

and $\mathfrak{F}$ is a linear topological mapping.

Proof. Let

$$\psi(\sigma) = \int_{R^n} \varphi(x)\, e^{i\sigma\cdot x}\, dx,$$

where $\varphi \in W^{\Omega,b}_{M,a}$. As in the proof of Theorem 12 we can extend ψ as an entire function $\psi(\sigma + i\tau)$, and as in the proof of Theorem 13 we may change the domain of integration, thus getting, for any $y \in R^n$,

$$\psi(\sigma + i\tau) = \int_{R^n} \varphi(x + iy)\, e^{i(x+iy)\cdot(\sigma+i\tau)}\, dx.$$

Hence

(4.11)

$$|\psi(\sigma + i\tau)| \leqslant C_{a'b'}\, \{\exp\,[-|y\sigma| + \Omega(b'y)]\} \int_{R^n} \{\exp\,[|x\tau| - M(a'x)]\}\, dx$$

for any $a' < a$, $b' > b$, provided we take $y_j = |y_j| \operatorname{sgn} \sigma_j$. As in the proof of Theorem 12 the integral is bounded by

$$C_\gamma \exp\,[\Omega^0(\tau/\gamma)] \qquad \text{for any } \gamma < a'.$$

On the other hand the expression outside the integral, on the right side of (4.11), is bounded by

$$C_{a'b'} \exp\,[-M^0(\sigma/b')],$$

as follows from the proof of Theorem 13. It follows that

$$|\psi(\sigma + i\tau)| \leqslant C'_{b'\gamma} \exp\,[-M^0(\sigma/b') + \Omega^0(\tau/\gamma)]$$

for any $b' > b$, $\gamma < a$. This proves that

$$\mathfrak{F}[W_{M,a}^{\Omega,b}] \subset W_{M^0,1/b}^{\Omega^0,1/a}.$$

Similarly,

$$\mathfrak{F}[W_{M^0,1/b}^{\Omega^0,1/a}] \subset W_{M,a}^{\Omega,b}.$$

Using (1.6) of Chap. 4, (4.10) follows. The continuity of $\mathfrak{F}$ and $\mathfrak{F}^{-1}$ follows from the inequalities which appear in the previous proof.

Corollary.

(4.12) $$\mathfrak{F}[W_M^\Omega] = W_{M^0}^{\Omega^0}$$

and $\mathfrak{F}$ *is a linear topological mapping.*

The most important examples of W spaces are the spaces for which

(4.13) $$M_j(x_j) = \frac{|x_j|^{p_j}}{p_j}, \qquad \Omega_j(y_j) = \frac{|y_j|^{q_j}}{q_j} \qquad (j = 1, \ldots, n).$$

We write (4.13) briefly in the form

(4.14) $$M(x) = x^p/p, \qquad \Omega(y) = y^q/q \qquad (x > 0,\ y > 0),$$

and use the following notation:

(4.15) $$W_{p,a} = W_{M,a}, \qquad W^{q,b} = W^{\Omega,b}, \qquad W_{p,a}^{q,b} = W_{M,a}^{\Omega,b}.$$

It is easily verified that M and Ω are conjugate to each other if and only if p and q are *conjugate* in the following sense:

$$\frac{1}{p} + \frac{1}{q} = 1 \qquad \left(\text{i.e., } \frac{1}{p_j} + \frac{1}{q_j} = 1 \text{ for } j = 1, \ldots, n\right).$$

Let $p > 1$, $r > 1$ and let q and s be the conjugates of p and r respectively. Then

(4.16) $$\mathfrak{F}[W_{p,a}] = W^{q,1/a}, \quad \mathfrak{F}[W^{q,b}] = W_{p,1/b}, \quad \mathfrak{F}[W_{p,a}^{r,b}] = W_{s,1/b}^{q,1/a},$$

(4.17) $$\mathfrak{F}[W_p] = W^q, \qquad \mathfrak{F}[W^q] = W_p, \qquad \mathfrak{F}[W_p^r] = W_s^q.$$

5. Nontriviality and Richness of W Spaces

A space Φ is said to be *nontrivial* if it contains nonzero functions. All the spaces $W_{M,a}$ are nontrivial since they all contain the space $K \equiv (D)$. By (4.8), all the spaces $W^{\Omega,b}$ are also nontrivial. As for $W^{\Omega,b}_{M,a}$, the situation is not quite as simple. From Theorem 7, Sec. 1, it follows that W^p_p is a nontrivial space if $1 < p < 2$. Using (4.17) it follows that W^p_p is nontrivial also if $2 < p < \infty$. W^2_2 is nontrivial since it contains e^{-z^2} (see the remark following Theorem 7). Summing up, we have the following:

Theorem 16. *All the spaces* W^p_p $(1 < p < \infty)$ *are nontrivial.*

As a trivial consequence we can state that W^{Ω}_M is nontrivial if

$$M_j(x_j) \leqslant \lambda(x_j)|x_j|^p, \quad \Omega_j(y_j) \geqslant \lambda(y_j)|y_j|^p \quad (j = 1, \ldots, n), \tag{5.1}$$

where $\lambda(t)$ (t real) is a continuous positive function bounded from above and from below by positive constants.

A continuous positive function $\lambda(t)$ (t real) is said to have a *slow variation at* ∞ if for any $\epsilon > 0$ there exist positive constants C_ϵ, C'_ϵ, and a $t_\epsilon > 0$ such that

$$C_\epsilon|t|^{-\epsilon} < \lambda(t) < C'_\epsilon|t|^{\epsilon} \qquad \text{for } |t| \geqslant t_\epsilon.$$

The following theorem is an extension of Theorem 7.

Theorem 17. *For any* $p > 1$ *and for any function* $\lambda(t)$ *of slow variation at* ∞ *there exists an entire function* $\varphi(z) \not\equiv 0$ *satisfying the inequality*

$$|\varphi(x + iy)| \leqslant C \exp\left[-\lambda(|x|)|x|^p + \gamma\lambda(|y|)|y|^p\right]$$

for some positive constants C, γ.

Corollary. *If* $\lambda(t)$ *is a function of slow variation at* ∞, *and if* M *and* Ω *satisfy* (5.1), *then* W^{Ω}_M *is a nontrivial space.*

Theorem 17 (and its corollary) will not be used in the future. The proof of Theorem 17 is given in Reference 46 [p. 124] and it will not be presented here.

It should be mentioned that Theorem 17 cannot be essentially improved. Indeed, we have the following:

Theorem 18. *If* $p > q$ *then* W^q_p *does not contain nonzero functions.*

Proof. Suppose $\varphi \in W^q_p$. Then

$$|\varphi(x + iy)| \leqslant C \exp\left[-a|x|^p + b|y|^q\right]$$

for some $a > 0$, $b > 0$. The function $\varphi(iz) = \varphi(ix - y)$ satisfies

$$|\varphi(ix - y)| \leqslant C \exp[-a|y|^p + b|x|^q].$$

Hence,

$$|\varphi(z)\varphi(iz)| \leqslant C^2 \{\exp[-a|x|^p + b|x|^q]\} \{\exp[-a|y|^p + b|y|^q]\}.$$

Since $p > q$, the right side is a bounded function which, in fact, tends to zero as $|x| + |y| \to \infty$. By Liouville's theorem it follows that

$$\varphi(z)\varphi(iz) \equiv 0,$$

from which one easily deduces that $\varphi(z) \equiv 0$.

If a space W_M^Ω contains a nonzero function φ, then φ is contained in some W_{M,a_0}^{Ω,b_0}. Hence $W_{M,a}^{\Omega,b}$ is a nontrivial space if $a \leqslant a_0$, $b \geqslant b_0$. Moreover, $W_{M,a}^{\Omega,b}$ is nontrivial whenever $b/a \geqslant b_0/a_0$. Indeed $\varphi(\lambda z)$ belongs to $W_{M,a}^{\Omega,b}$ if λ is a properly chosen positive vector. It follows that there exists a $\gamma \geqslant 0$ such that $W_{M,a}^{\Omega,b}$ is nontrivial if

$$b/a > \gamma \qquad (\text{i.e., } b_j/a_j > \gamma_j \text{ for } j = 1, \ldots, n)$$

and $W_{M,a}^{\Omega,b}$ is trivial if $b/a < \gamma$. In general, γ is not uniquely determined except in the case $n = 1$. For the sake of later reference we state the following:

Theorem 19. *$W_{p,a}^{p,b}$ is nontrivial if $b/a > \gamma$, for some $\gamma > 0$.*

It will be shown in this section that nontrivial W spaces are "sufficiently rich" in the sense of the following definition.

Definition. A fundamental space Φ, whose elements are continuous functions on R^n, is said to be *sufficiently rich* if whenever a locally integrable function $f(x)$ satisfies, for any $\varphi \in \Phi$,

$$\int_{R^n} |f(x)\varphi(x)|\, dx < \infty, \qquad \int_{R^n} f(x)\varphi(x)\, dx = 0,$$

then $f(x) = 0$ almost everywhere on R^n.

We shall prove an important property of sufficiently rich spaces.

Theorem 20. *Let Φ be a sufficiently rich fundamental space and let $N(x)$ be a continuous positive function on R^n such that, for some $1 \leqslant p < \infty$,*

$$\|\varphi\|_p \equiv \left[\int_{R_n} N(x)|\varphi(x)|^p\, dx\right]^{1/p} < \infty \tag{5.2}$$

for all $\varphi \in \Phi$. Then Φ is dense in the normed space E consisting of all the locally integrable functions g having a finite norm $\|g\|_p$.

Proof. Φ is a linear subspace of E. If $\Phi \neq E$ then (by Theorem 8c, Sec. 4, Chap. 1) there exists a functional f in E' such that $f \not\equiv 0$ and $f\cdot\varphi = 0$ for all $\varphi \in \Phi$. Each functional f in E' is known to have the form

$$f\cdot g = \int_{R^n} N(x)h(x)g(x)\,dx \qquad (g \in E) \tag{5.3}$$

for some locally integrable function $h(x)$ having a finite norm $\|h\|_q$, where $(1/p) + (1/q) = 1$. Note, by Hölder's inequality ((12.2) of Chap. 3), that

$$\int_{R^n} N(x)|h(x)g(x)|\,dx \leqslant \|h\|_q\|g\|_p < \infty. \tag{5.4}$$

Substituting $g = \varphi \in \Phi$ into (5.3) and using (5.4) and the fact that Φ is sufficiently rich it follows that $N(x)h(x) = 0$ almost everywhere, and from (5.3) we see that $f = 0$, which is a contradiction.

We shall now give sufficient conditions on a space Φ in order that it be sufficiently rich.

Theorem 21. *Let Φ be a fundamental space of continuous functions on R^n which satisfies the following properties:*

(a) *Φ is nontrivial;*
(b) *if $\varphi(x) \in \Phi$ then $\varphi(x+h) \in \Phi$ for any $h \in R^n$, and*
(c) *if $\varphi(x) \in \Phi$ then $\varphi(x)e^{i\sigma\cdot x} \in \Phi$ for any $\sigma \in R^n$.*

Then Φ is sufficiently rich.

Proof. Suppose that f is a locally integrable function and

$$\int_{R^n} |f(x)\varphi(x)|\,dx < \infty, \qquad \int_{R^n} f(x)\varphi(x)\,dx = 0,$$

for all $\varphi \in \Phi$. Let $\varphi_0 \in \Phi$ be such that $\varphi_0 \not\equiv 0$. Then $\varphi_0(x) \neq 0$ in a neighborhood of some point x^0. For any $y \in R^n$, the function

$$\varphi_y(x) = \varphi(x - y + x^0)$$

belongs to Φ and $\varphi_y(x) \neq 0$ for x in some neighborhood of y. Since $\varphi_y(x)e^{i\sigma\cdot x}$ belongs to Φ, we have

$$\int_{R^n} |f(x)\varphi_y(x)|\,dx < \infty, \qquad \int_{R^n} f(x)\varphi_y(x)\,e^{i\sigma\cdot x}\,dx = 0.$$

Thus, $f(x)\varphi_y(x)$ belongs to $L^1(R^n)$, and its Fourier transform is equal everywhere to 0. By a well-known uniqueness theorem [4, p. 59],

$$f(x)\varphi_y(x) = 0$$

almost everywhere. Hence $f(x) = 0$ almost everywhere in some neighborhood of y. Since this is true for any $y, f = 0$ almost everywhere on R^n.

W spaces possess the property (*b*). They also possess the property (c), as follows by Corollary 2 at the end of Sec. 3 and the remark following it. Hence:

Theorem 22. *Any nontrivial W space is a sufficiently rich space. In particular, $W_{p,a}^{p,b}$ is sufficiently rich whenever $b/a > \gamma$.*

PROBLEMS

1. Extend Theorems 1′, 2′ to $p \leqslant \frac{1}{2}$, $n = 1$.
2. Prove that if in Theorem 1 (1.2) is replaced by

$$|f(z)| \leqslant C_1 \exp [a|z|^h] \qquad (a > 0, h < p),$$

then the assertion

$$|f(z)| \leqslant C_1 \exp [a'|z|^h] \qquad (z_j \in G_j, j = 1, \ldots, n)$$

is valid, where $a' = a/(\cos h\varphi/2)$. [*Hint:* For $n = 1$ take

$$f_\epsilon(z) = f(z) \{\exp [-\epsilon z^q]\} \{\exp [-a'z^h]\}.]$$

3. For some functions $f(z)$ satisfying the assumptions of Theorem 2 μ may be $> 1 - (p - h)$. Prove, however, that if $n = 1$ and $f(z)$ is an entire function of (precise) order p satisfying (1.8) with $h < p$ (or with $h = p$ and $a < 0$) then $\mu \leqslant 1$. [*Hint:* Use Problem 2.]
4. Verify that the functions

$$M(x) = (x + 1) \log (x + 1) - x, \qquad \Omega(y) = e^y - y - 1$$

($x > 0$, $y > 0$, $n = 1$ are conjugate to each other. Note also that $M(x)$ is equivalent to $M_1(x) = x \log (x + 1)$.
5. Prove that W_M^Ω ($n = 1$) is a trivial space if for any positive numbers a, b

$$\lim_{x \to \infty} [\Omega(bx) - M(ax)] = -\infty.$$

6. Prove that if $\varphi_1 \in W_p^p$, $\varphi_2 \in W_p^p$, then $\varphi_1 * \varphi_2 \in W_p^p$.
7. Prove Theorem 12, Sec. 4, Chap. 3 and its corollary, for generalized functions over $W_{p,a}^{p,b}$. [*Hint:* if $\varphi \in W_{p,a}^{p,b}$, $\int_{-\infty}^{\infty} \varphi(x)\, dx = 0$ then $\varphi(x) = d\chi(x)/dx$ for some $\chi \in W_{p,a}^{p,b}$.]
8. Prove: if g is a generalized function over $W_{p,a}^{p,b}$ ($n = 1$) which is of function type $g(x)$, and if $g(x)$ is continuous and $D^r g = 0$ over $W_{p,a}^{p,b}$, then $g(x)$ is a polynomial of degree $< r$.

CHAPTER 6

FOURIER TRANSFORMS OF ENTIRE FUNCTIONS

In this chapter we shall find the structure of a special but very important class of generalized functions, namely the Fourier transforms of functions $f(\sigma)$ $(\sigma = (\sigma_1, \ldots, \sigma_n))$ which can be extended into entire functions $f(s)$ $(s = (s_1, \ldots, s_n),\ s = \sigma + i\tau)$ of finite order. In view of the applications in Chap. 7, we shall be interested also in the case where f depends on a real parameter t, say, $t_0 \leqslant t \leqslant t_1$. Thus we assume that $f = f(\sigma, t)$ can be extended into an entire function $f(s, t)$ for every value of the parameter t, and

$$(0.1) \qquad |f(s, t)| \leqslant C \exp [b|s|^p]$$

for all $s \in C^n$ and $t_0 \leqslant t \leqslant t_1$. C, b, p are positive constants independent of (s, t).

The Fourier transform $\tilde{f}(x, t)$ is a generalized function, over an appropriate space, for every value of t. Our investigation will be directed toward determining the functional behavior of $\tilde{f}(x, t)$ with respect to x and t *combined*.

In studying $\tilde{f}(x, t)$, we have to use different methods for different growth behavior of $f(s, t)$ and $f(\sigma, t)$, as $|s|$ and $|\sigma|$ tend to infinity. We shall distinguish between the different cases according to the following types of behavior:

(i) $f(s, t)$ is either of order $p > 1$ or of order $\leqslant 1$;
(ii) $f(\sigma, t)$ is either of "fast" decrease, or of "slow" increase, or of "fast" increase, as $|\sigma| \to \infty$.

For the sake of simplicity we shall deal, in most cases, first with f which is independent of t. The result for $f = f(s, t)$ then follows by trivial modifications.

1. Entire Functions of Order $\leqslant p$ and of Fast Decrease

We assume that

(1.1)

$$|f(s)| \leqslant C_1 \exp [b|s|^p], \qquad |f(\sigma)| \leqslant C_2 \exp [-a|\sigma|^h] \qquad (0 < h \leqslant p, a > 0).$$

By Theorem 2, Sec. 1, Chap. 5,

$$|f(\sigma + i\tau)| \leqslant C_3 \exp\,[-a'|\sigma|^h] \quad \text{in a region } |\tau_j| \leqslant K_j(1 + |\sigma_j|)^{\mu_j} \quad (j = 1, \ldots, n), \tag{1.2}$$

where a' is arbitrarily close to a, and $\mu_j \geqslant 1 - (p_j - h_j)$ $(j = 1, \ldots, n)$.

If (1.2) holds for μ, then it clearly holds also for any $\mu' < \mu$ (with a different K). Hence we may assume that either $\mu > 0$ or $\mu \leqslant 0$. We shall study separately the cases $\mu > 0$ and $\mu \leqslant 0$, and begin with the case $\mu \leqslant 0$. In evaluating

$$\tilde{f}(x) = \int_{R^n} e^{ix\cdot\sigma} f(\sigma)\, d\sigma$$

we may change the domain of integration from $\tau_j = 0$ $(1 \leqslant j \leqslant n)$ into $\tau_j = K_j(1 + |\sigma_j|)^{\mu_j} \operatorname{sgn} x_j$. Using (1.2) we then obtain

$$|\tilde{f}(x)| \leqslant A_1 \int_{R^n} \{\exp\,[-A_2(1 + |\sigma|)^\mu |x| - a'|\sigma|^h]\}\, d\sigma, \tag{1.3}$$

where the A_j are used, here and in the following, to denote positive constants depending only on b, p, h, a, a', K and μ.

If $n = 1$, we divide the integral in (1.3) into two parts by taking $|\sigma| < M$ for the first part and $|\sigma| \geqslant M$ for the second part, where

$$M = |x|^{1/(h-\mu)}.$$

We then get

$$\begin{aligned} |\tilde{f}(x)| &\leqslant A_1 \int_{|\sigma|<M} \{\exp\,[-A_2(1 + |\sigma|)^\mu |x|]\}\, d\sigma \\ &\qquad + A_1 \int_{|\sigma|\geqslant M} \{\exp\,[-a'|\sigma|^h]\}\, d\sigma \\ &\leqslant A_3 M \exp\,[-A_2(1 + M)^\mu |x|] + A_3 \exp\,[-a'' M^h] \\ &\leqslant A_4 \exp\,[-A_5 |x|^{h_1}] \end{aligned} \tag{1.4}$$

for any $a'' < a'$, where $h_1 = h/(h - \mu)$. If $n > 1$, we divide the integral (1.3) into 2^n integrals according to whether $|\sigma_j| < M_j$ or $|\sigma_j| \geqslant M_j$ $(j = 1, \ldots, n)$, where $M_j = |x_j|^{1/(h-\mu)}$, and then proceed similarly to (1.4).

Since for every polynomial $v(s)$, the function $v(s)f(s)$ satisfies the inequality (1.2), with μ, K independent of v and with a' replaced by any fixed vector $< a'$ and independent of v, and since $\mathfrak{F}(v(\sigma)f(\sigma))$ is $v(-iD)\tilde{f}(x)$ $(D = (D_1, \ldots, D_n)$, $D_j = \partial/\partial x_j)$, we obtain

$$|D_x^q \tilde{f}(x)| \leqslant B_q \exp\,[-\alpha |x|^{h_1}] \qquad (\alpha > 0, B_q > 0, 0 \leqslant |q| < \infty). \tag{1.5}$$

From the proof of (1.5) one easily infers the following theorem.

Theorem 1. *Let $D_t^j f(s, t)$ $(0 \leqslant j \leqslant j_0)$ be continuous functions of (s, t) $(s \in C^n, t_0 \leqslant t \leqslant t_1)$ and let each $D_t^j f(s, t)$ be an entire function of s, for* *fixed* t $(t_0 \leqslant t \leqslant t_1)$, *which satisfies* (1.2) *for some constants* $\mu \leqslant 0$, $a' > 0$, $C_3 > 0$, $K > 0$ *independent of* t. *Then the derivatives* $D_x^q D_t^j \tilde{f}(x, t)$ $(0 \leqslant |q| < \infty, 0 \leqslant j \leqslant j_0)$ *are all continuous functions of* (x, t) $(x \in R^n, t_0 \leqslant t \leqslant t_1)$ *satisfying*

$$|D_x^q D_t^j \tilde{f}(x, t)| \leqslant B_q \exp[-\alpha|x|^{h_1}] \qquad \left(h_1 = \frac{h}{h - \mu}\right), \tag{1.6}$$

where B_q, α *are positive constants independent of* t.

Consider next the case $\mu > 0$. Then, by Theorem 3, Sec. 1, Chap. 5, if $\mu \leqslant 1$, $n = 1$,

$$|f(\sigma + i\tau)| \leqslant C_3 \exp[-a'|\sigma|^h + b'|\tau|^{p/\mu}]. \tag{1.7}$$

We shall now assume that (1.7) holds for any n and that $p > \mu$. Then, f belongs to $W^{p/\mu,\alpha}$ where $\mu\alpha^{p/\mu} = pb'$. By (4.5), Chap. 5, $\tilde{f}$ belongs to $W_{(p/\mu)',1/\alpha}$, i.e., for $0 \leqslant |q| < \infty$,

$$|D_x^q \tilde{f}(x)| \leqslant B_q \exp[-\alpha'|x|^{p_1}] \qquad (p_1 = p/(p - \mu)), \tag{1.8}$$

where $\alpha' = 1/p_1\alpha_1^{p_1}$ and α_1 is any number larger than α.

From the proof of Theorem 13, Sec. 4, Chap. 5 it follows that if $f = f(s, t)$ satisfies (1.7), independently of t, and if it is continuous in (s, t), then for any $0 \leqslant |q| < \infty$, $D_x^q \tilde{f}(x, t)$ is a continuous function of (x, t) satisfying (1.8) with constants independent of t. For later references we state this in the following slightly more general form:

Theorem 1′. *Let* $D_t^j f(s, t)$ $(0 \leqslant j \leqslant j_0)$ *be continuous functions of* (s, t) $(s \in C^n, t_0 \leqslant t \leqslant t_1)$ *and let each* $D_t^j f(s, t)$ *be an entire function of* s, *for fixed* t, *satisfying* (1.7) *with constants* $p > \mu > 0$, $a' > 0$, $b' > 0$, $C_3 > 0$ *independent of* t. *Then, for all* $0 \leqslant |q| < \infty$ *the functions* $D_x^q D_t^j \tilde{f}(x, t)$ *are continuous functions in* (x, t) *for* $x \in R^n$, $t_0 \leqslant t \leqslant t_1$, *satisfying*

$$|D_x^q D_t^j \tilde{f}(x, t)| \leqslant B_q \exp[-\alpha'|x|^{p_1}] \qquad \left(p_1 = \frac{p}{p - \mu}\right) \tag{1.9}$$

where the B_q *and* α' *are positive constants independent of* t.

2. Entire Functions of Order $\leqslant 1$

Let $f(z)$ be an entire function of order $\leqslant 1$ and type $\leqslant b$, i.e., for every $\epsilon > 0$

$$|f(z)| \leqslant C_\epsilon \exp[(b + \epsilon)|z|]. \tag{2.1}$$

Since Taylor's series $\sum_p a_p z^p$ of $f(z)$ is uniformly convergent in bounded sets of C^n,

(2.2)
$$\int_{R^n} f(x)\varphi(x)\,dx = \sum_p a_p \int_{R^n} x^p\varphi(x)\,dx$$

for any $\varphi \in K$. We may consider (2.2) as an equation in K' and we can therefore write

(2.3)
$$f = \sum a_p x^p \qquad \text{in } K'.$$

By (2.3), Chap. 4,

$$\mathfrak{F}(x^p) = (2\pi)^n(-iD)^p\delta,$$

since $\mathfrak{F}(1) = (2\pi)^n\,\delta$. Taking the Fourier transforms of both sides of (2.3) we then obtain

(2.4)
$$\mathfrak{F}f = (2\pi)^n \sum a_p(-iD)^p\delta \qquad \text{in } Z'.$$

Applying both sides to any $\psi \in Z$ we have

(2.5)
$$(\mathfrak{F}f, \psi) = (2\pi)^n \sum a_p((-iD)^p\delta(\sigma), \psi(\sigma)) = (2\pi)^n \sum i^p a_p\, D^p\psi(0).$$

Let G be a closed bounded set in C^n and denote by $\mathring{Z}(G)$ the linear space of all the functions which are analytic in G (i.e., analytic in some neighborhood of G). A sequence $\{g_m\}$ in $\mathring{Z}(G)$ is said to be *convergent* to 0 if there exists a neighborhood G' of G such that all the g_m are analytic functions in G' and $g_m(z) \to 0$, as $m \to \infty$, uniformly with respect to z in G'.

Theorem 2. *The Fourier transform of an entire function of exponential type $\leqslant b$ is a continuous linear functional over Z which can be extended into a continuous linear functional over $\mathring{Z}(G_b)$ where*

$$G_b = \{s;\, |s_1| \leqslant b_1, \ldots, |s_n| \leqslant b_n\}.$$

Proof. We use the right side of (2.5) to define $\tilde{f}\cdot\psi$ for any $\psi \in \mathring{Z}(G_b)$. In order to estimate $\tilde{f}\cdot\psi$, we use Cauchy's inequalities for ψ and get

(2.6)
$$|D^p\psi(0)| \leqslant \frac{Cp!}{(b')^p} \qquad (0 \leqslant |p| < \infty)$$

for any b' $(> b)$ such that ψ is analytic in $G_{b'}$, where

(2.7)
$$C = \max_{s \in G_{b'}} |\psi(s)|.$$

The Cauchy inequalities for $f(z)$ give

$$|a_p| \leqslant \frac{\exp\,[(b + \epsilon)R]}{R^p}.$$

Taking R_j $(j = 1, \ldots, n)$ which minimize

$$\frac{\exp[(b_j + \epsilon_j)R_j]}{R_j^{p_j}}$$

(compare (1.25), Chap. 5) we get

$$|a_p| \leqslant C' \left(\frac{be\theta}{p}\right)^p \tag{2.8}$$

for any $\theta > 1$, where C' is a constant depending on θ, b, p. If we take $\theta < b'/b$, i.e., $\theta_j < b_j'/b_j$ $(j = 1, \ldots, n)$, then we have

$$|\tilde{f}\cdot\psi| = (2\pi)^n\,|\Sigma\, i^p a_p\, D^p\psi(0)| \leqslant (2\pi)^n\, CC'\,\Sigma \left(\frac{be\theta}{p}\right)^p p! \left(\frac{1}{b'}\right)^p \leqslant CC'C'',$$

where C'' depends only on $\theta b/b'$. From this calculation it follows that $\tilde{f}\cdot\psi$ is a finite number if $\psi \in \mathring{Z}(G_b)$. Furthermore, if $\psi_m \to 0$ in $\mathring{Z}(G_b)$ then $\tilde{f}\cdot\psi_m \to 0$. Indeed, b' can be taken independently of m and $C = C(m)$ (defined by (2.7) with $\psi = \psi_m$) tends to 0 as $m \to \infty$.

We shall now prove the converse of Theorem 2.

Theorem 3. *If a functional $g \in Z'$ can be extended into a continuous linear functional over $\mathring{Z}(G_b)$ then g is the Fourier transform of a functional in K' which is of function type $f(x)$, where $f(z)$ is an entire function of exponential type $\leqslant b$.*

Proof. Let $\psi \in \mathring{Z}(G_b)$. By (2.6),

$$\psi(s) = \sum_p c_p s^p \qquad \text{and} \qquad |c_p| \leqslant C\left(\frac{\theta}{b}\right)^p$$

for some $\theta < 1$, where C depends on θ. Since g is both a linear and a continuous functional over $\mathring{Z}(G_b)$ and since the partial sums of the Taylor series of $\psi(s)$ converge to $\psi(s)$ in the sense of $\mathring{Z}(G_b)$,

$$g\cdot\psi = \sum_p c_p(g\cdot s^p) = \sum_p c_p g_p,$$

where $g_p = g\cdot s^p$ are some complex numbers. Taking ψ with

$$c_p = \left(\frac{\alpha}{b}\right)^p$$

for any $0 < \alpha < 1$ (such ψ clearly belong to $\mathring{Z}(G_b)$), we conclude from $c_p g_p \to 0$ (as $|p| \to \infty$) that

$$|g_p| \leqslant C_1\left(\frac{b}{\alpha}\right)^p \qquad (0 \leqslant |p| < \infty) \tag{2.9}$$

for some constant C_1.

We proceed to calculate $f = \mathfrak{F}^{-1}g$. Taking $\varphi \in K$, $\psi = \mathfrak{F}\varphi$ and noting that $\mathfrak{F}^{-1}\bar{\varphi} = (2\pi)^{-n}\bar{\psi}$, we have

$$\begin{aligned}(f, \bar{\varphi}) &= (\mathfrak{F}^{-1}g, \bar{\varphi}) = (g, \mathfrak{F}^{-1}\bar{\varphi}) = (2\pi)^{-n}(g, \bar{\psi}) = (2\pi)^{-n} \sum \bar{c}_p g_p \\ &= (2\pi)^{-n} \sum_p \frac{D^p\bar{\psi}(0)}{p!} g_p = (2\pi)^{-n} \sum_p \int \frac{g_p}{p!} (-ix)^p \overline{\varphi(x)}\, dx.\end{aligned} \tag{2.10}$$

Hence, if we show that the power series

$$f(z) \equiv \frac{1}{(2\pi)^n} \sum_p \frac{g_p}{p!} (-iz)^p \tag{2.11}$$

is uniformly convergent in bounded sets of C^n, it would follow that we can change the order of summation and integration on the right side of (2.10) and thus conclude that the functional f is of function type $f(x)$. Now, both the uniform convergence (in bounded sets) of the series in (2.11) and the assertion that $f(z)$ is of exponential type $\leqslant b$ follow from the inequalities

$$(2\pi)^n|f(z)| \leqslant \sum_p \left|\frac{g_p}{p!} (-iz)^p\right| \leqslant \sum_p C_1 \left(\frac{b}{\alpha}\right)^p \frac{|z^p|}{p!} \leqslant C_1 \exp [(b/\alpha)|z|],$$

which hold for any $\alpha < 1$.

3. Entire Functions of Order $\leqslant$ 1 and of Slow Increase

We shall need a classical theorem of Paley-Wiener.

Theorem 4 (*Paley-Wiener*). *Let $f(z)$ be an entire function of exponential type $\leqslant b$ and assume that $f(x)$ belongs to $L^2(R^n)$. Then the Fourier transform $g(\sigma)$ of $f(x)$ also belongs to $L^2(R^n)$ and $g(\sigma) = 0$ almost everywhere outside*

$$G_b \equiv \{\sigma; |\sigma_1| \leqslant b_1, \ldots, |\sigma_n| \leqslant b_n\}.$$

The following theorem generalizes the Paley-Wiener theorem.

Theorem 5. *Let $f(z)$ be an entire function of exponential type $\leqslant b$ and let $f(x)$ be $O(|x|^q)$ for some $q \geqslant 0$, as $|x| \to \infty$. Considering f as a functional in S', its Fourier transform $\mathfrak{F}f$ is again in S' and its support lies in G_b.*

Proof. Since $\mathfrak{F}S = S$, $\mathfrak{F}f \in S'$. It remains to prove that the support of $\mathfrak{F}f$ lies in G_b. Let $\psi_\epsilon(\sigma)$ be a function in K which vanishes for $|\sigma| \geqslant \epsilon$. Then $\varphi_\epsilon(x) = \mathfrak{F}^{-1}\psi_\epsilon$ is an entire function of exponential type $\leqslant \epsilon$ and it converges to 0, as $|x| \to \infty$, faster than any power of $|x|^{-1}$. $\varphi_\epsilon(z)f(z)$ is again an entire function. It is of exponential type $\leqslant b + \epsilon$ and it decreases to 0, as $|x| \to \infty$, faster than any power of $|x|^{-1}$. Hence, by Theorem 4, $\mathfrak{F}(\varphi_\epsilon f)$ is a function which vanishes outside $G_{b+\epsilon}$.

Since φ_ϵ is a multiplier in S, we can apply Theorem 9, Sec. 4, Chap. 4 (with $\mathfrak{F}$ replaced by $\mathfrak{F}^{-1}$), and thus get

$$(2\pi)^n \varphi_\epsilon(x)f(x) = \mathfrak{F}^{-1}(\mathfrak{F}\varphi_\epsilon * \mathfrak{F}f),$$

or

$$(2\pi)^n \mathfrak{F}[\varphi_\epsilon(x)f(x)] = \mathfrak{F}\varphi_\epsilon * \mathfrak{F}f = \psi_\epsilon(\sigma) * \mathfrak{F}f.$$

Taking $\psi_\epsilon \to \delta$ in S' and using Theorem 8, Sec. 3, Chap. 4, we conclude that

$$\mathfrak{F}f = \delta * \mathfrak{F}f = \lim_{\epsilon \to 0} \psi_\epsilon * \mathfrak{F}f \qquad (\text{in } S')$$

has its support in G_b since each $\psi_\epsilon * \mathfrak{F}f$ has its support in $G_{b+\epsilon}$.

Theorem 5 is the converse of Theorem 2, Sec. 2, Chap. 4. Combining both theorems we deduce:

Corollary. *If $f(z)$ is an entire function of exponential type $\leqslant b$ and if $f(x)$ is bounded by $O(|x|^q)$ for some $q \geqslant 0$, as $|x| \to \infty$, then, for any $\epsilon > 0$,*

$$|f(x + iy)| \leqslant C_\epsilon(1 + |x|)^q \exp[(b + \epsilon)|y|].$$

We shall now assume that f depends on a parameter t and, for $0 \leqslant j \leqslant j_0$,

$$\text{(3.1)} \qquad \begin{aligned} |D_t^j f(s, t)| &\leqslant C_1 \exp[b|s|], \\ |D_t^j f(\sigma, t)| &\leqslant C_2(1 + |\sigma|)^h \qquad (h \text{ an integer} \geqslant 0), \end{aligned}$$

and we shall study more closely the structure of $\tilde{f}$.

Theorem 6. *Let $D_t^j f(s, t)$ $(0 \leqslant j \leqslant j_0)$ be continuous functions of (s, t) $(s \in C^n, t_0 \leqslant t \leqslant t_1)$ which, for each fixed t, are entire functions of s, satisfying* (3.1) *with constants C_1, C_2, b, h independent of t. Then, for any $\epsilon > 0$, there exist functions $f_k(x, t)$ $(k = 1, \ldots, N;$ N depending only on $h + n)$ which vanish if $|x_j| > b_j + \epsilon$ for some j, and which are continuous in (x, t) $(x \in R^n, t_0 \leqslant t \leqslant t_1)$ together with their first j_0 t-derivatives such that*

$$\text{(3.2)} \qquad \tilde{f}(x, t) = \sum_{k=1}^{N} P_k\left(\frac{\partial}{\partial x}\right) f_k(x, t) \qquad \text{in } S',$$

where $P_k(\partial/\partial x)$ are some polynomials in $\partial/\partial x$ of degree $\leqslant n + 2 + h$ $(\leqslant n + 1 + h$ if $n + 1 + h$ is even).

Proof. We introduce the function

$$\text{(3.3)} \qquad f_6(\sigma, t) = \frac{f(\sigma, t)}{(1 + |\sigma|^2)^{(h+\nu)/2}} \qquad (h + \nu \text{ is even}, \nu \geqslant n + 1).$$

Using the rule (2.3), Chap. 4, we obtain from (3.3)

$$\tilde{f}(x, t) = (1 - \Delta)^{(h+\nu)/2}\tilde{f}_0(x, t) \qquad \text{in } S', \tag{3.4}$$

where $\Delta = \Sigma \dfrac{\partial^2}{\partial x_i^2}$.

By Theorem 5, $\tilde{f}(x, t)$ vanishes if $|x_j| > b_j$ for some j. Hence, if $h(x)$ is a C^∞ function which is equal to 1 for $|x_j| < b_j$ ($j = 1, \ldots, n$) and to zero if $|x_j| > b_j + \epsilon$ for some j, then

(3.5)

$$(\tilde{f}(x, t), \psi(x)) = (\tilde{f}(x, t), h(x)\psi(x)) = ((1 - \Delta)^{(h+\nu)/2}\tilde{f}_0(x, t), h(x)\psi(x))$$

for any $\psi \in S$. Using the definition of a derivative of a generalized function we obtain from the right side of (3.5)

$$(\tilde{f}_0(x, t), (1 - \Delta)^{(h+\nu)/2}(h(x)\psi(x)))$$
$$= \sum_{k=1}^{N} \left(\tilde{f}_0(x, t), \left[P_{1k}\left(\frac{\partial}{\partial x}\right)h(x)\right]P_{2k}\left(\frac{\partial}{\partial x}\right)\psi(x)\right)$$
$$= \left(\sum_{k=1}^{N} P_k\left(\frac{\partial}{\partial x}\right)f_k(x, t), \psi(x)\right),$$

where P_{1k}, P_{2k} are of orders $\leqslant h + \nu$, $f_k(x, t) = \tilde{f}_0(x, t)P_{1k}(i\partial/\partial x)h(x)$, and $P_k(\partial/\partial x)$ is $P_{2k}(-\partial/\partial x)$. Since

$$\int_{R^n} |f_0(\sigma, t)|\, d\sigma$$

is convergent uniformly with respect to t, the $f_k(x, t)$ are continuous functions of (x, t), and the proof of the theorem for $j_0 = 0$ is completed.

The proof for any j_0 follows by noting that the $\int |D_t^j f_0(\sigma, t)|\, d\sigma$ is convergent uniformly with respect to t, so that $\tilde{f}_0(x, t)$, and, consequently, the $f_k(x, t)$ are continuous functions in (x, t) together with their first j_0 t-derivatives.

4. Entire Functions of Order $\leqslant p$ and of Slow Increase

We assume that

(4.1)

$$|f(s)| \leqslant C_1 \exp[b|s|^p], \quad |f(\sigma)| \leqslant C_2(1 + |\sigma|)^h \quad (h \text{ an integer} \geqslant 0, p > 1).$$

f will be considered as an element of S'.

By Theorem 2″, Sec. 1, Chap. 5,

$$|f(\sigma + i\tau)| \leqslant C_3(1 + |\sigma|)^h \quad \text{if } |\tau_j| \leqslant K_j(1 + |\sigma_j|)^{\mu_j},\ j = 1, \ldots, n, \tag{4.2}$$

where $\mu \geqslant 1 - p$.

Consider first the case $\mu < 0$. Introducing the function

$$f_0(\sigma) = \frac{f(\sigma)}{(H + |\sigma|^2)^{(h+\nu)/2}} \qquad (h + \nu \text{ even}, \nu \geqslant n + 1) \tag{4.3}$$

and using the rule (2.3), Chap. 4, we get

$$\tilde{f}(x) = (H - \Delta)^{(h+\nu)/2}\tilde{f}_0(x) \qquad \text{in } S'. \tag{4.4}$$

In order to evaluate $\tilde{f}_0(x) = \int e^{ix\cdot\sigma} f_0(\sigma)\, d\sigma$ we modify the domain of integration from $\tau_j = 0$ into $\tau_j = K_j(1 + |\sigma_j|)^{\mu_j} \operatorname{sgn} x_j$ $(1 \leqslant j \leqslant n)$. If H is sufficiently large (depending only on K) then, using (4.2), we obtain

$$|\tilde{f}_0(x)| \leqslant A_1 \int_{R^n} \frac{\exp[-A_2(1 + |\sigma|)^{\mu}|x|]}{(1 + |\sigma|)^{\nu}}\, d\sigma$$

$$\leqslant \frac{A_3}{|x|^{(\nu-n)/|\mu|}} \int_0^{\infty} \{\exp[-A_2 r]\} r^{(\nu-n+\mu)/|\mu|}\, dr,$$

where the middle term is evaluated by introducing polar coordinates and substituting $r = |x|(1 + |\sigma|)^{\mu}$, the A_i being appropriate positive constants. We thus obtain

$$|\tilde{f}_0(x)| \leqslant \frac{A_4}{(1 + |x|)^{(\nu-n)/|\mu|}}. \tag{4.5}$$

From the above proof we can easily deduce the following theorem:

Theorem 7. *Let $D_t^j f(s, t)$ $(0 \leqslant j \leqslant j_0)$ be continuous functions of (s, t) $(s \in C^n, t_0 \leqslant t \leqslant t_1)$ and assume that, for every fixed t, $D_t^j f(s, t)$ is an entire function of s satisfying (4.2) with constants $\mu < 0$, C_3, h, K independent of t. Then, for any $\nu \geqslant n + 2$ ($\geqslant n + 1$ if $h + n + 1$ is even),*

$$\tilde{f}(x, t) = (H - \Delta)^{(h+\nu)/2}\tilde{f}_0(x, t) \qquad \text{in } S', \tag{4.6}$$

where $D_t^j \tilde{f}_0(x, t)$ are continuous functions of (x, t) $(x \in R^n, t_0 \leqslant t \leqslant t_1)$ satisfying

$$|D_t^j \tilde{f}_0(x, t)| \leqslant \frac{A_5}{(1 + |x|)^{(\nu-n)/|\mu|}}, \tag{4.7}$$

and H, A_5 are independent of t. If $\mu = 0$ then the above assertion remains true except that instead of (4.7) we have

$$|D_t^j \tilde{f}_0(x, t)| \leqslant A_6 \exp[-A_7|x|] \tag{4.8}$$

and A_6, A_7 are positive constants independent of t.

The proof for $\mu = 0$ follows by slightly modifying the above calculations for $\mu < 0$.

Consider now the case $\mu > 0$. By Theorem 3′, Sec. 1, Chap. 5, we have, if $\mu \leqslant 1$ and $n = 1$,

$$|f(\sigma + i\tau)| \leqslant C_4(1 + |\sigma|)^h \exp [b'|\tau|^{p/\mu}]. \tag{4.9}$$

We shall assume that (4.9) is satisfied for any n, and that $p/\mu > 1$ (if $p/\mu \leqslant 1$ we have Theorem 6).

We shall prove that the Fourier transform of such a function f has the form

$$\tilde{f}(x) = \sum_{k=1}^{N} P_k\left(\frac{\partial}{\partial x}\right) f_k(x) \qquad \text{in } S', \tag{4.10}$$

where the $P_k(\partial/\partial x)$ are polynomials in $\partial/\partial x$ of orders $\leqslant h + n + 2$ ($\leqslant h + n + 1$ if this number is even), and the $f_k(x)$ are continuous functions satisfying

$$|f_k(x)| \leqslant C'_k \exp [-b''|x|^{p_1}] \qquad (p_1 = p/(p - \mu)) \tag{4.11}$$

for some $b'' > 0$ depending on p, b', h, μ.

First, we claim that $g_B(x) = \int_{-B}^{B} e^{ix\cdot\sigma} f(\sigma)\, d\sigma$ is convergent in S' to $\tilde{f}(x)$, as $B \to \infty$ (i.e., $B = (B_1, \ldots, B_n)$ and all the B_j tend to ∞). Indeed, if $\varphi \in S$, $\psi = \mathfrak{F}\varphi$, then

$$g_B \cdot \varphi = \int_{-B}^{B} f(\sigma) \left\{\int_{R^n} e^{ix\cdot\sigma}\varphi(x)\, dx\right\} d\sigma = \int_{-B}^{B} f(\sigma)\psi(\sigma)\, d\sigma$$

$$\to \int_{R^n} f(\sigma)\psi(\sigma)\, d\sigma = f \cdot \mathfrak{F}\varphi = \mathfrak{F}f \cdot \varphi.$$

Next, we write (if $h + n + 1$ is even)

$$g_B(x) = \int_{-B}^{B} \frac{(|\sigma|^2 + C|x|^{2\beta} + 1)^{(h+n+1)/2} f(\sigma)\, e^{ix\cdot\sigma}}{(|\sigma|^2 + C|x|^{2\beta} + 1)^{(h+n+1)/2}}\, d\sigma = \sum C_{km} I^B_{km0}$$

where $C > 0$, β is a positive integer and

$$I^B_{kmj} = \int_{-B}^{B} \frac{x^k \sigma^m f(\sigma) e^{ix\cdot\sigma}}{(|\sigma|^2 + C|x|^{2\beta} + 1)^{((h+n+1)/2)+j}}\, d\sigma.$$

It is easily seen that I^B_{kmj} can be written as $(-iD_x)^m I^B_{k0j}$ plus a linear combination of integrals $I^B_{k'm'j'}$ with $m' < m$ and $j' \geqslant j$. Hence, we can rewrite $g_B(x)$ in the form

$$\sum P_{kj}\left(\frac{\partial}{\partial x}\right) I^B_{k0j} \qquad \text{where } j \geqslant 0.$$

Since, as easily verified, $I^B_{k0j} \to I^\infty_{k0j}$ in S', as $B \to \infty$, the last sum converges to $\sum P_{kj}(\partial/\partial x) I^\infty_{k0j}$ and it remains to estimate I^∞_{k0j}. This is done by modifying the domain of integration from $\tau_i = 0$ into $\tau_i = \alpha_i|x_i|^{d_i} \operatorname{sgn} x_i$, where $d = (d_1, \ldots, d_n)$, $d = p_1\mu/p$. Taking $\beta > |d|$, $C > |\alpha|^2$, and making use of (4.9), we obtain

$$|I^{\infty}_{k0j}(x)| \leqslant C_4 \int_{R^n} \frac{|x|^k(1+|\sigma|)^h \exp[b'|\tau|^{p/\mu}] \exp[-|x\tau|]}{(|\sigma|^2 - |\tau|^2 + C|x|^{2\beta} + 1)^{(h+n+1)/2}}\, d\sigma$$
$$\leqslant C_5|x|^k \exp[b'\alpha^{(p/\mu)}|x|^{p_1} - \alpha|x|^{p_1}] \leq C_6 \exp[-(1/2)\alpha|x|^{p_1}]$$

provided α is sufficiently small (recall that $p/\mu > 1$). In the last calculation we have used the notation

$$\exp[bc|x|^q] = \exp[b_1c_1|x_1|^{q_1} + \cdots + b_nc_n|x_n|^{q_n}].$$

From the above proof we deduce the following result:

Theorem 7′. *Let $D^j_tf(s, t)$ $(0 \leqslant j \leqslant j_0)$ be continuous functions of (s, t) $(s \in C^n,\ t_0 \leqslant t \leqslant t_1)$ and assume that, for each fixed t, $D^j_tf(s, t)$ are entire functions satisfying* (4.9) *with constants $\mu > 0$, C_4, b', h independent of t $(\mu < p)$. Then there exist functions $f_k(x, t)$ $(1 \leqslant k \leqslant N)$ continuous in (x, t) $(x \in R^n,\ t_0 \leqslant t \leqslant t_1)$ together with their first j_0 t-derivatives and satisfying*

$$(4.12)\qquad |D^j_tf_k(x, t)| \leqslant C'_k \exp[-b''|x|^{p_1}] \qquad (p_1 = p/(p-\mu))$$

such that

$$(4.13)\qquad \tilde{f}(x, t) = \sum_{k=1}^{N} P_k\left(\frac{\partial}{\partial x}\right) f_k(x, t) \qquad \text{in } S'.$$

Here the $P_k(\partial/\partial x)$ are polynomials in $\partial/\partial x$ of degree $\leqslant h + n + 2$ ($\leqslant h + n + 1$ if this number is even) and C'_k, b'' are positive constants independent of t.

5. Entire Functions of Order $\leqslant p$ and of Mildly Fast Increase

We assume that

$$(5.1)\qquad |f(x)| \leqslant C_1 \exp[b_0|s|^p], \qquad |f(\sigma)| \leqslant C_2 \exp[a_0|\sigma|^h]$$
$$(0 < h < p,\ a_0 > 0).$$

The case $h = p$ will be considered in the next section. f may be considered as an element of $(W^{\beta,b}_{\beta,a})'$, for any $a > 0$, $b > 0$, $b/a > \gamma$ (recall Theorem 19, Sec. 5, Chap. 5), provided $\beta > h$.

By Theorem 2, Sec. 1, Chap. 5,

$$(5.2)\quad |f(\sigma + i\tau)| \leqslant C_3 \exp[a'|\sigma|^h] \quad \text{in some region } |\tau_j| \leqslant K_j(1 + |\sigma_j|)^{\mu_j}$$
$$(j = 1, \ldots, n),$$

where a' is positive (and can be taken arbitrarily close to a_0) and

$$\mu \geqslant 1 - (p - h).$$

We consider first the case $\mu \leqslant 0$. We need an auxiliary theorem which is an extension of the rule (2.3), Chap. 4:

Lemma 1. *Assume that Φ satisfies the properties (Φ_1) through (Φ_4) of Chap. 4, Sec. 1. Let $P(s)$ be an entire function such that $P(i\sigma)$ is a multiplier in Φ, and such that the series $P(i\sigma)\varphi(\sigma)$ $(\varphi \in \Phi)$ is convergent in the Φ topology, uniformly with respect to φ in bounded sets of Φ. Then, for any $u \in \Phi'$,*

$$\mathfrak{F}(P(i\sigma)u) = P(D)\mathfrak{F}(u). \tag{5.3}$$

Proof. Let $P_m(s)$ $(m = 0, 1, 2, \ldots)$ be partial sums of the power series expansion of $P(s)$. By (2.3), Chap. 4,

$$\mathfrak{F}(P_m(i\sigma)u) = P_m(D)\mathfrak{F}(u) \qquad \text{in } \tilde{\Phi}'. \tag{5.4}$$

For every $\varphi \in \Phi$ we have

$$(P_m(i\sigma)u, \varphi) = (u, P_m(i\sigma)\varphi) \rightarrow (u, P(i\sigma)\varphi) = (P(i\sigma)u, \varphi). \tag{5.5}$$

Hence, by the continuity of $\mathfrak{F}$, for any $\tilde{\varphi} \in \tilde{\Phi}$,

$$(\mathfrak{F}(P_m(i\sigma)u), \tilde{\varphi}) \rightarrow (\mathfrak{F}(P(i\sigma)u), \tilde{\varphi}) \qquad \text{as } m \rightarrow \infty. \tag{5.6}$$

Observe next that $P(D)$ is a bounded operator in $\tilde{\Phi} = \mathfrak{F}(\Phi)$. Indeed, since

$$P_m(D)\tilde{\varphi} = \mathfrak{F}(P_m(i\sigma)\varphi) \tag{5.7}$$

and since $P_m(i\sigma)\varphi \rightarrow P(i\sigma)\varphi$ in the Φ topology, uniformly with respect to φ in bounded sets of Φ, a similar statement holds for the transforms, namely (by (5.7)), $\{P_m(D)\tilde{\varphi}\}$ converges in the $\tilde{\Phi}$ topology to, say, g, and the convergence is uniform with respect to $\tilde{\varphi}$ varying in bounded sets of $\tilde{\Phi}$. Since $\{P_m(D)\tilde{\varphi}\}$ is then also pointwise convergent to g (by property 2, Sec. 1, Chap. 2 of fundamental spaces), $g = P(D)\tilde{\varphi}$ and $P(D)$ is therefore a bounded operator on $\tilde{\Phi}$. The same is true of $P(-D)$, since the mapping $\varphi(x) \rightarrow \varphi(-x)$ is a topological mapping of Φ onto itself. (Note that

$$\mathfrak{F}[\mathfrak{F}[\varphi(x)]] = (2\pi)^n \varphi(-x).)$$

Hence, by the definition (4.9) of Chap. 2,

$$(P_m(D)\tilde{u}, \tilde{\varphi}) \rightarrow (P(D)\tilde{u}, \tilde{\varphi}). \tag{5.8}$$

Applying both sides of (5.4) to $\tilde{\varphi}$, taking $m \rightarrow \infty$ and using (5.6), (5.8), the proof of (5.3) is completed.

In studying both cases $\mu \leqslant 0$ and $\mu > 0$ we need to introduce a certain entire function $f_\alpha(z)$ where $\alpha = m_0/k_0 > 0$ is a rational number (m_0, k_0 are positive integers) and z is one complex variable. We define

$$f_\alpha(z) = \sum_{j=0}^{\infty} \frac{z^{2m_0 j}}{(2k_0 j)!}. \tag{5.9}$$

Note that if ρ is a $(2k_0)$th root of unity (for example, $\rho = e^{i\pi/k_0}$) then

$$f_\alpha(z) = \frac{1}{2k_0} \sum_{h=1}^{2k_0} \exp\,[\rho^h z^{2m_0/2k_0}]. \tag{5.10}$$

$f_\alpha(z)$ is an entire function of order α and type 1. Since $\exp\,[z^{2m_0/2k_0}]$ dominates the other terms in the sum (5.10) when z is restricted to some region $|y| \leqslant \epsilon_0|x|$ ($\epsilon_0 > 0$ and sufficiently small), we conclude that

$$|f_\alpha(z)| \geqslant A_1 \exp\,[(1/2)|z|^\alpha] - A_0 \qquad \text{if } |y| \leqslant \epsilon_0|x|, \tag{5.11}$$

where the A_i are positive constants depending only on α.

Consider now the function

$$f_0(\sigma) = \frac{f(\sigma)}{H + f_\alpha(|\sigma|)} \qquad (|\sigma| = (\sigma_1^2 + \cdots + \sigma_n^2)^{1/2}), \tag{5.12}$$

where $H \geqslant 2A_0$. We have

$$[H + f_\alpha(|\sigma|)]f_0(\sigma) = f(\sigma). \tag{5.13}$$

By Corollary 1 to Theorem 10, Sec. 3, Chap. 5, $g(s) = H + f_\alpha[(s_1^2 + \cdots + s_n^2)^{1/2}]$ is a multiplier in $\tilde{\Phi} \equiv W_{\beta,a}^{\beta,b}$ $(b/a > \gamma)$ if $\beta > \alpha$. Furthermore, the partial sums $g_m(i\sigma)\tilde{\varphi}$ in the power series expansion of $g(i\sigma)\tilde{\varphi}$ converge to $g(i\sigma)\tilde{\varphi}$ in the sense of $\tilde{\Phi}$ uniformly with respect to $\tilde{\varphi}$ in bounded sets V of $\tilde{\Phi}$. Indeed, since $g_m(s)$ is bounded by $A \exp\,[|s|^\alpha]$ for some A independent of m, the sequence $\{(g_m - g)\tilde{\varphi}\}$ is bounded in $\tilde{\Phi}$ uniformly with respect to $\tilde{\varphi} \in V$, and is pointwise convergent to zero.

Applying Lemma 1 we conclude from (5.13) that

$$\tilde{f}(x) = \mathfrak{F}\{[H + f_\alpha(|\sigma|)]f_0(\sigma)\} = [H + f_\alpha(i\Delta^{1/2})]\tilde{f}_0(x) \qquad \text{in } \Phi' \tag{5.14}$$

$$(\Phi \equiv W_{\beta',1/b}^{\beta',1/a}),$$

where $\beta' = \beta/(\beta - 1)$ (we take $\beta > 1$).

We take $\alpha > h$ and observe that the functional $\tilde{f}_0$ is of function type, the function being the classical Fourier transform of $f_0(\sigma)$. (This follows from the definition of $\tilde{f}_0$ and from Plancherel's theorem.) It remains to estimate

$$\tilde{f}_0(x) = \int_{R^n} e^{ix\cdot\sigma} f_0(\sigma)\, d\sigma.$$

By (5.2), (5.12) it follows that we may modify the domain of integration from $\tau_j = 0$ to $\tau_j = K_j(1 + |\sigma_j|^{\mu_j}) \operatorname{sgn} x_j$ $(1 \leqslant j \leqslant n)$. To make sure that the denominator of $f_0(s)$ does not vanish when the domain of integration is modified, we take H sufficiently large, independently of x. We obtain

$$(5.15) \qquad |\tilde{f}_0(x)| \leqslant A_2 \int_{R^n} \{\exp[-A_3(1 + |\sigma|^\mu)|x|]\} \{\exp[-A_4|\sigma|^\alpha]\}\, d\sigma,$$

where the A_i are appropriate positive constants.

The integral in (5.15) can be estimated in the same way that the integral in (1.3) was estimated, and we get

$$(5.16) \qquad |\tilde{f}_0(x)| \leqslant A_5 \exp[-A_6|x|^{\alpha_1}] \qquad (\alpha_1 = \alpha/(\alpha - \mu)).$$

If f depends continuously on a parameter t and if the constants in (5.2) are independent of t, then from the uniform convergence of the integral of $\tilde{f}_0$ we conclude that $\tilde{f}_0$ is a continuous function of t. We state this result in a slightly more general form:

Theorem 8. *Let $D_t^j f(s, t)$ $(0 \leqslant j \leqslant j_0)$ be continuous functions of (s, t) $(s \in C^n, t_0 \leqslant t \leqslant t_1)$ which, for each fixed t, are entire functions of s satisfying (5.2) with constants $\mu \leqslant 0$, C_3, a', h, p, K independent of t. Then there exists a function $\tilde{f}_0(x, t)$ continuous in (x, t) $(x \in R^n, t_0 \leqslant t \leqslant t_1)$ together with its first j_0 t-derivatives, such that, for $0 \leqslant j \leqslant j_0$,*

$$(5.17) \quad |D_t^j \tilde{f}_0(x, t)| \leqslant A_5 \exp[-A_6|x|^{\alpha_1}] \qquad \left(A_6 > 0,\ \alpha_1 = \frac{\alpha}{\alpha - \mu}\right),$$

and

$$(5.18) \quad \tilde{f}(x, t) = [H + f_\alpha(i\Delta^{1/2})]\tilde{f}_0(x, t) \qquad \text{in } \Phi' \qquad (\Phi = W_{\beta',1/b}^{\beta',1/a})$$

for any given $\beta > \alpha > h$, $b/a > \gamma$ and α a rational number.

Consider next the case $\mu > 0$. By Theorem 3, Sec. 1, Chap. 5,

$$(5.19) \qquad |f(\sigma + i\tau)| \leqslant C_4 \exp[a'|\sigma|^h + b'|\tau|^{p/\mu}]$$

provided $\mu \leqslant 1$ and $n = 1$. We shall now assume that (5.19) holds for any n, and that $p > \mu$.

Introducing the function

$$\tilde{f}_B(x) = \int_{-B}^{B} e^{ix\cdot\sigma} f(\sigma)\, d\sigma,$$

we claim that if $B \to \infty$ then $\tilde{f}_B(x) \to \tilde{f}(x)$ in Φ', where $\Phi \equiv W_{\beta',1/b}^{\beta',1/a}$, provided $\beta > h$. Indeed, for every $\varphi \in \Phi$, $\tilde{\varphi} \in \tilde{\Phi} \equiv W_{\beta,a}^{\beta,b}$ and

$$\begin{aligned} \tilde{f}_B \cdot \varphi &= \int_{R^n} \tilde{f}_B(x)\varphi(x)\, dx = \int_{-B}^{B} f(\sigma) \left\{\int_{R^n} e^{ix\cdot\sigma}\varphi(x)\, dx\right\} d\sigma \\ &= \int_{-B}^{B} f(\sigma)\tilde{\varphi}(\sigma)\, d\sigma \to \int_{R^n} f(\sigma)\tilde{\varphi}(\sigma)\, d\sigma = f\cdot\tilde{\varphi} = \tilde{f}\cdot\varphi. \end{aligned}$$

We next claim that

$$\tilde{f}_B(x) = [H_k + f_\alpha(i\Delta^{1/2})]g_{B,k}(x), \tag{5.20}$$

where

$$g_{B,k}(x) = \int_{-B}^{B} \frac{f(\sigma)\, e^{ix\cdot\sigma}}{H_k + f_\alpha(|\sigma|)}\, d\sigma. \tag{5.21}$$

Here H_k is a sufficiently large constant to be determined later on, and $\alpha > h$. The proof of (5.20) follows by applying partial sums $f_{\alpha,m}(i\Delta^{1/2})$ of $f_\alpha(i\Delta^{1/2})$ to both sides of (5.21) and then taking $m \to \infty$.

We shall need the following lemma.

Lemma 2. *There exists a family of functions $\chi_k(x)$ having the following properties:*

(a) *The support of $\chi_k(x)$ lies in $k \leqslant |x| \leqslant k + 2$;*

(b) *$\chi_k(x)$ is a C^∞ function, and*

$$|D^q\chi_k(x)| \leqslant A_0 A^q q^{q\delta} \tag{5.22}$$

where A_0, A are constants independent of k, and δ is an arbitrarily given number > 1;

(c) *$\chi_k(x) \geqslant 0$, and $\sum_{k=0}^{\infty} \chi_k(x) \equiv 1$.*

In other words, the χ_k form a partition of unity and in addition they satisfy the inequalities (5.22).

The proof of the lemma is given in Sec. 7.

Multiplying both sides of (5.20) by $\chi_k(x)\varphi(x)$, for any $\varphi \in \Phi$, and integrating over R^n, then summing with respect to k, we obtain, upon using (c),

$$\tilde{f}_B \cdot \varphi = \sum_{k=0}^{\infty} \int_{R^n} \chi_k(x)\varphi(x)[H_k + f_\alpha(i\Delta^{1/2})]g_{B,k}(x)\, dx. \tag{5.23}$$

We shall prove that if

$$\alpha\delta < 1 \tag{5.24}$$

then

$$\begin{aligned} \int_{R^n} \chi_k(x)\varphi(x)[H_k + f_\alpha(i\Delta^{1/2})]g_{B,k}(x)\, dx \\ = \int_{R^n} g_{B,k}(x)[H_k + f_\alpha(i\Delta^{1/2})][\chi_k(x)\varphi(x)]\, dx \equiv I_{B,k}. \end{aligned} \tag{5.25}$$

Note that if $h < 1$, we can choose $\delta > 1$ and α sufficiently close to h such that (5.24) is satisfied.

In proving (5.25) we need the inequalities

$$(5.26) \qquad |D^q\varphi(x)| \leqslant A_1 A_2^q q! \exp[-A_3 k^{\beta'}] \qquad \text{if } k \leqslant |x| \leqslant k+2.$$

These inequalities follow by using the inequality (recall that $\varphi \in W_{\beta',1/b}^{\beta',1/a}$)

$$|\varphi(x+iy)| \leqslant C_0' \exp[-C_1'|x|^{\beta'}] \qquad (|y| \leqslant 1)$$

and Cauchy's inequalities with radii equal to $1/\sqrt{n}$.

Expanding $D^q(\chi_k\varphi)$ according to Leibniz' rule and using the inequalities (5.22), (5.26), we get

(5.27)

$$|D^q(\chi_k\varphi)| \leqslant \Sigma \binom{q}{r} |D^r\varphi|\,|D^{q-r}\chi_k| \leqslant A_0 A_1 (A + A_2)^q q^{q\delta} \exp[-A_3 k^{\beta'}].$$

Recalling the definition (5.9) of f_α we then obtain

$$(5.28) \qquad |f_\alpha(i\Delta^{1/2})(\chi_k\varphi)| \leqslant \sum_{j=1}^{\infty} \frac{A_4^j j^{2m_0\delta j}}{j^{2k_0 j}} \exp[-A_3 k^{\beta'}] \leqslant A_5 \exp[-A_3 k^{\beta'}],$$

since, by (5.24), $m_0\delta < k_0$.

The proof of (5.25) now follows easily. Indeed, if we replace in (5.25) $f_\alpha(i\Delta^{1/2})$ by partial sums $f_{\alpha,m}(i\Delta^{1/2})$ then (5.25) follows by integration by parts. Taking $m \to \infty$ and noting that

$$[H_k + f_{\alpha,m}(i\Delta^{1/2})]g_{B,k} \quad \text{and} \quad [H_k + f_{\alpha,m}(i\Delta^{1/2})]\,(\chi_k\varphi)$$

are uniformly convergent for $k \leqslant |x| \leqslant k+2$ as follows from the proofs of (5.20) and (5.28) respectively, the proof of (5.25) is completed.

Our next assertion is that, as $B \to \infty$,

$$(5.29) \qquad I_{B,k} \to \int_{R^n} g_k(x)[H_k + f_\alpha(i\Delta^{1/2})][\chi_k(x)\varphi(x)]\,dx,$$

where

$$(5.30) \qquad g_k(x) = \int_{R^n} \frac{f(\sigma)\,e^{ix\cdot\sigma}}{H_k + f_\alpha(|\sigma|)}\,d\sigma.$$

The proof follows by noting that $g_{B,k}(x) \to g_k(x)$ uniformly with respect to x, whereas the other factor in the integrand of $I_{B,k}$ is a bounded function with a compact support.

Our main purpose is to prove that

$$(5.31) \qquad \tilde{f}\cdot\varphi = \sum_{k=0}^{\infty} \int_{R^n} g_k(x)[H_k + f_\alpha(i\Delta^{1/2})][\chi_k(x)\varphi(x)]\,dx.$$

In view of $\tilde{f}_B \to \tilde{f}$, (5.23), (5.25) and (5.29), the assertion (5.31) would follow if we prove that, as $M \to \infty$,

$$(5.32)\qquad \sum_{k=M}^{\infty} \int_{R^n} g_{B,k}(x)[H_k + f_\alpha(i\Delta^{1/2})][\chi_k(x)\varphi(x)]\,dx \to 0,$$

$$(5.33)\qquad \sum_{k=M}^{\infty} \int_{R^n} g_k(x)[H_k + f_\alpha(i\Delta^{1/2})][\chi_k(x)\varphi(x)]\,dx \to 0.$$

The proofs of (5.32), (5.33) follow by using (5.28) and noting that g_k $g_{B,k}$ are uniformly bounded functions (independently of B, k), provided we take

$$(5.34)\qquad H_k \leqslant A_6 \exp{[A_7 k^{\beta'}]} \qquad \text{for some } A_7 < \frac{1}{\beta' b^{\beta'}},$$

so that

$$\sum_{k=M}^{\infty} \int_{R^n} H_k \chi_k(x)|\varphi(x)|\,dx \to 0 \qquad \text{as } M \to \infty.$$

We have thus proved (5.31). We shall write (5.31) also in the abbreviated form

$$(5.35)\qquad \tilde{f} = \sum_{k=0}^{\infty} \chi_k[H_k + f_\alpha(i\Delta^{1/2})]g_k.$$

We proceed to evaluate $g_k(x)$ for $k \leqslant |x| \leqslant k + 2$. Since $\alpha > h$, for any $\rho > 0$ we may change the domain of integration from $\tau_j = 0$ to $\tau_j = |x_j|^\rho \operatorname{sgn} x_j$ $(j = 1, \ldots, n)$ provided H_k is sufficiently large (so that the denominator does not vanish). We divide the new domain of integration into two parts. In the first one, say I_1, $|\tau| < \epsilon_1|\sigma|$ (for some ϵ_1 depending on ϵ_0 of (5.11)) and we can apply (5.11), (5.19) and obtain

$$(5.36)\qquad \left|\int_{I_1}\right| \leqslant B_0 \exp{[B_1|x|^{\rho p/\mu} - B_2|x|^{\rho+1}]} \leqslant B_3 \exp{[-B_4|x|^{\rho+1}]}$$

provided $H_k > A_0$ and

$$(5.37)\qquad \frac{\rho p}{\mu} < \rho + 1, \quad \text{i.e.,} \quad \rho = \frac{\mu}{p - \mu} - \epsilon \qquad \text{for some } \epsilon > 0.$$

(Recall that we assume that $p/\mu > 1$). Here and in the following, the B_j are used to denote positive constants.

For the second part I_2, $|\tau| \geqslant \epsilon_1|\sigma|$. Since f_α is of order α, and since $|\tau| \sim |x|^\rho$ and we are considering points x in $k \leqslant |x| \leqslant k + 2$, $|f_\alpha|$ is bounded by $B_5 \exp{[B_6 k^{\rho\alpha}]}$. We define

$$(5.38)\qquad H_k = B_7 \exp{[B_6 k^{\rho\alpha}]} \qquad (B_7 = A_0 + B_5)$$

and conclude, analogously to (5.36), that if $k \leqslant |x| \leqslant k + 2$ then

$$(5.39)\qquad \left|\int_{I_2}\right| \leqslant B_8 \exp{[-B_9|x|^{\rho+1}]}.$$

We shall assume that

$$\rho\alpha < \beta' \tag{5.40}$$

so that H_k, as defined by (5.38), satisfies (5.34).

Combining (5.36), (5.39) we conclude that

$$|g_k(x)| \leqslant B_{10} \exp\left[-B_{11}|x|^{\rho+1}\right] \qquad \text{if } k \leqslant |x| \leqslant k+2. \tag{5.41}$$

In view of (5.37) we can rewrite (5.40) in the form

$$\alpha < \frac{p-\mu}{\mu - \epsilon(p-\mu)}\,\beta'. \tag{5.42}$$

If this inequality is satisfied when α is replaced by h, then it is also satisfied by any α sufficiently close to h.

Coming back to $g_k(x)$, it is readily seen that if $f = f(s, t)$ depends continuously on t and if (5.19) holds for $f(s, t)$ with constants independent of t, then $g_k = g_k(x, t)$ are continuous functions of (x, t) and all the constants A_j, B_m can be taken to be independent of t. The following theorem is an obvious extension of the results we have proved for the case $\mu > 0$.

Theorem 8′. *Let $D_t^j f(s, t)$ $(0 \leqslant j \leqslant j_0)$ be continuous functions of (s, t) $(s \in C^n, t_0 \leqslant t \leqslant t_1)$ which, for every fixed t, are entire functions of s satisfying* (5.19) *with $\mu > 0$, a', b', h, p, C_4 independent of t $(\mu < p)$. Let $h < \beta \leqslant p$ and $\epsilon > 0$ be such that*

$$h < 1, \qquad h < \frac{p-\mu}{\mu - \epsilon(p-\mu)}\,\beta' \tag{5.43}$$

$$\left(\beta' = \frac{\beta}{\beta - 1},\ \beta > 1, \qquad \mu > \epsilon\,(p-\mu)\right).$$

Then there exist functions $g_k(x, t)$ $(k = 0, 1, 2, \ldots)$ such that $D_t^j g_k(x, t)$ $(0 \leqslant j \leqslant j_0)$ are continuous functions of (x, t) for $x \in R^n$, $t_0 \leqslant t \leqslant t_1$, and if $k \leqslant |x| \leqslant k+2$,

$$|D_t^j g_k(x, t)| \leqslant B' \exp\left[-B|x|^{\eta}\right] \qquad \left(\eta = \frac{p}{p-\mu} - \epsilon\right), \tag{5.44}$$

where $B' > 0$, $B > 0$ are independent of t, k, and such that

$$\tilde{f}(x, t) = \sum_{k=0}^{\infty} \chi_k(x)[H_k + f_\alpha(i\Delta^{1/2})]g_k(x, t) \qquad \text{over } W_{\beta', 1/b}^{\beta', 1/a} \tag{5.45}$$

for any positive vectors a, b $(b/a > \gamma)$, α any rational number in the interval $h < \alpha < \beta$ satisfying

$$\alpha < 1, \qquad \alpha < \frac{p-\mu}{\mu - \epsilon\,(p-\mu)}\,\beta', \tag{5.46}$$

and

(5.47) $$H_k = B_0' \exp [B_0 k^{\nu}] \qquad \left(\nu = \alpha\left(\frac{\mu}{p-\mu} - \epsilon\right)\right);$$

B_0', B_0 being positive constants independent of t, k.

6. Entire Functions of Order $\leqslant$ p and of Fast Increase

We finally consider the case where

(6.1) $$|f(s)| \leqslant C \exp [|bs|^p] \qquad (p > 1)$$

holds but no other assumptions are made on the behavior of $f(\sigma)$. We introduce the space $Z_{p,b}$ as the set of all entire functions $g(z)$ of order $\leqslant p$, and type $\leqslant b^p$. A sequence $\{g_\nu\}$ is said to converge to zero if $g_\nu(z) \to 0$ uniformly on bounded sets of C^n and

$$|g_\nu(z)| \leqslant C_1 \exp [|b'z|^p] \qquad \text{for any } b' > b,$$

where C_1 is independent of ν. A topology can be introduced in such a way that $Z_{p,b}$ becomes a perfect $Z\{M_p\}$ space. Details are left to the reader.

Let $f(s) = \Sigma\, a_m s^m$ be Taylor's series of f, and let

$$f_M(s) = \sum_{|m| \leqslant M} a_m s^m \qquad (|m| = m_1 + \cdots + m_n).$$

f and f_M may be considered as elements in K' and then $f_M \to f$ in the sense of K', as $M \to \infty$. It follows that $\tilde{f}_M \to \tilde{f}$ in the sense of Z'. Hence, for any $\varphi \in K$,

$$(\tilde{f}, \tilde{\varphi}) = \Sigma\, (\widetilde{a_m \sigma^m}, \tilde{\varphi}) = (2\pi)^n \Sigma\, (a_m(-i\, D_x)^m\, \delta(x), \tilde{\varphi}),$$

i.e.,

(6.2) $$\tilde{f}(x) = (2\pi)^n \sum_m (-1)^m\, i^m a_m\, D_x^m\, \delta(x) \qquad \text{in } Z'.$$

The functional on the right side can be extended, in a natural way, to all $\psi \in Z_{p',1/b_1}$ for any $b_1 > bp^{1/p}(p')^{1/p'}$. Indeed, writing

$$\psi(z) = \Sigma\, \frac{z^m \psi^{(m)}(0)}{m!}$$

and using Cauchy's inequalities with $|z_j| = r_j$ $(j = 1, \ldots, n)$ we get (for some choice of the r_j, depending on m; compare (2.8))

$$|\psi^{(m)}(0)| \leqslant C_1' m! \frac{\theta^m}{b_0^m} \left(\frac{ep'}{m}\right)^{m/p'}$$

if $|\psi(z)| \leqslant C_1 \exp [|z/b_0|^{p'}]$, where b_0 is any vector $< b_1$ and θ is any vector having components > 1.

Similarly,

$$|a_m| \leqslant C'(b')^m \left(\frac{ep}{m}\right)^{m/p} \qquad \text{for any } b' > b.$$

We then obtain

$$(6.3) \quad |(\tilde{f}, \psi)| \leqslant (2\pi)^n C' C_1' \Sigma \left(\frac{\theta b'}{b_0}\right)^m \left(\frac{ep}{m}\right)^{m/p} \left(\frac{ep'}{m}\right)^{m/p'} m! \leqslant AC'C_1',$$

where A depends only on p and on $\theta b'/b_0$.

We observe that if $\psi_\nu \to 0$ in $Z_{p',1/b_1}$ then the series expansion of $(\tilde{f}, \psi_\nu)$ is convergent uniformly with respect to ν, whereas each term in the series tends to 0 as $\nu \to \infty$. Using this remark one can show, by appropriately dividing the series expansion of $(\tilde{f}, \psi_\nu)$ into two parts, that if $\psi_\nu \to 0$ in $Z_{p',1/b_1}$ then $(\tilde{f}, \psi_\nu) \to 0$. Hence $\tilde{f}$ is a continuous functional over $Z_{p',1/b_1}$.

If we consider $f(s)$ as a functional over $W_{p,c}^{p,d}$, where $c^p > pb^p$, then by slightly modifying the proof of (6.2), noting that for $\varphi \in W_{p,c}^{p,d}$

$$(\tilde{f}, \tilde{\varphi}) = (f, \mathfrak{F}\tilde{\varphi}) = (\Sigma\, a_m\sigma^m, \mathfrak{F}\tilde{\varphi}) = \Sigma\, (a_m\sigma^m, \mathfrak{F}\tilde{\varphi}) = \Sigma\, (\widetilde{a_m\sigma^m}, \tilde{\varphi}),$$

we derive the equality in (6.2) over $W_{p',1/d}^{p',1/c}$ $((1/p) + (1/p') = 1)$.

If $f = f(s, t)$ depends continuously on t, then the $a_m = a_m(t)$ also depend continuously on t. We conclude:

Theorem 9. *Let $D_t^j f(s, t)$ $(0 \leqslant j \leqslant j_0)$ be continuous functions of (s, t) $(s \in C^n,\ t_0 \leqslant t \leqslant t_1)$ which, for each fixed t, are entire functions of s satisfying* (6.1) *with C, b independent of t. Then f can be considered as an element of $(W_{p,c}^{p,d})'$ where $c^p > pb^p$ and*

$$(6.4) \qquad \tilde{f}(x, t) = (2\pi)^n \Sigma\, (-1)^m i^m a_m(t)\, D_x^m\, \delta(x) \qquad \text{over } W_{p',1/d}^{p',1/c}.$$

The functions $a_m(t)$ have j_0 continuous derivatives for $t_0 \leqslant t \leqslant t_1$, and

$$(6.5) \qquad |D_t^j a_m(t)| \leqslant C'(b')^m \left(\frac{ep}{m}\right)^{m/p} \qquad (0 \leqslant j \leqslant j_0)$$

for any $b' > b$ and C' independent of t. The right side of (6.4) *gives an extension of $\tilde{f}$ into a continuous functional over $Z_{p',1/b_1}$, for any $b_1 > bp^{1/p}(p')^{1/p'}$.*

7. Proof of Lemma 2

Let $\{M_j\}$ be a monotone-increasing sequence of positive numbers $\geqslant 1$ and satisfying the inequalities

$$(7.1) \qquad \binom{q}{j} M_j M_{q-j} \leqslant AM_q \qquad (0 \leqslant j \leqslant q,\ 0 \leqslant q < \infty,\ A > 0).$$

For example,

$$M_j = \Gamma(\delta j + 1) \qquad (\delta \geqslant 1), \tag{7.2}$$

where $\Gamma(t)$ is the gamma function.

Let $u_1(x), \ldots, u_N(x)$ be C^∞ functions in a domain D of R^n and let $f(u_1, \ldots, u_N)$ be a C^∞ function in a domain D_0 of R^N which contains the points $(u_1(x), \ldots, u_N(x))$ as x varies in D. We assume that if $x \in D$,

$$|D^q u_i(x)| \leqslant A_1 A_2^q M_{q-2} \qquad (i = 1, \ldots, n;\ 1 \leqslant |q| < \infty), \tag{7.3}$$

where for simplicity we define

$$M_q = M_{|q|}, \quad M_{q-i} = M_{|q|-|i|}, \quad M_{-2} = M_{-1} = 1, \quad A^q = A^{|q|},$$

We finally assume that if $u \in D_0$,

$$|D^q f(u)| \leqslant A_3 A_4^q M_{q-2} \qquad (0 \leqslant |q| < \infty). \tag{7.4}$$

We shall prove:

Theorem 10. *Under the foregoing assumptions (in particular, (7.1), (7.3), (7.4)), the function $f[x] \equiv f(u_1(x), \ldots, u_N(x))$ satisfies, for any $x \in D$,*

$$|D_x^q f[x]| \leqslant A_5 A_6^q M_{q-2} \qquad (0 \leqslant |q| < \infty), \tag{7.5}$$

where A_5, A_6 depend only on A_1, A_2, A_3, A_4.

Proof. It suffices to consider only $|q| > 0$. $D_x^q f[x]$ is a linear combination with positive coefficients of expressions of the form

$$\frac{\partial^r f}{\partial u_{i_1} \cdots \partial u_{i_r}} u_{t_1}^{(s_1)} \cdots u_{t_\lambda}^{(s_\lambda)} \qquad (1 \leqslant s_j \leqslant |q|,\ 1 \leqslant |r| \leqslant q).$$

Hence if we replace $f(u)$ by a function $F(u)$ for which

$$|D^r f(u)| \leqslant D^r F(u) \qquad (1 \leqslant |r| \leqslant |q|), \tag{7.6}$$

and if we replace $u_j(x)$ by functions $v_j(x)$ such that

$$|D^k u_j(x)| \leqslant D^k v_j(x) \qquad (1 \leqslant |k| \leqslant |q|, j = 1, \ldots, n), \tag{7.7}$$

then it follows that

$$|D_x^q f[x]| \leqslant D^q F[x], \tag{7.8}$$

where $F[x] \equiv F(v_1(x), \ldots, v_N(x))$.

We have to estimate $D^q f[x]$ for each fixed $x = x^0$ in D. For simplicity we may assume that $x^0 = 0$, $u_j(0) = 0$ $(j = 1, \ldots, N)$.

We then take $F(v)$ to be

$$F(v) = A_3 \sum_{s=1}^{|q|} \frac{A_4^s M_{s-2}}{s!} w^s \qquad \text{where } w = v_1 + \cdots + v_N. \tag{7.9}$$

(7.6) is then clearly satisfied at $u = 0$. We next take

$$(7.10)\qquad v_j(x) = v(x) = A_1 \sum_{r=1}^{|q|} \frac{A_2^r M_{r-2}}{r!} \xi^r \qquad \text{where } \xi = x_1 + \cdots + x_n$$

and it is clear that (7.7) is satisfied at $x = 0$. It thus remains to estimate $D^q F[\xi]$ at $\xi = 0$.

Using (7.1) and the notation $(-1)! = 1$, we get, if $p > 2$,

$$\sum_{\substack{r+s=p \\ r\geqslant 1, s\geqslant 1}} \frac{M_{r-2}}{r!} \frac{M_{s-2}}{s!}$$
$$\leqslant \max_{\substack{r+s=p \\ r\geqslant 1, s\geqslant 1}} \frac{M_{r-2}}{(r-2)!} \frac{M_{s-2}}{(s-2)!} \left[\sum_{\substack{r+s=p \\ r\geqslant 2, s\geqslant 2}} \frac{1}{r(r-1)s(s-1)} + \frac{2}{(p-1)(p-2)} \right]$$
$$\leqslant AA_7 \frac{M_{p-3}}{(p-3)!} \frac{1}{p^2}.$$

Hence,

$$(7.11)\qquad v^2(x) \prec A_1^2 A_8 \sum_{r=2}^{|q|} \frac{r A_2^r M_{r-3}}{r!} \xi^r,$$

where the relation $\prec$ is defined as follows: if $w = \sum a_p \xi^p$, $z = \sum b_p \xi^p$ then $w \prec z$ if and only if $|a_p| \leqslant b_p$ for all $p \leqslant |q|$. Note that A_8 is a universal constant.

Proceeding step by step to evaluate v^3, v^4, and so on, we find that

$$(7.12)\qquad v^s(x) \prec A_1^s A_8^{s-1} \sum_{r=s}^{|q|} A_2^r \frac{r(r-1)\cdots(r-s+2) M_{r-s-1}}{r!} \xi^r.$$

Substituting (7.12) into (7.9) we get

$$F[x] \prec A_3 \sum_{s=1}^{|q|} \frac{A_4^s A_1^s N^s A_8^{s-1} M_{s-2}}{s!} \sum_{r=s}^{|q|} A_2^r \frac{r \cdots (r-s+2) M_{r-s-1}}{r!} \xi^r.$$

Hence the coefficient of $\xi^{|q|}$ is bounded by

$$A_3 A_9^q \sum_{s=1}^{|q|} \frac{M_{s-2}}{s!} \frac{M_{q-s-1}}{(|q|-s+1)!} \leqslant A_{10} A_9^q \frac{M_{q-2}}{q!}.$$

But this means that

$$D^q F(0) \leqslant A_{10} A_9^q M_{q-2}.$$

Recalling (7.8), the proof of the theorem is thereby completed.

Remark. From the above proof it follows that if we assume that (7.3), (7.4) hold only for $|q| \leqslant m$ then (7.5) holds for all $|q| \leqslant m$ and A_5, A_6 are *independent* of m.

In what follows we shall use only some special cases of Theorem 10.

Beside Theorem 10, we shall need the following well-known theorem of Carleman and Mandelbrojt [50, p. 65]:

Theorem 11. *Let* $\{M_j\}$ *be a monotone-increasing sequence of positive numbers and let*

$$T(r) = \underset{1 \leqslant j < \infty}{\text{l.u.b.}} \frac{r^j}{M_j}.$$

If

$$\int_1^\infty \frac{\log T(r)}{r^2}\, dr < \infty, \tag{7.13}$$

then there exists a C^∞ *function* $f(x) \not\equiv 0$ (x *one-dimensional*) *which is non-negative, has its support in the interval* $0 \leqslant x \leqslant 1$, *and satisfies the inequalities*

$$|f^{(j)}(x)| \leqslant M_j \qquad (0 \leqslant j < \infty). \tag{7.14}$$

From now on we take the M_j to be defined by (7.2), where $\delta > 1$. Then (7.13) is satisfied. Let x_0 be a point where $f \neq 0$. Set

$$g(x) = g_1(ax + b),$$

where

$$g_1(x) = f(x_0 + x^{2m+1}),$$

m is a positive integer, and a, b satisfy

$$a\epsilon_0 + b = -1 + \epsilon_0,$$

$$a(1 - \epsilon_0) + b = 1 - \epsilon_0,$$

where ϵ_0 is an arbitrary positive number $< \frac{1}{2}$. Then, for any sufficiently large m (depending on ϵ_0),

$$g(x) > 0 \qquad \text{if } \epsilon_0 \leqslant x \leqslant 1 - \epsilon_0,$$

$$g(x) = 0 \qquad \text{if } x \leqslant 0 \text{ and if } x \geqslant 1.$$

Clearly $g(x) \geqslant 0$ for $0 \leqslant x \leqslant 1$ and, by Theorem 10, g also satisfies the inequalities

$$|g^{(j)}(x)| \leqslant H_0 H^j M_{j-2} \qquad (H_0 > 0,\ H > 0). \tag{7.15}$$

Now let x be a real n-dimensional variable and define

$$G_k(x) = g\left(\frac{|x|^2 - k^2}{\alpha_k}\right) \quad (k \geqslant 1), \qquad G_0(x) = g\left(\frac{|x|^2 + \alpha}{\alpha_0}\right),$$

where $\alpha_k = 4k$ for all k sufficiently large, say $k \geqslant k_0$.

Using Theorem 10 we conclude that

$$|D^q G_k(x)| \leqslant H_1 H_2^q M_q \qquad (0 \leqslant |q| < \infty),$$

where H_1, H_2 are independent of k. Note further that by appropriately choosing k_0 and the α_k with $k \leqslant k_0$ we get, for $k \geqslant 1$,

$$G_k(x) > 0 \qquad \text{if } k + \tfrac{1}{2} \leqslant |x| \leqslant k + \tfrac{3}{2},$$

$$G_k(x) = 0 \qquad \text{if } |x| \leqslant k,\ |x| \geqslant k + 2,$$

provided ϵ_0 is sufficiently small. We take α, α_0 such that $G_0(x) > 0$ if $0 \leqslant |x| \leqslant \frac{3}{2}$ and $G_0(x) = 0$ if $|x| \geqslant 2$. We finally define

$$\chi_k(x) = \frac{G_k(x)}{\Sigma\, G_j(x)} \tag{7.16}$$

and note that, in view of Theorem 10, the χ_k satisfy (5.22). All the other requirements of Lemma 2 are obviously satisfied.

PROBLEMS

1. Prove that every continuous linear functional over $\mathring{Z}(G_b)$ has the form

$$(f, g) = \int_{G_{b+\epsilon}} g(s)\, d\mu(s) \qquad (\text{for } g \text{ analytic in } G_{b+\epsilon})$$

for any given $\epsilon > 0$, where μ is a completely additive complex measure, depending on ϵ. [*Hint:* Use F. Riesz' theorem.]

2. Let $f(z) = \Sigma\, c_\nu z^\nu$ be an entire function of exponential type $\leqslant b$, i.e.,

$$|f(z)| \leqslant C_\epsilon \exp\,[(b + \epsilon)|z|] \qquad (\text{for any } \epsilon > 0).$$

Prove that

$$(f, g) = \Sigma\, a_\nu f_\nu,$$

where $g = \Sigma\, a_\nu s^\nu$, $f_\nu = (-i)^\nu c_\nu \nu!$, is a continuous linear functional over $\mathring{Z}(G_{b+\epsilon})$, for any $\epsilon > 0$.

3. Prove that every entire function f of exponential type $\leqslant b$ can be represented in the form

$$f(z) = \int_{G_{b+\epsilon}} e^{iz\cdot s}\, d\mu(s) \qquad (\text{for any } \epsilon > 0),$$

where μ is a completely additive complex measure. [*Hint:* Note that $f_\nu = (f, s^\nu) = \int s^\nu\, d\mu(s)$.]

4. In Sec. 6 it was proved that if $f(s)$ is an entire function of order $\leqslant p$ $(p > 1)$ and type $\leqslant b^p$ then, considering f as an element of K', its Fourier transform g (which belongs to Z') can be extended into a continuous linear functional over $Z_{p',1/b_1}$. Prove the converse: if $g \in Z'$ can be extended into a continuous linear functional over $Z_{p',1/b_1}$ then $g = \mathfrak{F}f$ where $f(\sigma) \in K'$ and $f(s)$ is an entire function of order $\leqslant p$ and type $\leqslant b^p$. (The analogues of both statements for $p = 1$ were given in Theorems 2, 3, Sec. 2.)

CHAPTER 7

THE CAUCHY PROBLEM FOR SYSTEMS OF PARTIAL DIFFERENTIAL EQUATIONS

In this chapter we solve the Cauchy problem for any system of partial differential equations having coefficients independent of x, i.e., we solve the problem

$$(0.1) \qquad \frac{\partial u(x, t)}{\partial t} = P\left(t, i\frac{\partial}{\partial x}\right)u(x, t) + f(x, t) \qquad (x \in R^n, 0 < t \leqslant T),$$

$$(0.2) \qquad u(x, 0) = u_0(x) \qquad (x \in R^n),$$

where f and u_0 are given vectors and P is a matrix whose elements are any partial differential operators with coefficients depending on t. The uniqueness and existence of "weak" solutions (i.e. of generalized functions which satisfy (0.1), (0.2) in the sense of the topology of generalized functions) are proven for any given generalized functions f, u_0. It is further shown that if u_0 and f satisfy some differentiability and growth conditions, depending only on bounds on the asymptotic growth of the eigenvalues of $P(t, s)$ for both complex s and real s, then the "weak" solution is a solution in the usual sense. It is also shown that, roughly speaking, the conditions on u_0, f cannot be weakened. We also discuss convolution equations and difference-differential equations.

Extensive use is made of the material of Chaps. 1, 2, 4, 5, and, in particular, 6.

1. Systems of Partial Differential Equations and the Cauchy Problem

Let $u = (u_1, \ldots, u_N)$ be a vector whose components u_j are functions of (x, t), where $x = (x_1, \ldots, x_n)$ varies in R^n and t is a real variable which varies in an interval $0 \leqslant t \leqslant T$ or $0 < t \leqslant T$. Let P_{jk} $(j, k = 1, \ldots, N)$ be linear differential operators of orders $\leqslant p$ with coefficients depending on t, i.e.,

$$(1.1) \qquad P_{jk}\left(t, i\frac{\partial}{\partial x}\right) = \sum_{|\alpha| \leqslant p} P_\alpha^{jk}(t)\, D_x^\alpha \qquad (i = \sqrt{-1}),$$

where $D_x^\alpha = D_1^{\alpha_1} \cdots D_n^{\alpha_n}$ and $D_m = \partial/\partial x_m$. The factor i is introduced in order to get, later on, simpler formulas. We shall be concerned with the problem of finding a solution u of the system of linear partial differential equations

$$\frac{\partial u_j}{\partial t} = \sum_{k=1}^{N} P_{jk}\left(t, i\frac{\partial}{\partial x}\right) u_k + f_j \qquad (j = 1, \ldots, N) \tag{1.2}$$

under the *initial conditions*

$$u_j(x, 0) = u_{0j}(x) \qquad (j = 1, \ldots, N), \tag{1.3}$$

where $f_j(x, t)$ and $u_{0j}(x)$ are given functions. This problem is called a *Cauchy problem.*

The more general Cauchy problem

$$\frac{\partial^{m_j} u_j}{\partial t^{m_j}} = \sum_{k=1}^{N} P_{jk}\left(t, \frac{\partial}{\partial t}, i\frac{\partial}{\partial x}\right) u_k + f_j \qquad (j = 1, \ldots, N), \tag{1.4}$$

$$\frac{\partial^h u_j(x, 0)}{\partial t^h} = u_{hj}(x) \qquad (h = 0, \ldots, m_j - 1; j = 1, \ldots, N), \tag{1.5}$$

where

$$P_{jk} = \sum_{|\alpha| \leqslant p} \sum_{h=0}^{m_k - 1} P_{h\alpha}^{jk}(t)\, D_t^h\, D_x^\alpha,$$

can be reduced to the problem (1.2), (1.3). Indeed, substituting

$$u_{j1} = u_j, \quad u_{j2} = \frac{\partial u_j}{\partial t}, \quad \ldots, \quad u_{jm_j} = \frac{\partial^{m_j - 1} u_j}{\partial t^{m_j - 1}}, \tag{1.6}$$

we can write (1.4), (1.5) in the equivalent form

$$\begin{aligned} \frac{\partial u_{j1}}{\partial t} &= u_{j2}, \\ \frac{\partial u_{j2}}{\partial t} &= u_{j3}, \\ \frac{\partial u_{jm_j}}{\partial t} &= \sum_{k=1}^{N} \sum_{h=1}^{m_k} P_{jkh}\left(t, i\frac{\partial}{\partial x}\right) u_{kh} + f_j \qquad (j = 1, \ldots, N), \end{aligned} \tag{1.7}$$

$$u_{jh}(x, 0) = u_{h-1,j}(x) \qquad (h = 1, \ldots, m_j; j = 1, \ldots, N), \tag{1.8}$$

where P_{jkh} are differential operators of orders $\leqslant p$.

In particular, any differential equation

$$\frac{\partial^m u}{\partial t^m} + \sum_{|\alpha| \leqslant p} \sum_{j=0}^{m-1} a_{j\alpha}(t)\, D_t^j\, D_x^\alpha u = f(x, t) \tag{1.9}$$

with the initial conditions

$$\frac{\partial^h u(x, 0)}{\partial t^h} = u_{0h}(x) \qquad (h = 0, 1, \ldots, m - 1) \tag{1.10}$$

can be reduced to the form (1.2), (1.3).

It will be shown later on that the conditions under which the Cauchy problem (1.2), (1.3) has a solution depend on the eigenvalues of the matrix (P_{jk}), i.e., on the solutions $\lambda = \lambda(s, t)$ of the equation

$$\det (P_{jk}(t, s) - \lambda\delta_{jk}) = 0, \tag{1.11}$$

where $\delta_{jk} = 0$ if $j \neq k$ and $\delta_{jj} = 1$. It is therefore useful to note that the eigenvalues corresponding to the system (1.7) can be calculated directly from the system (1.4). In fact, we have the following theorem.

Theorem 1. *The eigenvalues* $\lambda = \lambda(s, t)$ *corresponding to* (1.7) *are the solutions of the equation*

$$\det (P_{jk}(t, \lambda, s) - \lambda^{m_j}\delta_{jk}) = 0. \tag{1.12}$$

In particular, the eigenvalues corresponding to the system (1.7) to which the equation (1.9) is reduced by the transformation (1.6) are the roots of the equation

$$\lambda^m + \sum_{|\alpha|\leqslant p} \sum_{j=0}^{m-1} a_{j\alpha}(t)\lambda^j s^\alpha = 0. \tag{1.13}$$

Proof. Note first that λ is an eigenvalue corresponding to (1.2) if and only if there exists a nonzero solution

$$u_j = C_j \exp [\lambda t - is\cdot x] \qquad (j = 1, \ldots, N;\ \text{the } C_j \text{ are constants})$$

of (1.2) with $f_j = 0$. Applying this remark to the system (1.7) we conclude that λ is an eigenvalue corresponding to (1.7) if and only if there exists a nontrivial solution

$$u_{ij} = C_{ij} \exp [\lambda t - is\cdot x]$$

of (1.7) with $f_j = 0$. Substituting u_{ij} into (1.7) we get the equations

$$\begin{aligned} \lambda C_{j1} &= C_{j2}, \\ \lambda C_{j2} &= C_{j3}, \\ &\;\vdots \\ \lambda C_{jm_j} &= \Sigma\, P_{jk}(t, \lambda, s)C_{k1} \qquad (j = 1, \ldots, N). \end{aligned} \tag{1.14}$$

Multiplying the first equation by λ^{m_j-1}, the second by λ^{m_j-2}, and so on, and adding, we obtain the equivalent equations

$$\lambda^{m_j}C_{j1} = \Sigma\, P_{jk}(t, \lambda, s)C_{k1} \qquad (j = 1, \ldots, N). \tag{1.15}$$

Since a nontrivial solution C_{j1} of (1.15) exists if and only if λ satisfies (1.12), the proof is completed.

For the sake of simplicity we shall first deal only with the case where the coefficients of P_{jk} in (1.1) do not depend on t, and $f_j \equiv 0$. In Sec. 11 we shall extend the results to the general case. We mention, however, that the methods of this chapter cannot be extended to differential equations with coefficients depending on (x, t). Some of the results have been proved by other methods (which are technically more involved than the methods of this chapter) to equations of hyperbolic and of parabolic type with coefficients depending on (x, t).

We shall deal in this chapter with vectors whose components are generalized functions. We therefore need to observe that the theory of generalized functions developed in the previous chapters remains true also for vector-valued generalized functions and for matrices whose elements are generalized functions. In particular, the analysis of the Fourier transforms of entire functions holds for matrices whose elements are entire functions. Quite often we shall refer to a vector-valued (generalized) function simply as a (generalized) function. Similarly, by a W space we shall mean the Cartesian product of N identical W spaces.

If $f = (f_1, \ldots, f_N)$, $\varphi = (\varphi_1, \ldots, \varphi_N)$, where the f_j and φ_j are generalized and test functions respectively, we use the notation

$$f \cdot \varphi = (f, \varphi) = \sum_{j=1}^{N} f_j \cdot \varphi_j.$$

The notation of Sec. 1, Chap. 2, will be freely used in this chapter.

We conclude this section by defining the concept of a solution of the Cauchy problem.

Definition 1. A function $u(x, t)$ is called a *classical solution* (or a *smooth solution*) of (1.2), (1.3) in $0 \leqslant t \leqslant T$ if u satisfies the following conditions:

(i) $u(x, t)$ is a continuous function of (x, t) for $x \in R^n$, $0 \leqslant t \leqslant T$;
(ii) all the derivatives of $u(x, t)$ which appear in (1.2) are continuous functions of (x, t) for $x \in R^n$, $0 < t \leqslant T$;
(iii) $u(x, t)$ satisfies (1.2) for $x \in R^n$, $0 < t \leqslant T$, and
(iv) $u(x, t)$ satisfies (1.3) for $x \in R^n$.

Definition 2. $u = u(x, t)$ is called a *generalized solution* in Φ' (or over Φ) of (1.2), (1.3) if u satisfies the following conditions:

(i) for every fixed t, $0 < t \leqslant T$, $u(x, t)$ is an element of Φ';
(ii) u satisfies (1.2) in the weak sense of Φ', i.e., for every $\varphi \in \Phi$, $0 < t \leqslant T$,

$$\frac{d}{dt}(u(x, t), \varphi(x)) \equiv \lim_{h \to 0} \left(\frac{u(x, t+h) - u(x, t)}{h}, \varphi(x) \right)$$
$$= \left(u(x, t), P^*\left(t, -i\frac{\partial}{\partial x}\right) \varphi(x) \right) + (f(x, t), \varphi(x))$$

(for $t = T$ we take only $h < 0$), where P^* is the transpose of P, and

(iii) u satisfies (1.3) in the following sense:

$$\lim_{t\to 0} (u(x, t), \varphi(x)) = (u_0(x), \varphi(x)) \qquad \text{for any } \varphi \in \Phi.$$

Recall that if $u(x, t)$ is a locally integrable function of x, for each fixed t, then the generalized function which u defines is given by

$$(u(x, t), \varphi(x)) = \int u(x, t)\varphi(x)\, dx,$$

where by $u(x, t)\varphi(x)$ we understand the expression $\sum_{j=1}^{N} u_j(x, t)\varphi_j(x)$ whenever

$$u(x, t) = (u_1(x, t), \ldots, u_N(x, t)) \quad \text{and} \quad \varphi(x) = (\varphi_1(x), \ldots, \varphi_N(x)).$$

2. Auxiliary Theorems on Functions of Matrices

Let $A = (a_{ij})$ be an $N \times N$ matrix and let $v = (v_1, \ldots, v_N)$ be a variable point in the complex space C^N. Considering A as a linear operator in C^N, we denote its norm by $\|A\|$, i.e.,

$$\|A\| = \operatorname*{l.u.b.}_{v \neq 0} \frac{|Av|}{|v|},$$

where $|v| = (\Sigma\, |v_j|^2)^{1/2}$. The reader will easily verify that

$$\max_{1 \leqslant k \leqslant N} \sum_{j=1}^{N} |a_{jk}|^2 \leqslant \|A\|^2 \leqslant \sum_{j,k=1}^{N} |a_{jk}|^2. \tag{2.1}$$

Let $f(z)$ be an entire function of one complex variable z, and let A be an $N \times N$ matrix. We define $f(A)$ by

$$f(A) = \sum_{m=0}^{\infty} c_m A^m \qquad \text{if } f(z) = \sum_{m=0}^{\infty} c_m z^m. \tag{2.2}$$

To prove convergence of the series of $f(A)$ we use the obvious inequality

$$\|AB\| \leqslant \|A\|\, \|B\|$$

$m - 1$ times and get

$$\|A^m\| \leqslant \|A\|^m.$$

It follows that

$$\Big\| \sum_{m=M}^{H} c_m A^m \Big\| \leqslant \sum_{m=M}^{H} |c_m|\, \|A\|^m \to 0$$

as $M \to \infty$. Recalling (2.1) we see that each (j, k)-element in the series of $f(A)$ is convergent; in other words, the series of $f(A)$ is convergent.

From the above proof it follows, incidentally, that the definition (2.2) is also valid when $f(z)$ is only assumed to be analytic in the disc $|z| \leqslant \|A\|$.

Let λ_j ($j = 1, \ldots, N$) be the eigenvalues of the matrix A, and denote by $H(A)$ the convex hull (in the complex plane) of the points $\lambda_1, \ldots, \lambda_N$. We shall prove the following:

Theorem 2. *For any entire function* $f(z)$,

$$\|f(A)\| \leqslant \sum_{j=0}^{N-1} 2^j \|A\|^j \max_{z \in H(A)} |f^{(j)}(z)|. \tag{2.3}$$

Proof. We first observe that it suffices to prove (2.3) in the case where all the eigenvalues λ_j are distinct from each other. Indeed, every matrix $A = (a_{ij})$ can be approximated by matrices $A_m = (a_{ij}^m)$ ($a_{ij}^m \to a_{ij}$ as $m \to \infty$) having simple eigenvalues. If we apply (2.3) to each A_m and note that $f(A_m) \to f(A)$ as $m \to \infty$, we obtain (2.3) for any matrix A.

Let $R(z)$ be the polynomial of degree $N - 1$ which satisfies

$$R(\lambda_j) = f(\lambda_j) \qquad (j = 1, \ldots, N).$$

As is well known (see, for instance, [19, Chap. 5]) it then follows that

$$R(A) = f(A). \tag{2.4}$$

We shall evaluate the coefficients of $R(z)$ by writing it in Newton's form

$$\begin{aligned} R(z) = b_1 + b_2(z - \lambda_1) + b_3(z - \lambda_1)(z - \lambda_2) + \cdots \\ + b_N(z - \lambda_1)(z - \lambda_2) \cdots (z - \lambda_{N-1}). \end{aligned} \tag{2.5}$$

Setting $f_j = f(\lambda_j)$, the b_j are then defined by

$$\begin{aligned} f_1 &= b_1, \\ f_2 &= b_1 + b_2(\lambda_2 - \lambda_1), \\ f_3 &= b_1 + b_2(\lambda_3 - \lambda_1) + b_3(\lambda_3 - \lambda_1)(\lambda_3 - \lambda_2), \\ &\cdot \\ &\cdot \\ &\cdot \\ f_N &= b_1 + b_2(\lambda_N - \lambda_1) + b_3(\lambda_N - \lambda_1)(\lambda_N - \lambda_2) + \cdots \\ &\qquad + b_N(\lambda_N - \lambda_1) \cdots (\lambda_N - \lambda_{N-1}). \end{aligned} \tag{2.6}$$

Introducing the notation

$$[f_j] = f_j, \qquad [f_{j_1 j_2}] = \frac{[f_{j_2}] - [f_{j_1}]}{\lambda_{j_2} - \lambda_{j_1}}, \qquad \ldots,$$

$$[f_{j_1 j_2 \cdots j_k}] = \frac{[f_{j_1 \cdots j_{k-2} j_k}) - [f_{j_1 \cdots j_{k-2} j_{k-1}}]}{\lambda_{j_k} - \lambda_{j_{k-1}}},$$

we claim that the solution of (2.6) is given by

$$b_h = [f_{12\ldots h}]. \tag{2.7}$$

Indeed, assuming, by induction, that (2.7) is true for $h \leqslant k$, we get

$$\begin{aligned} b_{k+1} &= \frac{f_{k+1} - f_1 - [f_{12}](\lambda_{k+1} - \lambda_1) - \cdots - [f_{12\ldots k}](\lambda_{k+1} - \lambda_1) \cdots (\lambda_{k+1} - \lambda_k)}{(\lambda_{k+1} - \lambda_1)(\lambda_{k+1} - \lambda_2) \cdots (\lambda_{k+1} - \lambda_k)} \\ &= \frac{[f_{1,k+1}] - [f_{12}] - [f_{123}](\lambda_{k+1} - \lambda_2) - \cdots - [f_{12\ldots k}](\lambda_{k+1} - \lambda_2) \cdots (\lambda_{k+1} - \lambda_k)}{(\lambda_{k+1} - \lambda_2) \cdots (\lambda_{k+1} - \lambda_k)} \\ &= \frac{[f_{12,k+1}] - [f_{123}] - \cdots - [f_{12\ldots k}](\lambda_{k+1} - \lambda_3) \cdots (\lambda_{k+1} - \lambda_k)}{(\lambda_{k+1} - \lambda_3) \cdots (\lambda_{k+1} - \lambda_k)} \\ &= \cdots = [f_{12\ldots k,k+1}]. \end{aligned}$$

In order to evaluate $[f_{12\ldots k}]$ we introduce the function

$$u_k(\lambda) = \int_0^1 \int_0^{t_1} \cdots \int_0^{t_{k-1}} f^{(k)}[\lambda_1 + (\lambda_2 - \lambda_1)t_1 + \cdots + (\lambda_k - \lambda_{k-1})t_{k-1} + (\lambda - \lambda_k)t_k]\, dt_1 \cdots dt_k.$$

Integrating with respect to t_k we get

$$\begin{aligned} u_k(\lambda) &= \frac{1}{\lambda - \lambda_k} \int_0^1 \int_0^{t_1} \cdots \int_0^{t_{k-2}} f^{(k-1)}[\lambda_1 + (\lambda_2 - \lambda_1)t_1 + \cdots + (\lambda_k - \lambda_{k-1})t_{k-1} + (\lambda - \lambda_k)t_k] \Big|_0^{t_{k-1}} dt_1 \cdots dt_{k-1} \\ &= \frac{u_{k-1}(\lambda) - u_{k-1}(\lambda_k)}{\lambda - \lambda_k}, \end{aligned}$$

from which we easily find, by induction, that

$$u_{k-1}(\lambda_k) = [f_{12\ldots k}].$$

Recalling (2.7) we conclude that

$$b_k = \int_0^1 \int_0^{t_1} \cdots \int_0^{t_{k-2}} f^{(k-1)}[\lambda_1 + (\lambda_2 - \lambda_1)t_1 + \cdots + (\lambda_k - \lambda_{k-1})t_{k-1}]\, dt_1 \cdots dt_{k-1}. \tag{2.8}$$

The argument of $f^{(k-1)}$ is

$$\tau \equiv (1 - t_1)\lambda_1 + (t_1 - t_2)\lambda_2 + \cdots + t_{k-1}\lambda_k = \sum_{j=1}^{k} c_j\lambda_j$$

and $c_j \geqslant 0$, $\Sigma\, c_j = 1$. Hence τ belongs to the convex hull $H(A)$ of the eigenvalues of A. It follows that

$$|b_k| \leqslant \max_{z \in H(A)} |f^{(k-1)}(z)|. \tag{2.9}$$

Noting that for any λ_j there exists a unit eigenvector x, i.e., $Ax = \lambda_j x$ and $|x| = 1$, we find that

$$|\lambda_j| \leqslant \|A\|. \tag{2.10}$$

It follows that $\|A - \lambda_j I\| \leqslant 2\|A\|$. Combining this inequality with (2.9), (2.5), and (2.4), the inequality (2.3) follows.

Consider now the special case where

$$f(z) = e^{tz}, \qquad A = P(s),$$

$0 \leqslant t \leqslant T$, and $P(s)$ is a matrix whose elements are polynomials in $s = (s_1, \ldots, s_n)$ of degrees $\leqslant p$. Let $\lambda_1(s), \ldots, \lambda_n(s)$ be the eigenvalues of $P(s)$ and set

$$\Lambda(s) = \max_{1 \leqslant j \leqslant N} \operatorname{Re} \{\lambda_j(s)\}. \tag{2.11}$$

Using (2.1) we get the inequality

$$\|P(s)\| \leqslant C_1|s|^p + C_2. \tag{2.12}$$

Since, further, $f^{(j)}(z) = t^j e^{tz}$ and

$$|e^{t\mu}| = |\exp[t \operatorname{Re} \mu]| \leqslant \exp[t\Lambda(s)] \qquad \text{if } \mu \in H(P(s)),$$

it follows from (2.3) that

$$\|\exp[tP(s)]\| \leqslant C_0(1 + t^{1/p} + t^{1/p}|s|)^{p(N-1)} \exp[t\Lambda(s)], \tag{2.13}$$

where C_0 is a constant independent of t.

As is well known, $\exp[t\lambda_j(s)]$ are the eigenvalues of $\exp[tP(s)]$. By the rule (2.10) we then have

$$\|\exp[tP(s)]\| \geqslant \max_{1 \leqslant j \leqslant N} |\exp[t\lambda_j(s)]| = \exp[t\Lambda(s)].$$

Combining this inequality with (2.13) we obtain the following theorem.

Theorem 3. *There exists a constant C_0 independent of t, s such that*

$$\exp[t\Lambda(s)] \leqslant \|\exp[tP(s)]\| \leqslant C_0(1 + t^{1/p} + t^{1/p}|s|)^{p(N-1)} \exp[t\Lambda(s)]. \tag{2.14}$$

Consider the algebraic equation

$$\lambda^N + P_1(s)\lambda^{N-1} + \cdots + P_N(s) = 0, \tag{2.15}$$

where the $P_j(s)$ are polynomials in $s = (s_1, \ldots, s_n)$ of degrees p_j, and set

$$p_0 = \max_{1 \leqslant j \leqslant N} \frac{p_j}{j}. \tag{2.16}$$

Let $\lambda_j(s)$ be the roots of Eq. (2.15). Introducing the functions

$$\Lambda(s) = \max_{1 \leqslant j \leqslant N} \operatorname{Re} \{\lambda_j(s)\}, \qquad \Lambda(r) = \max_{|s| \leqslant r} \Lambda(s),$$

$$\bar{\Lambda}(s) = \max_{1 \leqslant j \leqslant N} |\operatorname{Re} \{\lambda_j(s)\}|, \qquad \bar{\Lambda}(r) = \max_{|s| \leqslant r} \bar{\Lambda}(s),$$

$$I(s) = \max_{1 \leqslant j \leqslant N} |\operatorname{Im} \{\lambda_j(s)\}|, \qquad I(r) = \max_{|s| \leqslant r} I(s),$$

$$M(s) = \max_{1 \leqslant j \leqslant N} |\lambda_j(s)|, \qquad M(r) = \max_{|s| \leqslant r} M(s),$$

we shall prove the following theorem:

Theorem 4. *Let* $\Gamma(s)$ *be any one of the functions* $\Lambda(s)$, $\bar{\Lambda}(s)$, $I(s)$, $M(s)$. *Then there exist positive constants* α_1, α_2 *such that*

$$\Gamma(s) \leqslant \alpha_1(1 + |s|)^{p_0} \qquad \text{for all } s \in C^n, \tag{2.17}$$

$$\Gamma(s_k) \geqslant \alpha_2 |s_k|^{p_0} \qquad \text{for some sequence } \{s_k\},\ |s_k| \to \infty. \tag{2.18}$$

Before proving the theorem we give an application. Let $P(s)$ be a matrix whose elements are polynomials of degrees $\leqslant p$. Writing the characteristic equation in the form

$$\det (P(s) - \lambda I) \equiv (-1)^N \lambda^N + P_1(s)\lambda^{N-1} + \cdots + P_N(s) = 0,$$

we call the number p_0 defined by (2.16), where p_j is the degree of P_j, the *reduced order* of P. Combining (2.13) and Theorem 4 we get:

Theorem 5. *There exist constants* C, c *independent of* t, s *such that*

$$\|\exp [tP(s)]\| \leqslant C(1 + t^{1/p} + t^{1/p}|s|)^{p(N-1)} \exp [ct(1 + |s|)^{p_0}]. \tag{2.19}$$

It follows that if $0 \leqslant t \leqslant T$,

$$\|\exp [tP(s)]\| \leqslant C'(1 + t^{1/p}|s|)^{p(N-1)} \exp [ct(1 + |s|)^{p_0}], \tag{2.20}$$

$$C' = C(1 + T^{1/p})^{p(N-1)}.$$

Note that the weaker inequality

(2.21)

$$\|\exp [tP(s)]\| \leqslant C'' \exp [c't(1 + |s|)^p] \qquad (C'',\ c' \text{ are independent of } t,\ s)$$

can be obtained immediately from the definition of $\exp [tP]$ (see (2.2)) and the inequality (2.12).

Proof of Theorem 4. We first prove (2.17) for $\Gamma(s) = M(s)$, i.e.,

$$|\lambda_j(s)| \leqslant B_0(1 + |s|)^{p_0} \qquad (B_0 \text{ a constant}), \tag{2.22}$$

for $j = 1, \ldots, N$. Suppose (2.22) is not true. Then there exists a sequence $\{s_m\}$, $|s_m| \to \infty$ as $m \to \infty$, and some λ_j, say λ_1, such that

$$|\lambda_1(s_m)| \geqslant m(1 + |s_m|)^{p_0}.$$

From (2.15) we have

$$[\lambda_1(s_m)]^N \left\{1 + \frac{P_1(s_m)}{\lambda_1(s_m)} + \cdots + \frac{P_N(s_m)}{[\lambda_1(s_m)]^N}\right\} = 0.$$

It is therefore impossible that each term $P_j(s_m)/[\lambda_1(s_m)]^j$ will tend to 0 as $m \to \infty$. Hence for some subsequence of $\{s_m\}$, which is denoted again by $\{s_m\}$, and for some j,

$$|P_j(s_m)| \geqslant B_1|\lambda_1(s_m)|^j \geqslant B_2 m^j(1 + |s_m|)^{p_0 j} \qquad (B_1 > 0,\ B_2 > 0),$$

thus implying that $p_j > p_0 j$, which is a contradiction to (2.16).

We next prove (2.18) for $\Gamma(s) = M(s)$. If (2.18) is not true, then, for all $j = 1, \ldots, N$,

$$|\lambda_j(s)| = \eta_j(s)|s|^{p_0},$$

where $\eta_j(s) \to 0$ as $|s| \to \infty$. Since $P_h(s)$ is a sum of products

$$\lambda_{j_1}(s) \cdots \lambda_{j_h}(s),$$

it follows that

$$|P_h(s)| \leqslant \hat{\eta}_h(s)|s|^{hp_0},$$

where $\hat{\eta}_h(s) \to 0$ as $|s| \to \infty$. This implies that $p_h < hp_0$ for $h = 1, \ldots, N$, which is in contradiction with the definition (2.16) of p_0.

We shall need the following elementary lemma.

Lemma 1. *Let $Q_R(s)$, $Q_I(s)$ be the real and the imaginary parts of a polynomial $Q(s)$ of degree k. Then, for all r sufficiently large,*

$$\begin{aligned} B_3 r^k &\leqslant \max_{|s| \leqslant r} Q_R(s) \leqslant B_4 r^k, \\ B_5 r^k &\leqslant \max_{|s| \leqslant r} |Q_R(s)| \leqslant B_6 r^k, \\ B_7 r^k &\leqslant \max_{|s| \leqslant r} |Q_I(s)| \leqslant B_8 r^k \end{aligned} \tag{2.23}$$

for some positive constants $B_3, \ldots, B_8$.

Proof. The second half of each of the inequalities (2.23) is trivial. To prove the first half, we first observe that it suffices to consider the case where Q is a homogeneous polynomial. Now let α, β be solutions of

$$\alpha^k = i, \qquad \beta^k = -1.$$

Then,

$$Q(\alpha s) = \alpha^k Q(s) = iQ(s), \tag{2.24}$$

$$Q_R(\beta s) = \beta^k Q_R(s) = -Q_R(s). \tag{2.25}$$

From the first equality it follows that

$$\max_{|s| \leqslant r} |Q_R(s)| = \max_{|s| \leqslant r} |Q_I(s)|$$

thus proving that

$$\max_{|s| \leqslant r} |Q_R(s)| \geqslant \tfrac{1}{2} \max_{|s| \leqslant r} |Q(s)| \geqslant B_5 r^k.$$

A similar inequality holds for Q_I.

Equation (2.25) shows that

$$\max_{|s| \leqslant r} Q_R(s) = \max_{|s| \leqslant r} |Q_R(s)|$$

and the proof is thereby completed.

We return to the proof of Theorem 4. We have already proved the assertions (2.17), (2.18) for $\Gamma(s) = M(s)$. We shall now prove that if r is sufficiently large then

$$\frac{\bar{\Lambda}(r)}{\Lambda(r)} < B_9, \qquad \Lambda(r) > 0, \tag{2.26}$$

$$\frac{I(r)}{\bar{\Lambda}(r)} < B_{10}, \qquad \frac{\bar{\Lambda}(r)}{I(r)} < B_{11}. \tag{2.27}$$

If we next use the obvious inequalities

$$\max \{I(r), \bar{\Lambda}(r)\} \leqslant M(r) \leqslant I(r) + \bar{\Lambda}(r),$$

then the assertions (2.17), (2.18) follow also for $\Gamma = \Lambda, \bar{\Lambda}, I$.

To prove (2.26) denote by k the degree of the polynomial $P_1(s)$. Since $-P_1(s) = \lambda_1(s) + \cdots + \lambda_n(s)$, Lemma 1 yields, for all sufficiently large r,

$$B_3 r^k \leqslant \max_{|s| \leqslant r} \sum_j \operatorname{Re} \{\lambda_j(s)\} \leqslant B_4 r^k, \tag{2.28}$$

$$B_5 r^k \leqslant \max_{|s| \leqslant r} \left|\sum_j \operatorname{Re} \{\lambda_j(s)\}\right| \leqslant B_6 r^k. \tag{2.29}$$

It is also clear that

$$\max_{|s| \leqslant r} \sum_j \operatorname{Re} \{\lambda_j(s)\} \leqslant N\Lambda(r). \tag{2.30}$$

Suppose now that (2.26) is false. Then, since (by (2.28), (2.30)) $\Lambda(r) > 0$ for all sufficiently large r, there exist two sequences $\{s_m\}$ and $\{r_m\}$, $|s_m| \leqslant r_m$, and $|s_m| \to \infty$ as $m \to \infty$, such that

$$(2.31) \qquad -\frac{\operatorname{Re}\{\lambda_j(s_m)\}}{\Lambda(r_m)} \equiv \gamma_m \to \infty \qquad \text{as } m \to \infty,$$

for some j. Let λ_h satisfy $\operatorname{Re}\{\lambda_h(s)\} = \min_k \operatorname{Re}\{\lambda_k(s)\}$ and assume, for simplicity, that $h = 1$. Then (2.31) holds also for $j = 1$ and we have

$$(2.32) \qquad \begin{aligned} -\sum_{j=1}^{N} \operatorname{Re}\{\lambda_j(s_m)\} &= -\operatorname{Re}\{\lambda_1(s_m)\} - \sum_{j=2}^{N} \operatorname{Re}\{\lambda_j(s_m)\} \\ &\geqslant -\operatorname{Re}\{\lambda_1(s_m)\} - (N-1)\Lambda(r_m) \\ &= (\gamma_m - N + 1)\Lambda(r_m). \end{aligned}$$

Using the inequality (2.30) with $r = r_m$, we easily obtain from (2.32)

$$\max_{|s| \leqslant r_m} \left|\sum_j \operatorname{Re}\{\lambda_j(s)\}\right| \geqslant \frac{\gamma_m - N + 1}{N} \max_{|s| \leqslant r_m} \sum_j \operatorname{Re}\{\lambda_j(s)\},$$

which is impossible for large m, because of (2.28), (2.29). This completes the proof of (2.26).

We next prove the first inequality of (2.27). Let $\mu_j(s) = |\operatorname{Im}\{\lambda_j(s)\}|$ and arrange the μ_j in a decreasing order, i.e.,

$$\mu_1(s) \geqslant \mu_2(s) \geqslant \cdots \geqslant \mu_N(s).$$

If the first inequality of (2.27) is not true, then the function

$$(2.33) \qquad \max_{|s| \leqslant r} \frac{\mu_1(s)}{\bar{\Lambda}(r)} \equiv C_r$$

is an unbounded function of r. Let q be the largest integer for which

$$(2.34) \qquad \max_{|s| \leqslant r} \frac{\mu_1(s) \cdots \mu_q(s)}{C_r^q(\bar{\Lambda}(r))^q} \text{ does not tend to 0 as } r \to \infty.$$

q is $\geqslant 1$. We claim that if $q < N$ then

$$(2.35) \qquad \max_{|s| \leqslant r} \frac{\mu_1(s) \cdots \mu_{q-1}(s)\mu_{q+1}(s)}{C_r^q(\bar{\Lambda}(r))^q} \to 0 \qquad \text{as } r \to \infty.$$

Indeed, in the contrary case,

$$(2.36) \qquad \mu_1(s_m) \cdots \mu_{q-1}(s_m)\mu_{q+1}(s_m) > A C_{r_m}^q(\bar{\Lambda}(r_m))^q \qquad (A > 0)$$

for some sequences $\{s_m\}$ and $\{r_m\}$, $|s_m| \leqslant r_m$, $|s_m| \to \infty$. Since, if $|s| \leqslant r_m$,

$$\mu_j(s) \leqslant \mu_1(s) \leqslant C_{r_m}\bar{\Lambda}(r_m),$$

it follows from (2.36) that $\mu_{q+1}(s_m) \geqslant AC_{r_m}\bar{\Lambda}(r_m)$. Hence,

$$\frac{\mu_1(s_m) \cdots \mu_q(s_m)\mu_{q+1}(s_m)}{C_{r_m}^{q+1}(\bar{\Lambda}(r_m))^{q+1}} \geqslant A^{q+1},$$

thus contradicting the definition of q.

From (2.35) and from the assumption that the function C_r (in (2.33)) is unbounded it follows that each of the functions

$$\max_{|s| \leqslant r} \frac{\mu_{i_1}(s) \cdots \mu_{i_a}(s)\mu_{j_1}(s) \cdots \mu_{j_b}(s)[\operatorname{Re} \lambda_{h_1}(s)] \cdots [\operatorname{Re} \lambda_{h_c}(s)]}{C_r^q(\bar{\Lambda}(r))^q} \tag{2.37}$$

$$(i_1 < \cdots < i_a \leqslant q < j_1 \cdots < j_b;\ a + b + c = q,\ a < q)$$

tends to zero for any sequence $r = R_m \to \infty$ for which $C_{R_m} \to \infty$.

It will be necessary to know not only that C_r is an unbounded function of r but also that $C_r \to \infty$ as $r \to \infty$. Now, in Sec. 14 it will be shown (see (14.4)) that $\Lambda(r) = Ar^h + 0(r^{h'})$ for some $h' < h$, $A \neq 0$. The same asymptotic behavior can be proved, in the same way, for $I(r)$, $\bar{\Lambda}(r)$. Hence

$$C_r = A_0 r^{h_0} + \text{lower terms},$$

for some $A_0 \neq 0$. Since C_r is an unbounded non-negative function of r, $h_0 > 0$ and $A_0 > 0$. Hence, $C_r \to \infty$ as $r \to \infty$.

We conclude that each term in (2.37) tends to 0 as $r \to \infty$.

We now observe that the real and imaginary parts, P_{qR} and P_{qI}, of the polynomial $P_q(s)$, each consists of a linear combination of products

$$[\operatorname{Re} \lambda_{h_1}(s)] \cdots [\operatorname{Re} \lambda_{h_c}(s)][\operatorname{Im} \lambda_{k_1}(s)] \cdots [\operatorname{Im} \lambda_{k_d}(s)] \qquad (c + d = q).$$

In view of (2.34) and the last statement concerning (2.37), the maximum absolute value, for $|s| \leqslant r$, of the product $[\operatorname{Im} \lambda_1] \cdots [\operatorname{Im} \lambda_q]$ dominates the maximum absolute value, for $|s| \leqslant r$, of each of the other products, for some sequence $r = \rho_m \to \infty$. Since this product appears only in one of the polynomials P_{qR}, P_{qI}, we get a contradiction to Lemma 1.

The proof of the second inequality in (2.27) follows by applying the first inequality to the polynomial

$$(-i)^N\lambda^N + (-i)^{N-1}P_1(s)\lambda^{N-1} + \cdots + P_N(s)$$

whose roots $\lambda_j'(s)$ are equal to $i\lambda_j(s)$.

In view of the results of Sec. 14 (i.e., (14.4)) and the analogous asymptotic behavior for $\bar{\Lambda}$, I, M, we can restate Theorem 4 in the following form:

Theorem 4′. *For* $\Gamma = \Lambda, \bar{\Lambda}, I, M$ *the asymptotic formula*

$$\Gamma(r) = Br^{p_0} + 0(r^q) \qquad (r \to \infty) \tag{2.38}$$

is valid, where $q < p_0$, $B > 0$, *and* B, q *depend on the particular function* Γ.

3. Uniqueness of Solutions of the Cauchy Problem

We shall first prove a uniqueness theorem for an *abstract* Cauchy problem, where the differential operators are replaced by any linear operators.

Let Φ, Φ_1, E be linear topological spaces, or countable union spaces, such that

$$\Phi \subset \Phi_1 \subset E,$$

and assume that the topology of Φ is stronger than the topology of Φ_1 and that the topology of Φ_1 is stronger than the topology of E. Assume further that Φ is dense in E. It is clear that

$$E' \subset \Phi_1' \subset \Phi'.$$

Next let A_t be a continuous linear mapping from Φ_1 into Φ_1, for any value of t, $0 < t \leqslant T$. The adjoint A_t^* is then a continuous linear operator from Φ_1' into Φ_1'.

Theorem 6. *If for any t_0 in the interval $0 < t \leqslant T$ and for any $\varphi_0 \in \Phi$ there exists a function $\varphi(t)$, $0 \leqslant t \leqslant t_0$, such that $\varphi(t) \in \Phi_1$ and*

$$\frac{d\varphi(t)}{dt} = A_t\varphi(t) \qquad (0 \leqslant t \leqslant t_0), \qquad \varphi(t_0) = \varphi_0, \tag{3.1}$$

then there exists at most one solution $u(t)$ $(0 \leqslant t \leqslant T)$ in Φ_1' of

$$\frac{du(t)}{dt} = -A_t^*u(t) \qquad (0 < t \leqslant T), \qquad \lim_{t\to 0} u(t) = u_0, \tag{3.2}$$

where $u(t) \in E'$ for any $0 < t \leqslant T$ and u_0 is any given element in Φ_1'.

It should be clarified that by $d\varphi(t)/dt$ we understand the limit in Φ_1 of $[\varphi(t + h) - \varphi(t)]/h$ as $h \to 0$, and

$$\text{by } \frac{du(t)}{dt} \text{ we understand } \lim_{h\to 0} \frac{u(t + h) - u(t)}{h},$$

the limit being taken in the strong topology of Φ_1' (when $h \to 0$ along any sequence). $\lim_{t\to 0} u(t) = u_0$ is taken in the strong topology of Φ_1' (as $t \to 0$ along any sequence).

Proof. It suffices to prove that if $u_0 = 0$ then for any t_0, $0 < t_0 \leqslant T$, $u(t_0) = 0$. Consider the function $a(t) = (u(t),\varphi(t))$ in the interval $0 < t \leqslant t_0$, where $\varphi(t)$ is a solution of (3.1). Then

$$\frac{a(t + h) - a(t)}{h} = \left(\frac{u(t + h) - u(t)}{h}, \varphi(t + h)\right) + \left(u(t), \frac{\varphi(t + h) - \varphi(t)}{h}\right). \tag{3.3}$$

As $h \to 0$ along any sequence, the first term on the right side converges to $(du(t)/dt, \varphi(t))$ since

$$\left(\frac{1}{h}(u(t+h) - u(t), \varphi(t)\right) \to \left(\frac{du(t)}{dt}, \varphi(t)\right),$$

whereas $\varphi(t+h) - \varphi(t) \to 0$ in Φ_1 while $\{(1/h)(u(t+h) - u(t))\}$ is strongly convergent in Φ_1'.

Taking $h \to 0$ (along any sequence) in (3.3) we get

$$\frac{da(t)}{dt} = \left(\frac{du(t)}{dt}, \varphi(t)\right) + \left(u(t), \frac{d\varphi(t)}{dt}\right).$$

Substituting from (3.1), (3.2) we find that $da(t)/dt = 0$. Hence,

$$a(t_0) = a(t) = \text{const.} = 0,$$

since $a(\tau) \to 0$ as $\tau \to 0$. Having proved that $(u(t_0), \varphi_0) = 0$ for any $\varphi_0 \in \Phi$ and recalling that $u(t_0) \in E'$, $\bar{\Phi} = E$, it follows that $u(t_0) = 0$.

We shall use Theorem 6 in proving the uniqueness of solutions for the Cauchy problem

$$\frac{\partial u(x,t)}{\partial t} = P\left(i\frac{\partial}{\partial x}\right)u(x,t), \tag{3.4}$$

$$u(x, 0) = u_0(x), \tag{3.5}$$

where $P(s)$ is an $N \times N$ matrix whose elements are polynomials in $s = (s_1, \ldots, s_n)$ of degrees $\leqslant p$ (at least one polynomial is of degree p) and whose reduced order is p_0. We first consider the case $p_0 > 1$.

We shall define Φ, Φ_1 as W spaces and E as a normed space of locally integrable functions. Before introducing these spaces, however, let us try to solve the *adjoint problem* of (3.4), (3.5), namely,

$$\frac{\partial \varphi(x,t)}{\partial t} = -P^*\left(-i\frac{\partial}{\partial x}\right)\varphi(x,t) \qquad (0 \leqslant t \leqslant t_0), \tag{3.6}$$

$$\varphi(x, t_0) = \varphi_0(x), \tag{3.7}$$

where P^* is the transpose of P.

Taking, formally, the Fourier transforms of (3.6), (3.7) we obtain the system

$$\frac{\partial \psi(\sigma,t)}{\partial t} = -P^*(-\sigma)\psi(\sigma,t), \tag{3.8}$$

$$\psi(\sigma, t_0) = \psi_0(\sigma), \tag{3.9}$$

where $\psi_0(\sigma) = \mathfrak{F}[\varphi_0(x)]$. The solution of this system is given by

$$\psi(\sigma, t) = \{\exp[-(t - t_0)P^*(-\sigma)]\}\,\psi_0(\sigma). \tag{3.10}$$

Assume now that

$$\varphi_0(x) \in W_{p_0',1/b}^{p_0',1/a} \qquad \text{where } \frac{1}{p_0} + \frac{1}{p_0'} = 1 \tag{3.11}$$

and a, b are restricted by the inequality $b/a > \gamma$ (so that the space is nontrivial; see Theorem 19, Sec. 5, Chap. 5). Then

$$\psi_0(\sigma) \in W^{p_0,b}_{p_0,a} \tag{3.12}$$

By (2.20) we have

$$\|\exp[-(t-t_0)P^*(-s)]\| \leqslant C'_0(1+|s|)^{(N-1)p}\exp[c(t_0-t)|s|^{p_0}], \tag{3.13}$$

where C'_0 depends on T. By Theorem 10, Sec. 3, Chap. 5 it follows that

$$\exp[-(t-t_0)P^*(-\sigma)]$$

is a multiplier from $W^{p_0,b}_{p_0,a}$ into $W^{p_0,b+c_0}_{p_0,a-c_0}$, where

$$c_0 = (p_0cT_0)^{1/p_0} \qquad \text{provided } t_0 < T_0,\ a > c_0. \tag{3.14}$$

Hence $\psi(\sigma, t)$ as defined by (3.10) belongs to $W^{p_0,b+c_0}_{p_0,a-c_0}$.

We next show that $\psi(\sigma, t)$ satisfies (3.8) in the sense of $W^{p_0,b+c_0}_{p_0,a-c_0}$. Writing

$$\frac{\psi(\sigma, t+h) - \psi(\sigma, t)}{h} = \frac{1}{h}\{\exp[-(t+h-t_0)P^*(-\sigma)] - \exp[-(t-t_0)P^*(-\sigma)]\}\psi_0(\sigma), \tag{3.15}$$

we see that, as $h \to 0$, the right side converges properly to

$$-P^*(-\sigma)\{\exp[-(t-t_0)P^*(-\sigma)]\}\,\psi_0(\sigma) = -P^*(-\sigma)\psi(\sigma, t). \tag{3.16}$$

At the same time the right side of (3.15) is bounded in $W^{p_0,b+c_0}_{p_0,a-c_0}$ if, say, $|h| \leqslant h_0$. Indeed, its value for any complex value s is equal to

$$\frac{1}{h}\int_t^{t+h}\frac{\partial}{\partial\rho}\{[\exp[-(\rho-t_0)P^*(-s)]]\psi_0(s)\}\,d\rho = -\frac{1}{h}\int_t^{t+h}(\rho-t_0)P^*(-s)\{\exp[-(\rho-t_0)P^*(-s)]\}\psi_0(s)\,d\rho,$$

and the integrand is bounded by

$$C_\epsilon \exp[(-1/p_0)|(a-c_0-\epsilon)\sigma|^{p_0} + (1/p_0)|(b+c_0+\epsilon)\tau|^{p_0}]$$

for any $\epsilon > 0$, where C_ϵ is independent of ρ.

Since in W spaces, proper convergence plus boundedness is equivalent to convergence, it follows that the right side of (3.15) is convergent, in the sense of $W^{p_0,b+c_0}_{p_0,a-c_0}$, to the right side of (3.16). Hence (3.8) is satisfied in the same sense. (3.9) follows immediately from $\exp[0A] = I$ (= the identity matrix).

We now take the inverse Fourier transforms of (3.8), (3.9). Denoting $\mathfrak{F}^{-1}[\psi(\sigma, t)]$ by $\varphi(x, t)$ and using the rule (2.3) of Chap. 4 we see that the right side of (3.8) is transformed into the right side of (3.6). Since $\mathfrak{F}^{-1}$ is a continuous linear operator from $W_{p_0,a-c_0}^{p_0,b+c_0}$ onto $W_{p_0',1/(b+c_0)}^{p_0',1/(a-c_0)}$, we also have

$$\mathfrak{F}^{-1}\left[\frac{\partial\psi(\sigma, t)}{\partial t}\right] = \mathfrak{F}^{-1}\left\{\lim_{h\to 0}\left[\frac{\psi(\sigma, t+h) - \psi(\sigma, t)}{h}\right]\right\}$$
$$= \lim_{h\to 0} \mathfrak{F}^{-1}\left[\frac{\psi(\sigma, t+h) - \psi(\sigma, t)}{h}\right]$$
$$= \lim_{h\to 0} \frac{\varphi(x, t+h) - \varphi(x, t)}{h}$$
$$= \frac{\partial\varphi(x, t)}{\partial t}.$$

Introducing the spaces

$$(3.17) \qquad \Phi = W_{p_0',1/b}^{p_0',1/a}, \qquad \Phi_1 = W_{p_0',1/(b+c_0)}^{p_0',1/(a-c_0)},$$

we conclude that (3.6) is satisfied in the sense of Φ_1. (3.7) follows immediately from (3.9).

We now introduce the space E as the space of all measurable functions $g(x)$ on R^n having a finite norm

$$\|g\|_q = \left\{\int_{R^n} [\exp\,[q\beta|x|^{p_0'}]]g(x)|^q\,dx\right\}^{1/q} \qquad (1 \leqslant q < \infty).$$

If

$$(3.18) \qquad \beta = \frac{\theta}{p_0'(b + c_0)^{p_0'}} \qquad \text{for some } \theta < 1$$

then $\Phi \subset \Phi_1 \subset E$. Since Φ is a nontrivial W space, it follows from Theorems 22, 20 of Sec. 5, Chap. 5, that Φ is dense in E.

The operator $A_t = A = -P^*(-i\,(\partial/\partial x))$ is clearly a linear continuous operator from Φ_1 into Φ_1. We can therefore apply Theorem 6 and conclude that if $u_0(x) = 0$ and u is a solution in Φ_1' belonging to E', then $u(x, t) = 0$ for all $0 \leqslant t \leqslant T_0$. Recall that T_0 is taken sufficiently small so that $c_0 < a$ (see (3.14)).

We can now repeat the same argument, replacing t by $t + T_0$, and thus conclude that $u(x, t) = 0$ for $T_0 \leqslant t \leqslant 2T_0$. Proceeding step by step we find that $u(x, t) = 0$ for all $0 \leqslant t \leqslant T$.

Noting that b and c_0 can be taken a priori to be arbitrarily small, so that β is arbitrarily large, and using the form of the continuous linear functionals of E', we can sum up our results as follows.

Theorem 7. *Let $p_0 > 1$ and let β be any positive number. Then there exists at most one function $u(x, t)$, measurable in x (for each fixed t), which is a generalized solution over $\Phi_1 \equiv W_{p_0',1/(b+C_0)}^{p_0',1/(a-C_0)}$ of (3.4), (3.5) for $0 \leqslant t \leqslant T$,* *and which satisfies*

$$\int_{R^n} \{\exp[-\beta r|x|^{p_0'}]\}\, |u(x,t)|^r < A \qquad \text{for some } 1 < r < \infty, \tag{3.19}$$

or

$$|u(x,t)| \leqslant A \exp[\beta|x|^{p_0'}] \qquad \text{almost everywhere (for each } t), \tag{3.20}$$

where A is a constant independent of t.

It should be remarked that the concept of a solution u as used in Theorem 6 is stronger than the concept of generalized solution as defined in Sec. 1; in Theorem 6 the equations are satisfied in the strong topology of Φ_1' whereas in Sec. 1 the equations are satisfied in the sense of the weak topology. Since, however Φ_1 is a perfect space, the two definitions are equivalent (in view of Theorem 21, Sec. 6, Chap. 1). A similar remark applies to all the other spaces that will be used in the future.

If $r = 1$ and (3.19) holds for *any* $\beta > 0$, then we can modify the proof of Theorem 7, taking $\Phi = \Phi_1 = E = W_{p_0'}^{p_0'}$, and again establish uniqueness.

If $p_0 \leqslant 1$, then, by replacing p_0 by $1 + \epsilon$, for any $\epsilon > 0$, we conclude:

Corollary. *If $p_0 \leqslant 1$, then the assertion of Theorem 7 remains valid provided we replace p_0' by an arbitrary positive number.*

If

$$\|\exp[tP(\sigma)]\| \leqslant A_1(1 + |\sigma|^h) \qquad \text{for some } h > 0,\ A_1 > 0, \tag{3.21}$$

then more can be said about the case $p_0 \leqslant 1$. Indeed, we then can take $\mathfrak{F}[\Phi] = \mathfrak{F}[\Phi_1] = Z$. By the corollary to Theorem 5, Sec. 3, Chap. 6, it follows that

$$\|\exp[tP(s)]\| \leqslant A_2(1 + |\sigma|^h) \exp[d_0 t|\tau|]$$

and $\exp[tP(\sigma)]$ is therefore a multiplier in Z. The proof of Theorem 7 can now be easily modified, taking $E = \Phi = \Phi_1 = K$. We thus arrive at the following theorem:

Theorem 7′. *If $p_0 \leqslant 1$ and* (3.21) *is satisfied then there exists at most one distribution which is a generalized solution of* (3.4), (3.5) *in the sense of K'.*

Thus, no restrictions are made on the growth of the solution.

We next consider the question of uniqueness for classical solutions. We first prove the following:

Theorem 8. *Any classical solution $u(x, t)$ of* (3.4), (3.5) *which satisfies the inequality* (3.19) *or* (3.20) *is a generalized solution in Φ_1' provided t is sufficiently small (more precisely, $t \leqslant T_0$, where T_0, β are related by* (3.18), (3.14)).

Proof. Let α_j be the functions of K defined by (12.14), Chap. 3. Then, for any $\varphi \in \Phi_1$,

$$\int_{R^n} \left[P\left(i \frac{\partial}{\partial x}\right) u(x, t) \right] [\alpha_j(x)\varphi(x)]\, dx$$
$$= \int_{R^n} u(x, t) \left[P^*\left(-i \frac{\partial}{\partial x}\right) (\alpha_j(x)\varphi(x)) \right] dx$$

as follows by integration by parts.

Integrating both sides with respect to t we obtain

$$\int_{R^n} u(x, t)\alpha_j(x)\varphi(x)\, dx - \int_{R^n} u_0(x)\alpha_j(x)\varphi(x)\, dx$$
$$= \int_0^t \int_{R^n} u(x, t) \left[P^*\left(-i \frac{\partial}{\partial x}\right) (\alpha_j(x)\varphi(x)) \right] dx\, dt. \tag{3.22}$$

Taking $j \to \infty$ and using Hölder's inequality ((12.2), Chap. 3) and (3.19) (or using (3.20) directly) one can easily see that the inner integral on the right side of (3.22) is convergent to

$$\int_{R^n} u(x, t) \left[P^*\left(-i \frac{\partial}{\partial x}\right) \varphi(x) \right] dx$$

uniformly with respect to t, provided $t \leqslant T_0$ where T_0, β are related by (3.18), (3.14). Similarly one derives the convergence of the integrals on the left side of (3.22), as $j \to \infty$. We thus get

$$\int_{R^n} u(x, t)\varphi(x)\, dx - \int_{R^n} u_0(x)\varphi(x)\, dx$$
$$= \int_0^t \int_{R^n} u(x, t) \left[P^*\left(-i \frac{\partial}{\partial x}\right) \varphi(x) \right] dx\, dt, \tag{3.23}$$

or

$$(u(x, t), \varphi(x)) - (u_0(x), \varphi(x)) = \int_0^t \left(u(x, t), P^*\left(-i \frac{\partial}{\partial x}\right) \varphi(x)\right) dt.$$

It follows that

$$\lim_{h \to 0} \left(\frac{u(x, t + h) - u(x, t)}{h}, \varphi(x) \right) = \left(u(x, t), P^*\left(-i \frac{\partial}{\partial x}\right) \varphi(x) \right),$$

i.e., (3.4) is satisfied in the sense of generalized solutions. Taking $t \to 0$ in (3.23) we find that (3.5) is also satisfied in the sense of generalized solutions. Hence, $u(x, t)$ is a generalized solution of (3.4), (3.5).

From Theorems 7, 8 we derive the uniqueness of classical solutions for $0 \leqslant t \leqslant T_0$. Proceeding step by step we obtain the following theorem:

Theorem 9. *Let $p_0 > 1$ and let β be any positive number. Then there exists at most one classical solution of* (3.4), (3.5) ($0 \leqslant t \leqslant T$) *satisfying* (3.19) *or* (3.20) *for some constant A independent of t.*

If $p_0 \leqslant 1$, then the assertion of the theorem remains true provided we replace p_0' by any positive number. For the sake of later reference we also state:

Theorem 9′. *If* $p_0 \leqslant 1$ *and* (3.21) *is satisfied, then there exists at most one classical solution of* (3.4), (3.5).

We conclude this section with two remarks concerning improvements of Theorems 7, 9.

Remark 1. Theorem 7, 9 can be improved by replacing (3.20) by

$$|u(x, t)| \leqslant A \exp [\beta\lambda(x)|x|^{p_0'}] \tag{3.24}$$

and similarly for (3.19), where $\lambda(x) = (\lambda_1(x_1), \ldots, \lambda_n(x_n))$, $\lambda_j(x_j) = \lambda_j(-x_j)$ and the λ_j are positive functions of slow variation at ∞ (see Chap. 5, Sec. 5) such that $|x_j|^{p_0'}\lambda_j(x_j)$ satisfy the properties of the functions $M_j(x_j)$ which were assumed in the definition of W spaces in Chap. 5. Indeed, by Theorem 17, Sec. 5, Chap. 5, space W_M^Ω, where

$$M_j(x_j) = \frac{|x_j|^{p_0'}}{p_0'} \lambda_j(x_j), \qquad \Omega_j(y_j) = \frac{|y_j|^{p_0'}}{p_0'} \lambda_j(y_j),$$

is nontrivial, and all the previous considerations can be extended in an obvious manner by replacing the spaces $W_{p_0',a_1}^{p_0',b_1}$ by W_{M,a_1}^{Ω,b_1}.

As an example we can take

$$\lambda_j(x_j) = [\log (1 + |x_j|)]^m \qquad (m > 0).$$

Remark 2. Theorems 7, 9 are not, in general, true if p_0' is replaced by $p_0' + \delta$ for some $\delta > 0$. We give a counterexample for the equation

$$\frac{\partial^q u}{\partial t^q} = \frac{1}{a} \frac{\partial^p u}{\partial x^p}. \tag{3.25}$$

Here $p_0 = p/q$ and we suppose that $p > q$. The function

$$u(x, t) = \sum_{m=0}^{\infty} a^m \frac{f^{(qm)}(t)}{(pm)!} x^{pm} \tag{3.26}$$

is, formally, a solution of (3.25). Let $f(t)$ be a nonzero C^∞ function which vanishes if $t < 0$ and if $t > 1$ and which satisfies

$$|f^{(m)}(t)| \leqslant C^m m^{(1+\epsilon)m} \qquad (\epsilon > 0).$$

By Theorem 11, Sec. 7, Chap. 6, such a function exists for any given $\epsilon > 0$. Taking ϵ such that

$$(1 + \epsilon)q < p,$$

we find that $u(x, t)$ is then a nonzero classical solution of (3.25) and it vanishes on $t = 0$ together with all its derivatives. Finally,

$$|u(x, t)| \leqslant \sum_{m=1}^{\infty} C_1^m \frac{x^{pm}}{m^{pm-(1+\epsilon)qm}} \leqslant C_2 \exp\,[C_3|x|^{p_0'+\delta}],$$

where

$$p_0' + \delta = \frac{p}{p - (1+\epsilon)q}; \qquad \delta \to 0 \text{ as } \epsilon \to 0.$$

4. Existence of Generalized Solutions

In this section we prove that for any P there exists a generalized solution of the Cauchy problem

$$\frac{\partial u(x, t)}{\partial t} = P\left(i\frac{\partial}{\partial x}\right) u(x, t), \tag{4.1}$$

$$u(x, 0) = u_0(x), \tag{4.2}$$

for $0 \leqslant t \leqslant T$. We begin by taking, formally, the Fourier transforms of both sides of (4.1), (4.2), thereby getting

$$\frac{\partial v(\sigma, t)}{\partial t} = P(\sigma)v(\sigma, t), \tag{4.3}$$

$$v(\sigma, 0) = v_0(\sigma), \tag{4.4}$$

where $v_0(\sigma) = \mathfrak{F}[u_0(x)]$. One is led to define

$$v(\sigma, t) = Q(\sigma, t)v_0(\sigma), \tag{4.5}$$

where

$$Q(s, t) = \exp\,[tP(s)]. \tag{4.6}$$

As in Sec. 3 we take

$$\begin{aligned} \Phi &= W_{p_0', 1/b}^{p_0', 1/a}, \qquad & \Phi_1 &= W_{p_0', 1/(b+c_0)}^{p_0', 1/(a-c_0)}, \\ \tilde{\Phi} &= W_{p_0, a}^{p_0, b}, \qquad & \tilde{\Phi}_1 &= W_{p_0, a-c_0}^{p_0, b+c_0}, \end{aligned} \tag{4.7}$$

where $b/a > \gamma$, $p_0 > 1$, but we now define c_0 slightly differently than in (3.14), namely,

$$c_0 = (p_0 cT)^{1/p_0}. \tag{4.8}$$

Recall that c appears in the inequality (2.20), i.e.,

$$\|Q(s, t)\| \leqslant C'(1 + t^{1/p}|s|)^{(N-1)p} \exp\,[ct(1 + |s|)^{p_0}]. \tag{4.9}$$

Since $Q(\sigma, t)$ is a multiplier from $\tilde{\Phi}$ into $\tilde{\Phi}_1$ (by Theorem 10, Sec. 3, Chap. 5), $Q(\sigma, t)$ is also a multiplier from $\tilde{\Phi}_1'$ into $\tilde{\Phi}'$. It follows that $v(\sigma, t)$ belongs to $\tilde{\Phi}'$ if $v_0(\sigma) \in \tilde{\Phi}_1'$, i.e., if $u_0(x) \in \Phi_1'$. We claim that if $u_0 \in \Phi_1'$ then $v(\sigma, t)$ is a solution of (4.3), (4.4) in the sense of $\tilde{\Phi}'$. To prove it let

$\psi \in \tilde{\Phi}$. We then have

$$\left(\frac{v(\sigma, t + h) - v(\sigma, t)}{h}, \psi(\sigma)\right)$$
$$= \left(\frac{1}{h}[\exp[(t + h)P(\sigma)] - \exp[tP(\sigma)]]v_0(\sigma), \psi(\sigma)\right)$$
$$= (v_0(\sigma), \frac{1}{h}[\exp[(t + h)P^*(\sigma)] - \exp[tP^*(\sigma)]]\psi(\sigma))$$
$$\rightarrow (v_0(\sigma), P^*(\sigma)[\exp[tP^*(\sigma)]]\psi(\sigma))$$
$$= (P(\sigma)[\exp[tP(\sigma)]]v_0(\sigma), \psi(\sigma)) = (P(\sigma)v(\sigma, t), \psi(\sigma)),$$

where in passing to the limit we have made use of the results of Sec. 3, i.e., the calculation following (3.15) with t_0 replaced by 0 and with $-P^*(-\sigma)$ replaced by $P^*(\sigma)$. We have thus proved that

$$\frac{\partial v(\sigma, t)}{\partial t} \equiv \lim_{h \to 0} \frac{v(\sigma, t + h) - v(\sigma, t)}{h}$$

exists in the weak topology of $\tilde{\Phi}'$ and that it is equal to $P(\sigma)v(\sigma, t)$. As for (4.4), if $t \rightarrow 0$ then

$$([\exp[tP(\sigma)]]v_0(\sigma), \psi(\sigma)) = (v_0(\sigma), [\exp[tP^*(\sigma)]]\psi(\sigma)) \rightarrow (v_0(\sigma), \psi(\sigma)),$$

since, as is easily verified, $\{[\exp[tP^*(\sigma)]]\psi(\sigma); 0 < t \leqslant T\}$ is a bounded set in $\tilde{\Phi}_1$ and

$$[\exp[tP^*(\sigma)]]\psi(\sigma) \rightarrow \psi(\sigma) \qquad \text{as } t \rightarrow 0$$

properly, hence also in the sense of $\tilde{\Phi}_1$.

We have thus proved that $v(\sigma, t)$ is a generalized solution of (4.3), (4.4). Taking the inverse Fourier transforms of both sides of (4.3), (4.4) and recalling that $\mathfrak{F}^{-1}$ is a continuous linear mapping we conclude, using the analogue of (2.3), Chap. 4, when $\mathfrak{F}$ is replaced by $\mathfrak{F}^{-1}$, that

$$u(x, t) \equiv \mathfrak{F}^{-1}[v(\sigma, t)]$$

is a generalized solution of (4.1), (4.2). On the other hand, from the explicit form (4.5) of $v(\sigma, t)$ we find, using Theorem 9′, Sec. 4, Chap. 4, that $\mathfrak{F}^{-1}v = (\mathfrak{F}^{-1}Q) * u_0$, i.e.,

$$u(x, t) = G(x, t) * u_0(x), \tag{4.10}$$

where

$$G(x, t) = \mathfrak{F}^{-1}[Q(\sigma, t)] = \mathfrak{F}^{-1}[\exp[tP(\sigma)]]. \tag{4.11}$$

$G(x, t)$ is called the *Green matrix* of the Cauchy problem (4.1), (4.2). Note that G belongs to Φ' and it convolves Φ into Φ_1.

We have thus proved the following:

Theorem 10. *Let* $p_0 > 1$. *For any* $u_0 \in \Phi_1'$ *there exists a generalized solution* $u(x, t)$ *in* Φ' *of* (4.1), (4.2) *for* $0 \leqslant t \leqslant T$, *provided* $c_0 < a$. *The solution is given by* (4.10), (4.11).

Note that uniqueness has been proved for solutions in Φ_1' lying in the space E' which is a linear subspace of the space Φ' in which the generalized solution (4.10) lies. If however we take $\Phi = \Phi_1 = W_{p_0'}^{p_0'}$ then for any $u_0 \in \Phi'$ the solution also belongs to Φ' and, by the second remark following Theorem 7, u is the unique solution in Φ'.

Note also that for any given T we have $c_0 < a$ if a is sufficiently large. The effect of increasing T is, roughly speaking, that the space Φ_1', to which the initial values $u_0(x)$ belong, becomes smaller.

If $p_0 \leqslant 1$ then Theorem 10 remains true provided p_0' is replaced by any positive number. We also have:

Theorem 10′. *If* $p_0 \leqslant 1$ *and* (3.21) *is satisfied then for any* $u_0 \in K'$ *there exists a generalized solution* $u(x, t)$ *in* K' *of* (4.1), (4.2), $(0 \leqslant t \leqslant T)$. *The solution is unique (by Theorem 7′) and it is given by* (4.10), (4.11).

5. Lemmas on Convolutions

In proving that the generalized solutions of Theorems 10, 10′ are classical solutions, for suitable initial values, we shall need several auxiliary results concerning convolutions. For the sake of clarity, we bring them all in this section.

The first lemma establishes the fact that if the generalized function (4.10) is a smooth function having an appropriate growth at ∞ then it is also a smooth solution.

Lemma 2. *If the convolution* (4.10) *is a functional in* Φ' $(\Phi \equiv W_{p_0',1/b}^{p_0',1/a})$ *of function type* $u(x, t)$, *where* $u(x, t)$ *is continuous in* (x, t) $(x \in R^n$, $0 \leqslant t \leqslant T)$, *and if* $u(x, t)$ *and all its derivatives appearing in* (4.1) *are continuous functions of* (x, t) $(x \in R^n, 0 < t \leqslant T)$ *bounded by* $A \exp[\nu|x|^{p_0'}]$ *for some constants* A, ν *independent of* t, *and* $\nu < 1/p_0'b^{p_0'}$, *then* $u(x, t)$ *is a classical solution of* (4.1), (4.2).

Proof. The convolution $u = G * u_0$ is a generalized solution in Φ'. Since $u(x, t)$ and its classical derivatives are bounded by $A \exp[\nu|x|^{p_0'}]$ and $\nu < 1/p_0'b^{p_0'}$, whereas each $\varphi \in \Phi$ and each of its derivatives is bounded by const. $\exp[-\delta|x|^{p_0'}]$ for any $\delta < 1/p_0'b^{p_0'}$ (hence for some $\delta > \nu$), we can integrate by parts the expression

$$\left(u(x, t), P^*\left(-i\frac{\partial}{\partial x}\right)\varphi(x)\right) = \int_{R^n} u(x, t)P^*\left(-i\frac{\partial}{\partial x}\right)\varphi(x)\,dx$$

and obtain $\int_{R^n}\left[P\left(i\frac{\partial}{\partial x}\right)u(x, t)\right]\varphi(x)\,dx.$

The expression

$$(u(x, t), \varphi(x)) = \int_{R^n} u(x, t)\varphi(x)\, dx$$

can be differentiated with respect to t $(0 < t \leqslant T)$ and the result is

$$\int u_t(x, t)\varphi(x)\, dx,$$

since the latter integral is uniformly convergent with respect to t.

Now, by definition of a generalized solution we have, for every $\varphi \in \Phi$,

$$\left(u(x, t), P^*\left(-i\frac{\partial}{\partial x}\right)\varphi\right) = \frac{d}{dt}(u(x, t), \varphi(x)).$$

Using the above remarks we then obtain

$$\left(\left[P\left(i\frac{\partial}{\partial x}\right) - \frac{\partial}{\partial t}\right] u(x, t), \varphi(x)\right) = 0.$$

Since Φ is sufficiently rich (see Theorem 22, Sec. 5, Chap. 5), we conclude that $u(x, t)$ satisfies (4.1) in the classical sense.

Next, (4.2) in the generalized sense means that, for every $\varphi \in \Phi$,

$$\lim_{t \to 0} (u(x, t), \varphi(x)) = (u_0(x), \varphi(x)). \tag{5.1}$$

Since $u(x, t)$ is a continuous function for $x \in R^n$, $0 \leqslant t \leqslant T$ and is bounded by $A \exp [\nu|x|^{p_0'}]$ with $\nu < 1/p_0'b^{p_0'}$, we easily see that one can pass to the limit $t \to 0$ inside the integral $(u(x, t), \varphi(x))$. We then obtain $(u(x, 0), \varphi(x)) = (u_0(x), \varphi(x))$. Employing again the richness property of Φ, the proof of (4.2), in the classical sense, follows.

By a similar argument one can easily prove the following:

Lemma 2′. *If the convolution* (4.10) *is a functional in* K' *of function type* $u(x, t)$, *where* $u(x, t)$ *is a continuous function of* (x, t) $(x \in R^n$, $0 \leqslant t \leqslant T)$ *and if* $u(x, t)$ *and all its derivatives appearing in* (4.1) *are continuous functions of* (x, t) $(x \in R^n, 0 < t \leqslant T)$, *then* $u(x, t)$ *is a classical solution of* (4.1), (4.2).

In what follows we shall use the notation $\circledast$ to denote a convolution taken in the classical sense. Thus, in particular, $g \circledast h = h \circledast g$.

In the following lemmas of this section we assume that Φ is any fundamental space satisfying the requirements (Φ_1) through (Φ_4) of Chap. 4, Sec. 1, and that Φ also admits a differentiable translation.

Lemma 3. *Let* $g(x)$, $h(x)$ *be functionals in* Φ' *of function type and suppose that* g *convolves with* Φ *and that the classical convolution* $g(x) \circledast h(x)$ *exists as a function* $k(x)$ *and* $k(x)$ *belongs to* Φ' (*i.e., the integral* $\int k(x)\varphi(x)\, dx$

is a continuous functional on Φ). *If, for every* $\varphi \in \Phi$,

$$\int_{R^n} \int_{R^n} |h(x)g(y - x)\varphi(y)| \, dx \, dy < \infty, \tag{5.2}$$

then $g * h = g \circledast h$.

Proof. By definition of a convolution

$$(g * h, \varphi) = (h, g * \varphi) = \left(h(x), \int g(\xi)\varphi(x + \xi) \, d\xi\right)$$
$$= \int h(x) \left\{ \int g(\xi)\varphi(x + \xi) \, d\xi \right\} dx.$$

Substituting in the inner integral of the last expression $\xi = \eta - x$ and changing the order of integration, which is permitted by Fubini's theorem and the assumption (5.2), we obtain

$$\int \left\{ \int h(x)g(\eta - x) \, dx \right\} \varphi(\eta) \, d\eta = (g \circledast h, \varphi).$$

Since $g \circledast h = h \circledast g$ we conclude:

Corollary. *Under the assumptions of Lemma 3 and the assumption that* h *convolves with* Φ, $g * h = h * g$.

When there may be confusion between generalized derivatives $D^\alpha g$ and usual (or classical) derivatives $D^\alpha g$ we enclose the latter in brackets; similarly for any differential operator $R(D)$.

Lemma 4. *Let* $u(x) = \sum_{j=1}^{N} R_j(D)g_j(x)$ *in* Φ' *and assume that*

(i) $R_j(D)$ *are linear differential operators of orders* $\leqslant q$, *and*
(ii) $g_j(x) \in \Phi'$ *and are of function type, and they convolve with* Φ.

Let $u_0(x) \in \Phi'$ *be of function type and assume that*

(iii) $u_0(x)$ *has* q *continuous derivatives for* $x \in R^n$ *and*

$$R_j(D)u_0(x) = [R_j(D)u_0(x)];$$

(iv) $g_j(x) \circledast R_j(D)u_0$ *exist and belong to* Φ', *and, for every* $\varphi \in \Phi$

$$\int \int |g_j(x)R_j(D_x)u_0(y - x)\varphi(y)| \, dx \, dy < \infty. \tag{5.3}$$

Then

$$u(x) * u_0(x) = \sum_{j=1}^{N} g_j(x) \circledast R_j(D)u_0(x) \qquad \text{in } \Phi'. \tag{5.4}$$

Proof. By Lemma 3 we have

$$g_j * R_j(D)u_0 = g_j \circledast R_j(D)u_0 \qquad \text{in } \Phi'. \tag{5.5}$$

Next, by Theorem 6, Sec. 3, Chap. 4, if $f_0 \in \Phi'$ and if it convolves with Φ then, for any linear differential operator $P(D)$ of finite order,

$P(D)f_0 \in \Phi'$ and convolves with Φ, and

$$(5.6) \qquad P(D)(f_0 * f) = P(D)f_0 * f = f_0 * P(D)f \qquad \text{in } \Phi'.$$

Applying this result in our case we get

$$(5.7) \qquad g_j * R_j(D)u_0 = R_j(D)g_j * u_0 \qquad \text{in } \Phi'.$$

Combining (5.5), (5.7) and summing over j, (5.4) follows.

Lemma 4′. *If in Lemma 4 the g_j are not assumed to convolve with Φ but, instead, $u(x)$ is assumed to convolve with Φ and*

$$(5.3') \qquad \int \int |g_j(x)u_0(y - x)\varphi(y)|\, dx\, dy < \infty,$$

then the assertion (5.4) *is still valid.*

Proof. Take for simplicity $N = 1$, $R_j = R$, $g_j = g$. Then

$$\begin{aligned}(u * u_0, \varphi) &= (u_0, u * \varphi) = (u_0, (R(D)g, \tau_\xi\varphi)) = (u_0, (g, R(-D)\tau_\xi\varphi)) \\ &= (g, (u_0, R(-D)\tau_\xi\varphi)) = (g, (R(D)u_0, \tau_\xi\varphi)) \\ &= (g, [(R(D)u_0], \tau_\xi\varphi)) \\ &= (g \circledast [R(D)u_0], \varphi),\end{aligned}$$

where the changes in the order of integration are justified by (5.3), (5.3′).

Lemma 4″. *If in Lemma 4′ $u(x)$ convolves Φ into Φ_1 and if $u_0 \in \Phi_1'$, then* (5.4) *is still valid.*

The proof is the same as that of Lemma 4′.

We shall give an extension of Lemma 4 to the case where the $R_j(D)$ are of infinite order. It will not be used in the future.

Lemma 5. *Let $u(x) = \sum_{j=1}^{N} R_j(D)g_j(x)$ in the sense of Φ' and assume that:*

(i) *$R_j(-D)$ is a power series in D, and $R_j(-D)\varphi$ is convergent in the Φ topology, uniformly with respect to φ in bounded sets of Φ (hence $R_j(D)f$ is a bounded operator in Φ');*
(ii) *$g_j(x) \in \Phi'$ and are of function type, and they convolve with Φ.*

Let $u_0(x) \in \Phi'$ be of function type and assume that

(iii) *$u_0(x)$ is infinitely differentiable in $x \in R^n$;*
(iv) *$R_j(D)u_0(x)$ (which belong to Φ') coincide with $[R_j(D)u_0(x)]$;*
(v) *$g_j(x) \circledast R_j(D)u_0(x)$ are functions belonging to Φ' and, for every $\varphi \in \Phi$ and $0 \leqslant |q| < \infty$,*

$$(5.8) \qquad \begin{aligned}&\int \int |g_j(x)R_j(D_x)u_0(y - x)\varphi(y)|\, dx\, dy < \infty, \\ &\int \int |g_j(x)D_x^q u_0(y - x)\varphi(y)|\, dx\, dy < \infty.\end{aligned}$$

Then

$$(5.9) \qquad u(x) * u_0(x) = \sum_{j=1}^{N} g_j(x) \circledast R_j(D)u_0(x) \qquad \text{in } \Phi'.$$

Proof. Since (5.5) remains true, we have only to establish (5.7). Note, by definition, that

(5.10) if g convolves with Φ, then $g * \varphi$ depends continuously on φ, in the Φ topology.

Let $R_{jm}(D)$ $(m = 1, 2, \ldots)$ be partial sums of the power series $R_j(D)$. We shall prove that $R_j(D)g_j$ convolves with Φ and, as $m \to \infty$,

$$(5.11) \qquad R_{jm}(D)g_j * \varphi \to R_j(D)g_j * \varphi \qquad \text{in the } \Phi \text{ topology.}$$

We begin with the identity (see (4.10), Chap. 2)

$$(R_j(D)g_j, \varphi) = (g_j, R_j(-D)\varphi).$$

Replacing $\varphi(\xi)$ by $\varphi(\xi + x)$ we obtain

$$(5.12) \qquad (R_j(D)g_j(\xi), \varphi(x + \xi)) = (g_j(\xi), R_j(-D)\varphi(x + \xi)).$$

$R_j(-D)$ is a bounded operator in Φ and is the strong limit of the $R_{jm}(-D)$. Since each $R_{jm}(-D)$ commutes with any translation, the same is true of $R_j(-D)$, and we may therefore write (5.12) in the form

$$(5.13) \qquad R_j(D)g_j * \varphi = g_j * R_j(-D)\varphi.$$

Since g_j convolves with Φ, it follows that the right side of (5.13) belongs to Φ and depends continuously on φ, in the Φ topology. Hence, $R_j(D)g_j$ convolves with Φ.

Next we have

$$(5.14) \qquad \begin{aligned} R_{jm}(D)g_j * \varphi &= (R_{jm}(D)g_j(\xi), \varphi(x + \xi)) \\ &= (g_j(\xi), R_{jm}(-D)\varphi(x + \xi)) \\ &= g_j * R_{jm}(-D)\varphi. \end{aligned}$$

As $m \to \infty$, $R_{jm}(-D)\varphi \to R_j(-D)\varphi$ in the Φ topology. By the property (5.10) it then follows that the right side of (5.14) tends to the right side of (5.13), and the statement (5.11) is thereby established.

We are now ready to proceed with the proof of (5.7). By (5.6),

$$(5.15) \qquad g_j * R_{jm}(D)u_0 = R_{jm}(D)g_j * u_0.$$

Applying each side to φ and taking $m \to \infty$, we obtain on the right side

$$(5.16) \qquad \begin{aligned} (R_{jm}(D)g_j * u_0, \varphi) &= (u_0, R_{jm}(D)g_j * \varphi) \to (u_0, R_j(D)g_j * \varphi) \\ &= (R_j(D)g_j * u_0, \varphi), \end{aligned}$$

where use has been made of (5.11).

On the left side we obtain

$$(5.17)\quad \begin{aligned}(g_j * R_{jm}(D)u_0, \varphi) &= (R_{jm}(D)u_0, g_j * \varphi) \to (R_j(D)u_0, g_j * \varphi)\\ &= (g_j * R_j(D)u_0, \varphi).\end{aligned}$$

Since the final expressions in (5.16) and (5.17) are equal for all $\varphi \in \Phi$, the proof of (5.7) (and thereby the proof of the lemma) is completed.

Lemma 5′. *If in Lemma 5 the $g_j(x)$ are not assumed to convolve with Φ but, instead, $u(x)$ is assumed to convolve with Φ, then the assertion* (5.9) *is still valid.*

Lemma 5″. *If in Lemma 5′ $u(x)$ convolves Φ into Φ_1 and if $u_0 \in \Phi_1'$, then* (5.9) *is still valid. ((iv) is assumed with Φ' replaced by Φ_1'.)*

Proofs are left to the reader.

6. Existence Theorems for Parabolic Systems

We recall the notation

$$\Lambda(s) = \max_{1\leqslant j\leqslant N} \operatorname{Re}\{\lambda_j(s)\},$$

where the $\lambda_j(s)$ are the eigenvalues of the matrix $P(s)$. The system (4.1) is called a *parabolic system* (in the sense of Shilov) if for all $\sigma \in R^n$

$$(6.1)\qquad \Lambda(\sigma) \leqslant -C|\sigma|^h + C_1 \qquad (C > 0,\ h > 0).$$

h is called the *exponent* of parabolicity. It is obvious that $h \leqslant p_0$. For simplicity we take h to be a scalar even though all the results given below remain true if h is any vector with positive components.

Let $P(s) = (P_{jk}(s))$ and denote by $P_{jk}^0(s)$ the *principal part* of the polynomial $P_{jk}(s)$; i.e., $P_{jk}^0(s)$ consists only of those terms of $P_{jk}(s)$ which are of degree p. Let $\lambda_1^0(s), \ldots, \lambda_N^0(s)$ be the eigenvalues of the matrix

$$P^0(s) = (P_{jk}^0(s)).$$

The system (4.1) is called a *parabolic system in the sense of Petrowski* if for all $\sigma \in R^n$, $|\sigma| = 1$,

$$(6.2)\qquad \Lambda^0(\sigma) \equiv \max_{1\leqslant j\leqslant N} \operatorname{Re}\{\lambda_j^0(\sigma)\} \leqslant -\delta \qquad (\delta > 0).$$

We shall prove the following:

Theorem 11. *The system* (4.1) *is parabolic in the sense of Petrowski if and only if it is parabolic (in the sense of Shilov) and $h = p = p_0$.*

Proof. Suppose that the system (4.1) is parabolic in the sense of

Petrowski. For any $\sigma \in R^n$ ($\sigma \neq 0$), let $\sigma^0 = \sigma/|\sigma|$. Defining

$$P^1(\sigma) = P(\sigma) - P^0(\sigma)$$

we have

$$\det(P(\sigma) - \lambda I) = \det(P^0(\sigma) + P^1(\sigma) - \lambda I)$$
$$= |\sigma|^{Np} \det(P^0(\sigma^0) + \epsilon(\sigma) - \lambda^* I),$$

where $\lambda^* = \lambda/|\sigma|^p$. The roots of the last determinant approach the roots of $\det(P^0(\sigma^0) - \lambda_0 I)$, as $|\sigma| \to \infty$, since $\epsilon(\sigma) = 0(1/|\sigma|)$. Hence,

$$\mathrm{Re}\,\{\lambda_j^*\} = \mathrm{Re}\left\{\frac{\lambda_j(\sigma)}{|\sigma|^p}\right\} < -\frac{\delta}{2}$$

for all $|\sigma|$ sufficiently large. This implies (6.1) with $h = p$. Since

$$h \leqslant p_0 \leqslant p,$$

$p = p_0$, and the proof is completed. The proof of the second part of the theorem is quite similar and is left to the reader.

The theory of Petrowski parabolic systems with *variable* coefficients has been studied in great detail in the literature. In particular the Cauchy problem has been solved in terms of classical solutions under very general assumptions on the smoothness of the coefficients of the system and of the initial data $u_0(x)$. On the other hand, no analogous theory has been developed for systems which are parabolic only in the sense of (6.1). In this section we shall solve the Cauchy problem (4.1), (4.2) in terms of classical solutions for parabolic systems in the sense of (6.1).

We first prove the following theorem.

Theorem 12. *For any parabolic system,* $p_0 > 1$.

Proof. If $p_0 \leqslant 1$ then the polynomial

$$(-1)^{N-1}\,\mathrm{Re}\,\{P_1(s)\} = \mathrm{Re}\,\{\lambda_1(s) + \cdots + \lambda_N(s)\}$$

is of degree $\leqslant 1$ and hence it is of the form $\Sigma\, a_i\sigma_i + \Sigma\, b_i\tau_i + d$. But then $\Lambda(\sigma)$ cannot tend to $-\infty$ as $|\sigma| \to \infty$, thus contradicting (6.1).

We next apply (2.14) and (2.19) and get the inequalities:

$$\|Q(\sigma, t)\| \leqslant B(1 + t^{1/p}|\sigma|)^{(N-1)p} \exp[-c't|\sigma|^h] \qquad (c' > 0), \tag{6.3}$$

$$\|Q(s, t)\| \leqslant B(1 + t^{1/p}|s|)^{(N-1)p} \exp[ct|s|^{p_0}], \tag{6.4}$$

where $Q(s, t) = \exp[tP(s)]$ and B is some constant depending continuously on T ($0 \leqslant T < \infty$). Using Theorem 2, Sec. 1, Chap. 5, we get

$$\|Q(\sigma + i\tau, t)\| \leqslant E_0 \exp[-a'|\sigma|^h] \quad \text{if } |\tau_j| \leqslant K(1 + |\sigma_j|)^\mu \quad (1 \leqslant j \leqslant n), \tag{6.5}$$

where E_0, a' are positive constants depending on t. μ may also depend on t, but we can always decrease $\mu(t)$ so as to obtain a number μ independent of t and $\geqslant 1 - (p_0 - h)$. (It will be shown in Sec. 14 that if $\mu_0(t) = \text{l.u.b. } \mu(t)$, for all the possible $\mu(t)$ for a fixed t, then $\mu_0(t)$ is independent of t; the same remark applies to all the other types of differential systems.)

Since

$$\frac{\partial Q(s, t)}{\partial t} = P(s)Q(s, t), \tag{6.6}$$

each function $\partial^j Q(s, t)/\partial t^j$ satisfies an equality similar to (6.5), with E_0 replaced by another constant E_j.

Noting that the classical inverse Fourier transform of Q coincides with the inverse Fourier transform of Q as a generalized function over any Φ ($\Phi \subset S$), and applying Theorems 1, 1′ of Sec. 1, Chap. 6, we conclude that

$$|D_x^q D_t^j G(x, t)| \leqslant C_{qj} \exp[-\rho|x|^{h_1}] \qquad \left(h_1 = \frac{h}{h - \mu}\right) \qquad \text{if } \mu \leqslant 0, \tag{6.7}$$

$$|D_x^q D_t^j G(x, t)| \leqslant C'_{qj} \exp[-\rho'|x|^{p_1}] \qquad \left(p_1 = \frac{p_0}{p_0 - \mu}\right) \qquad \text{if } 0 < \mu \leqslant 1. \tag{6.8}$$

provided

$$\|Q(\sigma + i\tau, t)\| \leqslant E' \exp[-a'|\sigma|^h + b'|\tau|^{p_0/\mu}] \qquad \text{if } 0 < \mu \leqslant 1. \tag{6.9}$$

(In view of Theorems 3, 4, Sec. 2, (6.9) is false if $\mu > 1$.) We recall that (6.9) is satisfied if $n = 1$. It can also be proved without difficulty (see Problem 1) that (6.9) with $\mu = 1$ is satisfied for parabolic systems in the sense of Petrowski.

The constants C_{qj}, C'_{qj}, ρ, ρ' depend on t; C_{qj}, C'_{qj} and ρ' tend to ∞ as $t \to 0$ and ρ (if $\mu < 0$) tends to 0 as $t \to 0$. By looking at the proofs of Theorems 2, 3, Sec. 1, Chap. 5, and of Theorems 1, 1′, Sec. 1, Chap. 6, these constants can be evaluated. Thus one finds that

$$|D_x^q G(x, t)| \leqslant \frac{A_q}{t^{(n+q+\nu)/h}} \exp[-\beta t^{-\mu/(h-\mu)}|x|^{h_1}] \qquad \left(h_1 = \frac{h}{h - \mu}\right) \qquad \text{if } \mu \leqslant 0, \tag{6.10}$$

$$|D_x^q G(x, t)| \leqslant \frac{A'_q}{t^{(n+q+\nu)/h}} \exp[-\beta_0 t^{-\mu/(p_0-\mu)}|x|^{p_1}] \qquad \left(p_1 = \frac{p_0}{p_0 - \mu}\right) \qquad \text{if } \mu > 0, \tag{6.11}$$

where $\nu = (N - 1)(p - h)$, and A_q, A'_q, β, β_0 are positive constants independent of t. It is assumed that the coefficients a', b' in (6.9) are of the same dimension as those obtained in proving (6.9) for $n = 1$.

Let $u_0(x)$ be a measurable (vector-) function which satisfies, almost everywhere,

$$|u_0(x)| \leqslant B_1 \exp [\gamma_1 |x|^{h_1}] \qquad \text{if } \mu \leqslant 0, \tag{6.12}$$

$$|u_0(x)| \leqslant B_2 \exp [\gamma_2 |x|^{p_1}] \qquad \text{if } \mu > 0. \tag{6.13}$$

Then, if $\gamma_1 < \beta t_0^{-\mu/(h-\mu)}$, $t_0 \leqslant t \leqslant T$, and $\mu \leqslant 0$, the convolution (4.10), is the function $G \circledast u_0$. This follows by a modified version of Lemma 3, in which g convolves Φ into Φ_1, $h \in \Phi'_1$ and the other assumptions are unchanged; Φ and Φ_1 are the spaces defined in (4.7). The function $G(x, t) \circledast u_0(x)$ is a uniformly convergent integral and it is therefore continuous in (x, t) for $x \in R^n$, $t_0 \leqslant t \leqslant T$. Furthermore, in view of our estimates (6.10), (6.11), and the analogous estimates on $D_x^q D_t^j G(x, t)$, it follows that the integral can be differentiated any number of times with respect to (x, t) if $t > t_0$, and the differentiation may be performed by differentiating $G(x - \xi, t)$ in the integrand. Hence $u(x, t)$ is a C^∞ function for $t_0 \leqslant t \leqslant T$.

To prove that $u(x, t)$ is a classical solution of (4.1) for $t_0 \leqslant t \leqslant T$, we proceed to estimate $D_x^q u(x, t)$. We have

$$\begin{aligned} |D_x^q u(x, t)| &= \left| \int_{R^n} D_x^q G(x - \xi, t) u_0(\xi)\, d\xi \right| \\ &\leqslant H_1 \int_{R^n} [\exp [-\alpha |x - \xi|^{h_1} + \gamma_1 |\xi|^{h_1}]]\, d\xi \\ &= H_1 \int_{R^n} [\exp [-\alpha |\xi|^{h_1} + \gamma_1 |x - \xi|^{h_1}]]\, d\xi \end{aligned} \tag{6.14}$$

provided $\mu \leqslant 0$. H_1 depends on t_0 where $t_0 \leqslant t \leqslant T$, and $\alpha = \beta t_0^{-\mu/(h-\mu)}$. We shall need the following lemma.

Lemma 6. *For any given $\alpha' > 0$, $\epsilon > 0$, there exists a $\gamma' > 0$ such that, for all x, ξ in R^n,*

$$-\alpha' |\xi|^{h_1} + \gamma' |x - \xi|^{h_1} \leqslant \beta' |x|^{h_1} \qquad \text{if } \beta' = \gamma' + \epsilon. \tag{6.15}$$

If $\alpha' \to 0$ (∞) then $\gamma' \to 0$ (∞).

Proof. Dividing both sides of (6.15) by $|\xi|$ and denoting $x/|\xi|$ again by x, we get the equivalent inequality

$$-\alpha' + \gamma' |x - e|^{h_1} \leqslant \beta' |x|^{h_1} \qquad \text{where } e \text{ is any unit vector.}$$

This inequality is satisfied for all sufficiently large $|x|$, say for $|x| \geqslant R$. If $|x| < R$ and γ' is sufficiently small, then the left side is $\leqslant 0$ and the inequality is also satisfied. From the above proof of (6.15) follows the second assertion of the lemma.

Writing $\exp[-\alpha|\xi|^{h_1}] = \exp\left[-\frac{\alpha}{2}|\xi|^{h_1}\right]\exp\left[-\frac{\alpha}{2}|\xi|^{h_1}\right]$ in (6.14), and applying (6.15) with $\alpha' = \alpha/2$, $\gamma_1 = \gamma'$, we conclude that, for $t_0 \leqslant t \leqslant T$,

$$|D_x^q u(x, t)| \leqslant B_q' \exp[\beta'|x|^{h_1}]. \tag{6.16}$$

A similar inequality holds for $D_t u$. Hence, applying a part of Lemma 2 it follows that $u(x, t)$ satisfies (4.1) in the classical sense, if $t_0 \leqslant t \leqslant T$.

In case $\mu > 0$, the calculation is similar, except that h_1 is replaced by p_1 and α is now $\beta_0 T^{-\mu/(p_0-\mu)}$. Furthermore, γ_2 must be restricted by the requirement that $u_0 \in \Phi_1'$. Note that we can take $\alpha' = \theta\alpha$ for any $0 < \theta < 1$, and that Lemma 6 remains true with $\gamma' = \theta\alpha'$, provided ϵ is sufficiently large.

For systems parabolic in the sense of Petrowski it can be proved (see the remark following Theorem 13′) that if $u_0(x)$ is a continuous function, then $u(x, t)$ is also a continuous function for $0 \leqslant t \leqslant T$ (and not only for $t_0 < t \leqslant T$ for any $t_0 > 0$). But for general parabolic systems this need not be true. The best bound (independent of t) on Q, for general parabolic systems, seems to be $\|Q(\sigma, t)\| = 0[(1 + |\sigma|)^k]$ for some $k > 0$. Using this bound it can be shown that $u(x, t)$ is continuous for $0 \leqslant t \leqslant T$ if we assume some growth conditions on $u_0(x)$ and on some of its derivatives. This will be done in Sec. 8. We sum up the results so far obtained.

Theorem 13. *Let $\mu \leqslant 0$. If $u_0(x)$ satisfies (6.12) for any $\gamma_1 > 0$ (B_1 depending on γ_1) then there exists a function $u(x, t)$ which satisfies (4.1) for $0 < t \leqslant T$ in the classical sense and (4.2) in the generalized sense over Φ (Φ is defined in (4.7)). u satisfies the inequalities*

$$|D_x^q u(x, t)| \leqslant B_1 B' \exp[\gamma_1'|x|^{h_1}] \qquad (0 < t \leqslant T; 0 \leqslant |q| \leqslant p), \tag{6.17}$$

where B' depends on γ_1' and on t_0 if $t_0 \leqslant t \leqslant T$ (for any $t_0 > 0$) and γ_1' is any given positive number.

Theorem 13′. *Let $0 < \mu \leqslant 1$ and assume that $n = 1$ or that (6.9) is satisfied. If $u_0(x)$ satisfies (6.13) for some $\gamma_2 < \beta_0 T^{-\mu/(p_0-\mu)}$, $\gamma_2 < \bar{\beta}_0 T^{-1/(p_0-1)}$ (where β_0, $\bar{\beta}_0$ depend only on P), then there exists a function $u(x, t)$ which satisfies (4.1) for $0 < t \leqslant T$ in the classical sense and (4.2) in the generalized sense over Φ. $u(x, t)$ satisfies the inequalities*

$$|D_x^q u(x, t)| \leqslant B_2 B'' \exp[\gamma_2'|x|^{p_1}] \qquad (0 < t \leqslant T; 0 \leqslant |q| \leqslant p), \tag{6.18}$$

where B'' depends on γ_2' and on t_0 if $t_0 \leqslant t \leqslant T$ (for any $t_0 > 0$) and γ_2' depends on γ_2, T. ($\gamma_2' \to \gamma_2$ if $T \to t_0$.)

If $p = h = p_0$, then, by Theorem 11, Problem 1 and (6.10),

$$|G(x, t)| \leqslant \frac{A_0}{t^{n/p}} \exp\left[-\beta_0 t^{-1/(p-1)}|x|^{p/(p-1)}\right]. \tag{6.19}$$

One then gets, for $u_0(x)$ satisfying (6.13) with $\gamma_2 < \beta_0 T^{-1/(p-1)}$,

$$|u(x, t)| \leqslant B_2 B'' \exp\left[\gamma_2'|x|^{p_1}\right],$$

where B'' is independent of t, $0 < t \leqslant T$. Hence u belongs to the class E' in which the uniqueness of solutions has been proved in Theorem 7, Sec. 3. It can furthermore be shown that $\int G(x, t)\, dx \to 1$ as $t \to 0$ ($\int G(x, t)\, dx \equiv 1$ when $u \equiv (1, \ldots, 1)$ satisfies $P(iD)u = 0$). Writing

$$u(x, t) - u_0(x) = \int G(x - \xi, t)[u_0(\xi) - u_0(x)]\, d\xi + o(1)$$

where $o(1) \to 0$ as $t \to 0$, and dividing the integral into two parts, I_1 with $|\xi - x| < \delta$ and I_2 with $|\xi - x| > \delta$ ($\delta \to 0$ as $t \to 0$), one can readily verify that if $u_0(x)$ is a continuous function then (4.2) is satisfied in the classical sense.

7. An Existence Theorem for Hyperbolic Systems

The system (4.1) is called a *hyperbolic system* if

$$|\Lambda(s)| \leqslant C_0|s| + C_1, \qquad \Lambda(\sigma) \leqslant C_2. \tag{7.1}$$

For one hyperbolic equation, written as a system (see (1.6), (1.7)), the above definition coincides with the definition of Gårding for equations with constant coefficients [20]. Since the degree of the polynomial $P_1(s)$, which is the coefficient of λ^{N-1} in the polynomial $\det (P(s) - \lambda I)$, is $\leqslant p_0 = 1$, and since

$$(-1)^{N-1} \operatorname{Re}\{P_1(\sigma)\} = \operatorname{Re}\{\lambda_1(\sigma)\} + \cdots + \operatorname{Re}\{\lambda_N(\sigma)\} \leqslant NC_2,$$

it follows that $(-1)^{N-1} \operatorname{Re}\{P_1(\sigma)\} \equiv \text{const.} = C'$. We then find that

$$\operatorname{Re}\{\lambda_j(\sigma)\} = C' - \sum_{k \neq j} \operatorname{Re}\{\lambda_k(\sigma)\} \geqslant C' - (N-1)C_2.$$

Hence, for some constant C_3,

$$C_3 \leqslant \operatorname{Re}\{\lambda_j(\sigma)\} \leqslant C_2 \qquad (j = 1, \ldots, N). \tag{7.2}$$

There are various different definitions of the concept of hyperbolic equations and of hyperbolic systems. The most common definition for one equation is the following: the equation has the form

$$L\left(\frac{\partial}{\partial t}, i\frac{\partial}{\partial x}\right) \equiv a_0 \frac{\partial^p u}{\partial t^p} + \sum_{|\alpha| \leqslant p-j} \sum_{j=0}^{p-1} a_{j\alpha} D_t^j (iD_x)^\alpha u = f \qquad (a_0 \neq 0),$$

and the roots of

$$L_0(\lambda, \sigma) \equiv a_0\lambda^p + \sum_{|\alpha|=p-j} \sum_{j=0}^{p-1} a_{j\alpha}\lambda^j\sigma^\alpha = 0 \qquad (\sigma \text{ real}, \sigma \neq 0)$$

are pure imaginary (i.e., $C_3 = C_2 = 0$ in (7.2)) and distinct from each other. For equations hyperbolic in the last sense, and also for systems which are hyperbolic in a similar sense, the Cauchy problem has been solved even in the case of *variable* coefficients.

The definition (7.1), which will be adopted in this book, is of course more general, but we do not consider coefficients which depend on x.

By (2.14) and (6.6) we get

$$(7.3) \qquad \begin{aligned} \|Q(\sigma, t)\| &= 0(|\sigma|^{(N-1)p}), & \left\|\frac{\partial}{\partial t} Q(\sigma, t)\right\| &= 0(|\sigma|^{Np}), \\ \|Q(s, t)\| &= 0(\exp[bt|s|]), & \left\|\frac{\partial Q(s, t)}{\partial t}\right\| &= 0(\exp[bt|s|]). \end{aligned}$$

Applying Theorem 6, Sec. 3, Chap. 6, we find that for every $\epsilon > 0$,

$$(7.4) \qquad G(x, t) = \sum_{k=1}^{M} P_k\left(\frac{\partial}{\partial x}\right) f_k(x, t) \qquad \text{in } S',$$

where $f_k(x, t)$, $\partial f_k(x, t)/\partial t$ are continuous functions of (x, t) ($x \in R^n$, $0 \leqslant t \leqslant T$) which vanish identically if $|x_j| > b_j t + \epsilon$ for some j, $j = 1, \ldots, n$. The P_k are linear differential operators of orders $\leqslant Np + n + 2$.

Since (7.4) holds in S', it also holds in K'. Applying Lemma 4 with $\Phi = K$ we conclude that if $u_0(x)$ has $Np + n + 2$ continuous derivatives then

$$(7.5) \qquad G(x, t) * u_0(x) = \sum_{k=1}^{M} \int_{R^n} f_k(x - \xi, t) P_k\left(\frac{\partial}{\partial \xi}\right) u_0(\xi)\, d\xi \qquad \text{over } K,$$

and the right side is a continuous function of (x, t) ($x \in R^n$, $0 \leqslant t \leqslant T$) together with its first t-derivative.

Similarly to (7.4) we can write another representation for $G(x, t)$ of the form $\sum \hat{P}_k\left(\frac{\partial}{\partial x}\right) g_k(x, t)$, where $g_k(x, t)$ are only continuous functions of (x, t) (vanishing if $|x_j| > b_j t + \epsilon$ for some j) but the orders of the P_k are $\leqslant (N - 1)p + n + 2$. Analogously to (7.5) we get

$$\begin{aligned} G(x,t) * u_0(x) &= \sum_k \int_{R^n} g_k(x - \xi, t)\hat{P}_k\left(\frac{\partial}{\partial \xi}\right) u_0(\xi)\, d\xi \\ &= \sum_k \int_{R^n} g_k(\xi, t)\hat{P}_k\left(\frac{\partial}{\partial x}\right) u_0(x - \xi)\, d\xi \equiv k(x, t) \qquad \text{in } K'. \end{aligned}$$

It is clear that if $u_0(x)$ is a continuous function together with all its first $Np + n + 2$ derivatives, then $k(x, t)$ is a continuous function of (x, t) together with its first p x-derivatives.

By Lemma 2′ we then obtain the following:

Theorem 14. *Let $u_0(x)$ be any function continuous for $x \in R^n$ together with its first $Np + n + 2$ derivatives ($Np + n + 1$ if this is an even number). Then there exists a unique classical solution of* (4.1), (4.2) *in any interval* $0 \leqslant t \leqslant T$.

The uniqueness follows from Theorem 9′, Sec. 3.

8. Existence Theorems for Correctly Posed Systems

The system (4.1) is called a *correctly posed system* if

$$\Lambda(\sigma) \leqslant C_0. \tag{8.1}$$

Both parabolic and hyperbolic systems are correctly posed systems. From (2.14) we conclude that

$$\|Q(\sigma, t)\| = 0(|\sigma|^h) \quad \text{for some } h \leqslant (N-1)p. \tag{8.2}$$

We shall take h to be an integer and assume that $p_0 > 1$. The (hyperbolic) case $p_0 \leqslant 1$ has already been treated in Sec. 7.

By Theorem 2″, Sec. 1, Chap. 5, we have

$$\|Q(\sigma + i\tau, t)\| \leqslant A_1(1 + |\sigma|)^h \qquad \text{in some region} \tag{8.3}$$
$$|\tau_j| \leqslant K(1 + |\sigma_j|)^\mu \qquad (1 \leqslant j \leqslant n),$$

where $\mu \geqslant 1 - p_0$ and K, A_1 depend only on P and T (if $0 \leqslant t \leqslant T$). μ can be taken independently of t. By (6.6) we also have, if $\mu \leqslant 1$,

$$\left\| \frac{\partial}{\partial t} Q(\sigma + i\tau, t) \right\| \leqslant A_2(1 + |\sigma|)^{h+p}. \tag{8.4}$$

Here and throughout this section, A_j will be used to denote constants depending only on P and T.

Let $\mu < 0$. Then, applying Theorem 7, Sec. 4, Chap. 6, we get

$$G(x, t) = (H - \Delta)^{(h+p+\nu)/2} f(x, t) \qquad \text{in } S' \qquad (\nu \geqslant n + 2), \tag{8.5}$$

where $f(x, t)$, $\partial f(x, t)/\partial t$ are continuous functions of (x, t) ($x \in R^n$, $0 \leqslant t \leqslant T$) and

$$|f(x, t)| + \left| \frac{\partial}{\partial t} f(x, t) \right| \leqslant \frac{A_3}{(1 + |x|)^{(\nu - n)/|\mu|}}. \tag{8.6}$$

Assume that $u_0(x)$ is continuously differentiable together with its first $h + p + \nu$ derivatives and that, for some $\gamma \geqslant 0$,

$$|D_x^q u_0(x)| \leqslant B_1(1 + |x|)^\gamma \qquad (|q| \leqslant h + p + \nu). \tag{8.7}$$

By Lemma 4″ (with Φ, Φ_1 defined in (4.7) (take $a > c_0$)),

$$\begin{aligned} u(x, t) &= G(x, t) * u_0(x) \\ &= \int_{R^n} f(x - \xi, t)(H - \Delta)^{(h+p+\nu)/2} u_0(\xi)\, d\xi \qquad \text{in } \Phi'. \end{aligned} \tag{8.8}$$

The integral can be estimated by substituting $x - \xi = \xi'$ and using (8.6), (8.7). We conclude that

$$|u(x, t)| \leqslant A_4(1 + |x|)^\gamma \qquad \text{provided } \frac{\nu - n}{|\mu|} \geqslant \gamma + n + 1. \tag{8.9}$$

We can also differentiate (8.8) with respect to t and obtain the same inequality for $\partial u/\partial t$.

To deal with $D_x^q u$, we apply Theorem 7, Sec. 4, Chap. 6, in a slightly different manner, considering $Q(s, t)$ to be only continuous, not t-differentiable, and obtain, analogously to (8.5),

$$G(x, t) = (H' - \Delta)^{(h+\nu)/2} f'(x, t) \qquad \text{in } S' \qquad (\nu \geqslant n + 2), \tag{8.5′}$$

where $f'(x, t)$ is only known to be continuous in (x, t) $(x \in R^n, 0 \leqslant t \leqslant T)$, and

$$|f'(x, t)| \leqslant \frac{A_5}{(1 + |x|)^{(\nu - n)/|\mu|}}. \tag{8.6′}$$

Proceeding similarly to (8.8) we conclude that, for all $|q| \leqslant p$, $D_x^q u(x, t)$ are continuous functions of (x, t) $(x \in R^n, 0 \leqslant t \leqslant T)$, and

$$|D_x^q u(x,t)| \leqslant A_6(1 + |x|)^\gamma \qquad \text{provided } \frac{\nu - n}{|\mu|} \geqslant \gamma + n + 1.$$

Using Lemma 2, we derive the following theorem.

Theorem 15. *Let $\mu < 0$. If $u_0(x)$ is continuous together with its first $Np + \nu$ derivatives $(\nu \geqslant n + 2)$ and if*

$$|D_x^q u_0(x)| \leqslant B_1(1 + |x|)^\gamma \qquad \text{for } |q| \leqslant Np + \nu, \tag{8.10}$$

where γ satisfies $\gamma \geqslant 0$, $\gamma + n + 1 \leqslant (\nu - n)/|\mu|$, then there exists in $0 \leqslant t \leqslant T$ a unique classical solution $u(x, t)$ of (4.1), (4.2) satisfying

$$|D_x^q u(x, t)| \leqslant B_1 B_1'(1 + |x|)^\gamma \qquad (0 \leqslant t \leqslant T; 0 \leqslant |q| \leqslant p), \tag{8.11}$$

where B_1' is a constant independent of $u_0(x)$. The solution is unique even within the larger class of functions $v(x)$ satisfying $|v(x)| = 0(\exp [\epsilon|x|^{p_0'}])$ for any $\epsilon > 0$.

If $0 \leqslant \mu \leqslant 1$, then by applying Theorems 7 (for $\mu = 0$) and 7′ (for $\mu > 0$) of Chap. 6, Sec. 4, we obtain

$$G(x, t) = \sum_{k=1}^{M} P_k\left(\frac{\partial}{\partial x}\right) f_k(x, t) \qquad \text{in } S', \tag{8.12}$$

where

$$|f_k(x, t)| + \left|\frac{\partial}{\partial t} f_k(x, t)\right| \leqslant A_7 \exp\left[-A_8 |x|^{p_1}\right] \qquad \left(p_1 = \frac{p_0}{p_0 - \mu}\right) \tag{8.13}$$

and the P_k are of orders $\leqslant h + p + n + 2$. If $n > 1$, $\mu > 0$, we have to assume that

$$\|Q(\sigma + i\tau, t)\| \leqslant A_9(1 + |\sigma|)^h \exp\left[A_{10}|\tau|^{p_0/\mu}\right]. \tag{8.14}$$

From (8.12), (8.13) we can derive the continuity of $u(x, t)$ and $\partial u(x, t)/\partial t$ in (x, t) ($x \in R^n$, $0 \leqslant t \leqslant T$), and the bound $0(\exp[\gamma_1'|x|^{p_1}])$ on u and u_t, provided a similar bound (with γ_1' replaced by some γ_1) holds for the derivatives of u_0. The calculation is based on Lemma 4″ and on estimates of classical convolutions of the type given previously in the parabolic case.

To derive the continuity of $D_x^q u$ ($|q| \leqslant p$) and the necessary bounds on it we use a different representation for $G(x, t)$ where now the f_k are only continuous functions (and not t-differentiable) and the P_k are of orders $\leqslant h + n + 2$.

Using Lemma 2, we conclude:

Theorem 15′. *Let $0 \leqslant \mu \leqslant 1$ and suppose that $n = 1$ or (if $n > 1$) that (8.14) holds. If $u_0(x)$ is continuous together with its first $Np + n + 2$ derivatives and if*

$$|D_x^q u_0(x)| \leqslant B_2 \exp\left[\gamma_1 |x|^{p_1}\right] \qquad \left(p_1 = \frac{p_0}{p_0 - \mu},\ |q| \leqslant Np + n + 2\right), \tag{8.15}$$

then there exists a unique classical solution in $0 \leqslant t \leqslant T$ of (4.1), (4.2) which satisfies the inequalities

$$|D_x^q u(x, t)| \leqslant B_2 B_2' \exp\left[\gamma_1' |x|^{p_1}\right] \qquad (0 \leqslant t \leqslant T;\ 0 \leqslant |q| \leqslant p), \tag{8.16}$$

where B_2' is a constant independent of $u_0(x)$. T depends on P and γ_1 (if $\gamma_1 \to 0$ then $T \to \infty$) and γ_1' depends on P, γ_1, and T ($\gamma_1' \to \gamma_1$ if $T \to 0$).

9. Existence Theorems for Mildly Incorrectly Posed Systems

The system (4.1) is called an *incorrectly posed system* if

$$\Lambda(\sigma) \leqslant C_0|\sigma|^h + C_1 \qquad (C_0 > 0,\ h > 0). \tag{9.1}$$

If $h < p_0$, we call it a *mildly* incorrectly posed system. We assume throughout this section that $p_0 > 1$.

Using (2.14) and Theorem 2, Sec. 1, Chap. 5, we get

$$\|Q(\sigma + i\tau, t)\| \leqslant C_2 \exp [b_0|\sigma|^h] \qquad \text{if } |\tau_j| \leqslant K(1 + |\sigma_j|)^\mu \quad (1 \leqslant j \leqslant n) \tag{9.2}$$

where $b_0 = b'T$, b' is independent of t and C_2 depends on T. μ can be taken independently of t.

Assume $\mu \leqslant 0$. By Theorem 8, Sec. 5, Chap. 6,

$$G(x, t) = [H + f_\alpha(i\Delta^{1/2})]f(x, t) \qquad \text{in } (W_{\beta',1/b}^{\beta',1/a})' \tag{9.3}$$

for any $\beta > \alpha > h$, $b/a > \gamma$, and $f(x, t)$ is a continuous function of (x, t) $(x \in R^n, 0 \leqslant t \leqslant T)$ satisfying

$$|f(x, t)| \leqslant C_3 \exp [-a'|x|^{\alpha_1}] \qquad \left(a' > 0, \alpha_1 = \frac{\alpha}{\alpha - \mu}\right). \tag{9.4}$$

Note that a' depends on T, and $a' \to \infty$ or 0 as $T \to 0$ or ∞. The function $f_\alpha(z)$ is defined by (5.9), Chap. 6. We shall take $\beta = p_0$ so that $h < \alpha < p_0$.

Let $u_0(x)$ be a C^∞ function, satisfying

$$|D_x^q u_0(x)| \leqslant B_1 B^q q^{q\alpha'} \exp [c|x|^{h_1}] \qquad \left(h_1 = \frac{h}{h - \mu}; 0 \leqslant |q| < \infty\right) \tag{9.5}$$

for some $\alpha' < 1/h$. Then, if $\alpha = m_0/k_0$,

$$\begin{aligned} |[H + f_\alpha(i\Delta^{1/2})]u_0(x)| &\leqslant B_1 \left(H + \sum_{j=0}^{\infty} \frac{(nB)^{2m_0j}(2m_0j)^{2m_0\alpha'j}}{(2k_0j)!}\right) \exp [c|x|^{h_1}] \\ &\leqslant C_4 \exp [c|x|^{h_1}] \end{aligned} \tag{9.6}$$

if $m_0\alpha' < k_0$, i.e., if $\alpha' < 1/\alpha$. Since $\alpha' < 1/h$, by taking α sufficiently close to h we achieve the inequality $\alpha' < 1/\alpha$ and (9.6) is then valid.

We wish to apply Lemma 5″, Sec. 5, with Φ and Φ_1 defined by (4.7). Of all the assumptions made in the lemma, the only one which needs to be verified is the first inequality in (5.8). (Note that the first statement in (v) follows from the inequalities derived in the proof of (5.8) given below.) The proof of (5.8) proceeds as follows: by (9.4), (9.6),

$$\begin{aligned} &\int\int |f(x, t)[H + f_\alpha(i\Delta_x^{1/2})]u_0(x - y)\varphi(y)|\, dx\, dy \\ &\leqslant C_5 \int \left\{\int \{\exp [-a'|x|^{\alpha_1}]\} \{\exp [c|x - y|^{h_1}]\} \{\exp [-c_0|y|^{p_0'}]\}\, dx\right\} dy, \end{aligned} \tag{9.7}$$

where c_0 is any number $< 1/p_0'b^{p_0'}$. Noting that $\alpha_1 > h_1$, we can estimate the inner integral in the same way that we estimated the integral (6.14). We obtain the bound

$$C_6 \int \{\exp [c_1|y|^{h_1}]\} \{\exp [-c_0|y|^{p_0'}]\} \, dy,$$

which is convergent, since $h_1 \leqslant 1 < p_0'$.

We can now apply Lemma 5″ and conclude that, in Φ',

$$\begin{aligned} G(x, t) * u_0(x) &= \int f(x - \xi, t)[H + f_\alpha(i\Delta_\xi^{1/2})]u_0(\xi) \, d\xi \\ &= \int f(\xi, t)[H + f_\alpha(i\Delta_x^{1/2})]u_0(x - \xi) \, d\xi \equiv k(x, t). \end{aligned}$$

If we differentiate $k(x, t)$ with respect to x any number of times, by differentiating the integrand, we obtain a uniformly convergent integral in (x, t) $(x \in R^n, 0 \leqslant t \leqslant T)$, bounded by

$$(9.8) \qquad C_7 \int \{\exp [-a'|\xi|^{\alpha_1}]\} \{\exp [c|x - \xi|^{h_1}]\} \, d\xi \leqslant C_8 \exp [c'|x|^{h_1}].$$

To prove that $\partial u/\partial t$ exists and is a continuous function of (x, t) $(x \in R^n, 0 \leqslant t \leqslant T)$ we note (by (6.6)) that $\partial Q/\partial t$ satisfies the same inequality (9.2) which Q satisfies. Hence we can find for G a representation of the form (9.3) but with both f and f_t continuous functions satisfying (9.4). We can now proceed as before and find the bound (9.8) on $\partial u/\partial t$. In the same way we can show that u is infinitely differentiable.

Applying, finally, Lemma 2, Sec. 5, we obtain the following:

Theorem 16. *Let $\mu \leqslant 0$, $0 < h < p_0$. If $u_0(x)$ is a C^∞ function satisfying* (9.5) *for some constants $B_1 > 0$, $B > 0$, $c > 0$, $\alpha' < 1/h$, then there exists a unique classical solution of* (4.1), (4.2) *in $0 \leqslant t \leqslant T$ which satisfies*

$$(9.9) \qquad |D_x^q u(x, t)| \leqslant B_1' \exp [c'|x|^{h_1}] \qquad \left(h_1 = \frac{h}{h - \mu}; \quad 0 \leqslant t \leqslant T; 0 \leqslant |q| \leqslant p\right).$$

The solution is an infinitely differentiable function of (x, t) for $x \in R^n$, $0 \leqslant t \leqslant T$. T depends on P and c $(T \to \infty$ if $c \to 0)$ and c' depends on P, c, and T $(c' \to c$ if $T \to 0)$.

Assume next that $0 < \mu \leqslant 1$ and $h < 1$.

By (2.14) and Theorem 3, Sec. 1, Chap. 5, we get

$$(9.10) \qquad \|Q(\sigma + i\tau, t)\| \leqslant \hat{C} \exp [d'|\sigma|^h + d''|\tau|^{p_0/\mu}]$$

for some constants $\hat{C}$, d', d'' depending on T, provided $n = 1$. We now assume that n is any positive integer and that (9.10) is satisfied.

By Theorem 8′, Sec. 5, Chap. 6 (with $\beta = p_0$),

$$G(x, t) = \sum_{k=0}^{\infty} \chi_k(x)[H_k + f_\alpha(i\Delta^{1/2})]g_k(x, t), \tag{9.11}$$

$$H_k = B_0' \exp{[B_0 k^\nu]} \qquad \left(\nu = \alpha\left(\frac{\mu}{p_0 - \mu} - \epsilon\right)\right),$$

where the $\chi_k(x)$ vanish outside $k \leqslant |x| \leqslant k + 2$ and satisfy the inequalities

$$|D^q \chi_k(x)| \leqslant A_0 A^q q^{q\delta} \qquad (\text{for any } \delta > 1), \tag{9.12}$$

and the $g_k(x, t)$ are continuous functions satisfying

$$|g_k(x, t)| \leqslant C' \exp{[-C|x|^\eta]} \qquad \text{if } k \leqslant |x| \leqslant k + 2 \qquad \left(\eta = \frac{p_0}{p_0 - \mu} - \epsilon\right). \tag{9.13}$$

h is assumed to satisfy the inequalities

$$h < \frac{p_0 - \mu}{\mu - \epsilon(p_0 - \mu)} \frac{p_0}{p_0 - 1}, \qquad h < 1, \tag{9.14}$$

and α is a rational number $> h$ and sufficiently close to h so that it satisfies the same inequalities and, in addition, δ is taken such that $\delta\alpha < 1$.

Let the initial values $u_0(x)$ be C^∞ and assume that

$$|D_x^q u_0(x)| \leqslant B_2 B_3^q q^{q\alpha'} \exp{[d|x|^{\bar{p}}]} \qquad \left(\bar{p} < p_1 \equiv \frac{p_0}{p_0 - \mu}; 0 \leqslant |q| < \infty\right). \tag{9.15}$$

Using (9.12), (9.15), and Leibniz' rule one easily finds that

$$|D_\xi^q[\chi_k(\xi)u_0(x - \xi)]| \leqslant \begin{cases} B_2'(B_3')^q q^{q\alpha''} \exp{[d|x - \xi|^{\bar{p}}]} & \text{if } k \leqslant |\xi| \leqslant k + 2, \\ 0 \quad \text{if } |\xi| < k \text{ and if } |\xi| > k + 2, \end{cases} \tag{9.16}$$

where $\alpha'' = \max\{\alpha', \delta\}$. We now assume that $\alpha' < 1/h$ and take α such that $\alpha' < 1/\alpha$. By a calculation similar to (9.6) we obtain

$$\begin{aligned} &|g_k(\xi, t)[H_k + f_\alpha(i\Delta_\xi^{1/2})]\,[\chi_k(\xi)u_0(x - \xi)]| \\ &\leqslant \begin{cases} B_4 [\exp{[-C''|\xi|^\eta]}]\,[\exp{[d|x - \xi|^{\bar{p}}]}] & \text{if } k \leqslant |\xi| \leqslant k + 2, \\ 0 \quad \text{if } |\xi| < k \text{ and if } |\xi| > k + 2, \end{cases} \end{aligned} \tag{9.17}$$

provided $\nu < \eta$. Integrating with respect to ξ on R^n, and summing with respect to k we find that if $\eta > \bar{p}$ then

$$\sum_{k=0}^{\infty} \int |g_k(\xi, t)[H_k + f_\alpha(i\Delta_\xi^{1/2})][\chi_k(\xi)u_0(x - \xi)]|\, d\xi \leqslant B_5 \exp{[d_0|x|^{\bar{p}}]}. \tag{9.18}$$

Note that for $\epsilon > 0$ sufficiently small the inequalities $\nu < \eta$, $\bar{p} < \eta$ are satisfied.

We now claim that

$$\{\sum_{k=0}^{\infty} \chi_k[(H_k + f_\alpha(i\Delta^{1/2}))g_k]\} * u_0 = K(x, t) \tag{9.19}$$

where

$$K(x, t) = \sum_{k=0}^{\infty} \int_{R^n} g_k(\xi, t)[H_k + f_\alpha(i\Delta_\xi^{1/2})][\chi_k(\xi)u_0(x - \xi)]\, d\xi. \tag{9.20}$$

The proof is given in two steps. In the first step we prove the analogous statement only for finite partial sums $\sum_{k=0}^{m}$. This is done by modifying the proof of Lemma 5″, Sec. 5. The second step consists in taking $m \to \infty$ and using the definition of convolution (compare the proof of (10.5) below). Details are left to the reader.

We have thus proved that

$$G(x, t) * u_0(x) = K(x, t) \qquad \text{in } \Phi' \tag{9.21}$$

and, as is seen from (9.20), (9.18), K is a continuous function in (x, t) $(x \in R^n, 0 \leqslant t \leqslant T)$ and it satisfies the inequality

$$|K(x, t)| \leqslant B_6 \exp [d_0|x|^{\bar{p}}]. \tag{9.22}$$

From (9.20) it also follows that each x-derivative of K is a continuous function in (x, t) $(x \in R^n, 0 \leqslant t \leqslant T)$ bounded by the right side of (9.22) (with a different B_6). To show that t-derivatives (and mixed derivatives) of $K(x, t)$ are continuous functions of (x, t), we use a representation (9.11) with g_k whose t-derivatives of any given order are continuous functions in $x \in R^n$, $0 \leqslant t \leqslant T$.

Noting finally that the first inequality in (9.14) is satisfied if $h < 1$, we can sum up our results in the following theorem:

Theorem 16′. *Let $0 < \mu \leqslant 1$, $0 < h < 1$ and assume that $n = 1$ or (if $n > 1$) that (9.10) is satisfied. If $u_0(x)$ is a C^∞ function satisfying (9.15) for some $\alpha' < 1/h$, $\bar{p} < p_1$ (where $p_1 = p_0/(p_0 - \mu)$) and for some constants $B_2 > 0$, $B_3 > 0$, $d > 0$, then there exists a unique classical solution of (4.1), (4.2), in any interval $0 \leqslant t \leqslant T$, which satisfies*

$$|D_x^q u(x, t)| \leqslant B_7 \exp [d_1|x|^{\bar{p}}] \qquad (0 \leqslant t \leqslant T; 0 \leqslant |q| \leqslant p) \tag{9.23}$$

for some d_1 depending on P, T, and d. The solution is an infinitely differentiable function of (x, t) for $x \in R^n$, $0 \leqslant t \leqslant T$.

Remark. If $0 \leqslant \alpha' < 1$ and $u_0(x)$ satisfies (9.15), then the Taylor series for $u_0(x + iy)$ about $y = 0$ is convergent for all y so that $u_0(z)$ is an entire function. Furthermore,

$$|u_0(x + iy)| \leqslant A_0 \exp [d|x|^{\bar{p}} + c|y|^{1/(1-\alpha')}] \tag{9.24}$$

for some $c > 0$. Conversely, if $u_0(z)$ is an entire function satisfying (9.24) (where $0 \leqslant \alpha' < 1$) then by using Cauchy's inequalities it follows that

$$|D_x^q u_0(x)| \leqslant A_1 A_2^q q^{q\alpha'} \exp [d'|x|^{\bar{p}}] \tag{9.25}$$

for some constants A_1, A_2, d', provided $\bar{p} \leqslant 1/(1 - \alpha')$.

10. An Existence Theorem for Incorrectly Posed Systems

We have

$$\|Q(s, t)\| \leqslant A \exp [cT|s|^{p_0}] \qquad (0 \leqslant t \leqslant T;\ p_0 > 1), \tag{10.1}$$

where c is independent of T, and no other assumptions are made on the growth of $\|Q(\sigma, t)\|$. Set $b^{p_0} = cT$ and let

$$b p_0^{1/p_0} (p_0')^{1/p_0'} < b_1, \qquad (\gamma + 1) b p_0^{1/p_0} (p_0')^{1/p_0'} < 2b_1. \tag{10.2}$$

We shall prove:

Theorem 17. *If $u_0(x)$ can be extended into an entire function of order $\leqslant p_0'$ and type $\leqslant (2b_1)^{-p_0'}$ then there exists a unique classical solution of (4.1), (4.2) for $0 \leqslant t \leqslant T$ which satisfies, for $x \in R^n$, $0 \leqslant t \leqslant T$,*

$$|D^q u(x, t)| \leqslant C' \exp [d|x|^{p_0'}] \qquad \left(d = \frac{1}{b_1^{p_0'}};\ 0 \leqslant t \leqslant T;\ 0 \leqslant |q| \leqslant p\right). \tag{10.3}$$

Proof. By Theorem 9, Sec. 6, Chap. 6, we have

$$G(x, t) = (2\pi)^n \Sigma\, (-1)^m i^m a_m(t)\, D_x^m \delta \qquad \text{over } \Phi \equiv W_{p_0', a_0}^{p_0', b_0}$$

provided $b_0^{p_0} < 1/(p_0 c T)$. Note that this space Φ can be taken to be the same space Φ over which $G * u_0$ is a generalized solution (see Theorem 10, Sec. 4) and u_0 can be considered to be an element of Φ_1' (here we use the second inequality of (10.2)). Introducing

$$G_M(x, t) = (2\pi)^n \sum_{|m| \leqslant M} (-1)^m i^m a_m(t)\, D_x^m \delta,$$

one easily verifies that G_M and G convolve Φ into Φ_1 and that for any $\varphi \in \Phi$

$$G_M * \varphi \to G * \varphi \qquad \text{in the } \Phi_1 \text{ topology.}$$

Hence

$$(G * u_0, \varphi) = (u_0, G * \varphi) = \lim_{M\to\infty} (u_0, G_M * \varphi) = \lim_{M\to\infty} (G_M * u_0, \varphi),$$

i.e.,

$$G * u_0 = \lim_{M\to\infty} G_M * u_0 \qquad \text{in } \Phi'. \tag{10.4}$$

Since the functional $D^m\delta * u_0$ is of function type $D^m u_0(x)$, it follows that

$$\begin{aligned}(G * u_0)(x) &= (2\pi)^n \lim_{M\to\infty} \sum_{|m|\leqslant M} (-1)^m i^m a_m(t)\, D^m u_0(x) \\ &= (2\pi)^n \sum_{|m|<\infty} (-1)^m i^m a_m(t)\, D^m u_0(x).\end{aligned} \tag{10.5}$$

The asserted properties of $u(x, t)$ now follow by looking at the right side of (10.5) and making use of (6.5), Chap. 6.

11. Nonhomogeneous Systems with Time-Dependent Coefficients

In this section we shall show how to extend all the previous results concerning the uniqueness and existence of solutions of the Cauchy problem to the more general systems

$$\frac{\partial u(x, t)}{\partial t} = P\left(t, i\frac{\partial}{\partial x}\right) u(x, t) + f(x, t), \tag{11.1}$$

$$u(x, 0) = u_0(x), \tag{11.2}$$

where $f(x, t)$ is a given function and where the coefficients of the matrix P are continuous functions of t.

To prove uniqueness we proceed as in Sec. 3, except that the matrix $\exp [tP(s)]$ is now replaced by the solution $Q(s, t)$ of the system

$$\frac{\partial Q(s, t)}{\partial t} = P(t, s)Q(s, t), \tag{11.3}$$

$$Q(s, 0) = I \qquad (I = \text{the unit matrix}). \tag{11.4}$$

We shall prove the following:

Theorem 18. *If the coefficients of the polynomials in $P(t, s)$ are continuous functions of t $(0 \leqslant t \leqslant T)$, then there exists a unique solution of* (11.3), (11.4) *and it satisfies the inequality*

$$\|Q(s, t)\| \leqslant A\,(1 + t + t|s|^p) \exp [at|s|^p] \tag{11.5}$$

where A, a are constants independent of t.

Before proving this theorem let us note that the uniqueness theorems of Sec. 3 can be proved for the present systems, by obvious modifications in the corresponding proofs, if we make use of Theorem 18 and if p_0 is replaced by p.

Proof of Theorem 18. The existence and uniqueness of Q follow from the theory of ordinary differential equations. It therefore remains to prove (11.5). Writing (11.3), (11.4) in the integrated form

$$Q(s, t) = I + \int_0^t P(\tau, s)Q(s, \tau)\, d\tau$$

we get

$$\|Q(s, t)\| \leqslant n + \int_0^t \|P(\tau, s)\|\, \|Q(s, \tau)\|\, d\tau. \tag{11.6}$$

We now need the following elementary lemma, the proof of which is left to the reader.

Lemma 7. *If φ, ψ, χ are real-valued continuous functions and $\chi > 0$ and if*

$$\varphi(t) \leqslant \psi(t) + \int_\alpha^t \chi(\tau)\varphi(\tau)\, d\tau \qquad (\alpha < t \leqslant \beta),$$

then

$$\varphi(t) \leqslant \psi(t) + \int_\alpha^t \chi(\tau)\psi(\tau)\, \{\exp\left[\int_\tau^t \chi(\rho)\, d\rho\right]\}\, d\tau.$$

Using the lemma, it follows from (11.6) that

$$\|Q(s, t)\| \leqslant n + n\int_0^t \|P(\tau, s)\| \exp\left[\int_\tau^t \|P(\rho, s)\|\, d\rho\right] d\tau.$$

Since $\|P(\tau, s)\| \leqslant C_1(1 + |s|^p)$ if $0 \leqslant \tau \leqslant T$, we obtain the asserted inequality (11.5).

We next consider the question of existence, taking first the case where $f \equiv 0$. The reader will easily verify that the existence of a generalized solution follows by slightly modifying the proof of Theorem 10, Sec. 4. The existence of smooth solutions also follows as in the case of constant coefficients, provided we assume that $Q(\sigma, t)$ has a prescribed growth as $|\sigma| \to \infty$ and $u_0(x)$ belongs to an appropriate class. For instance, if

$$\|Q(\sigma, t)\| \leqslant A_1(1 + t^{1/p}|\sigma|)^{(N-1)p} \exp\left[-a_1 t|\sigma|^h\right] \qquad (a_1 > 0), \tag{11.7}$$

then the solution $G(x, t) * u_0(x)$ is a smooth solution under the same assumptions on $u_0(x)$ as those made in the parabolic case (provided p_0 is replaced by p). Similarly, if

$$\|Q(\sigma, t)\| \leqslant A_2(1 + |\sigma|)^h, \tag{11.8}$$

then we have a smooth solution under the same assumptions on $u_0(x)$ as those made in the case of correctly posed systems (provided p_0 is replaced by p).

If $f \not\equiv 0$ then the generalized solution of (11.1), (11.2) is given by

$$u(x, t) = G(x, t) * u_0(x) + \int_0^t G(x, t - \tau) * f(x, \tau)\, d\tau \tag{11.9}$$

provided $f(x, t)$ belongs to Φ_1' (for each fixed t) and it varies (in Φ_1') continuously with t.

To prove this statement we first consider the transformed problem and find the solution

$$v(\sigma, t) = Q(\sigma, t)v_0(\sigma) + \int_0^t Q(\sigma, t - \tau)\tilde{f}(\sigma, \tau)\, d\tau. \tag{11.10}$$

Next we take the inverse Fourier transform and notice that in view of the continuity of $\mathfrak{F}^{-1}$ we have $\mathfrak{F}^{-1} \int = \int \mathfrak{F}^{-1}$. Hence, $\mathfrak{F}^{-1}v$ is equal to the right side of (11.9). The proof that $\mathfrak{F}^{-1}v$ is a solution of (11.1), (11.2) is similar to the proof in the case $f \equiv 0$.

The analysis of the second term on the right side of (11.9) is similar to the analysis of the first convolution, as given in Secs. 6–10. Hence, if $f(x, t)$ satisfies the same inequalities as $u_0(x)$, uniformly with respect to t, and if all the concerned x-derivatives of f are continuous functions of (x, t), then all the existence theorems of smooth solutions for the homogeneous system (4.1), (4.2) are valid also for the nonhomogeneous system (11.1), (11.2).

12. Systems of Convolution Equations

Consider the system of convolution equations

$$\frac{\partial u_j(x, t)}{\partial t} = \sum_{k=1}^{N} f_{jk}(x) * u_k(x, t) \qquad (j = 1, \ldots, N) \tag{12.1}$$

for $0 < t \leqslant T$, where the f_{jk} belong to some Φ' and convolve Φ into Φ_1, Φ and Φ_1 being fundamental spaces. Assume further that the $\tilde{f}_{jk}$ are functionals in $\tilde{\Phi}'$ of function type $\tilde{f}_{jk}(\sigma)$ and that the $\tilde{f}_{jk}(s)$ are entire functions. We want to prove the uniqueness of solutions of (12.1) under the initial conditions

$$u_j(x, 0) = u_{0j}(x) \qquad (j = 1, \ldots, N). \tag{12.2}$$

This will be done by using Theorem 6, Sec. 3. We therefore have to consider the adjoint problem

$$\frac{\partial \varphi_j(x, t)}{\partial t} = -\sum_{k=1}^{N} f_{kj}(x) * \varphi_k(x, t) \qquad (0 \leqslant t \leqslant t_0), \tag{12.3}$$

$$\varphi_j(x, t_0) = \varphi_{0j}(x) \qquad (j = 1, \ldots, N). \tag{12.4}$$

Taking the Fourier transforms, we get

$$\frac{\partial \psi_j(\sigma, t)}{\partial t} = -\sum_{k=1}^{N} \tilde{f}_{kj}(\sigma)\psi_k(\sigma, t), \tag{12.5}$$

$$\psi_j(\sigma, t_0) = \psi_{0j}(\sigma). \tag{12.6}$$

Writing this system in a matrix form

$$(12.7)\qquad \frac{\partial \psi(\sigma, t)}{\partial t} = -F(\sigma)\psi(\sigma, t), \qquad \psi(\sigma, t_0) = \psi_0(\sigma),$$

we can now proceed as in Sec. 3, provided $F(s)$ is an entire function of finite order.

Let us consider one example in detail. Assume that

$$(12.8)\qquad |\tilde{f}_{jk}(\sigma + i\tau)| \leqslant C(1 + |\sigma|)^h \exp [b_0|\tau|].$$

Using the inequality $\alpha\beta \leqslant \frac{\alpha^p}{p} + \frac{\beta^r}{r} \left(\frac{1}{p} + \frac{1}{r} = 1,\ \alpha > 0,\ \beta > 0\right)$, we get

$$|\tilde{f}_{jk}(\sigma + i\tau)| \leqslant C_1[(1 + |\sigma|)^{hp} + \exp [b_0 r|\tau|]].$$

It follows that if $0 \leqslant t \leqslant t_0$,

$$(12.9)\qquad \| \exp [(t_0 - t)F(s)]\| \leqslant \exp [(t_0 - t)\|F(s)\|]$$
$$\leqslant C_2 \{\exp [C_3(t_0 - t)|\sigma|^{hp}]\} \{\exp [C_3(t_0 - t) \exp [b_0 r|\tau|]]\}.$$

At this point we introduce the spaces

$$(12.10)\qquad \Phi = W_{M,a}^{\Omega,b}, \qquad \Phi_1 = W_{M,a'}^{\Omega,b},$$

where

$$(12.11)\qquad \frac{1}{a'} = \frac{1}{a} + b_0 r, \quad M_j(x_j) = (x_j + 1) \log (x_j + 1) - x_j, \quad \Omega_j(y_j) = \frac{y_j^{q'}}{q'},$$

if $x_j > 0$, $y_j > 0$, and a, b are any positive vectors. Then

$$(12.12)\qquad \tilde{\Phi} = W_{M^0,1/b}^{\Omega^0,1/a}, \qquad \tilde{\Phi}_1 = W_{M^0,1/b}^{\Omega^0,(1/a)+b_0 r},$$

where M^0, Ω^0 are the conjugate functions of Ω and M respectively, i.e. (see Problem 4, Chap. 5),

$$(12.13)\qquad \Omega_j^0(\tau_j) = \exp [\tau_j] - \tau_j - 1, \qquad M_j^0(\sigma_j) = \frac{\sigma_j^q}{q} \qquad \left(\frac{1}{q} + \frac{1}{q'} = 1\right)$$

if $\tau_j > 0$, $\sigma_j > 0$.

Taking $q > hp$ and using (12.9) one can easily verify that $\exp [(t_0 - t)F(\sigma)]$ is a multiplier from $\tilde{\Phi}$ into $\tilde{\Phi}_1$. Taking now the inverse Fourier transforms we find that $\varphi = \mathfrak{F}^{-1}\psi$ satisfies (12.3), (12.4) in the sense of Φ_1, provided $\varphi_0 \in \Phi$. Furthermore, for any $\delta > 0$,

$$(12.14)\qquad |\varphi(x, t)| \leqslant C_\delta \exp [-(a' - \delta)|x| \log |x|].$$

Introducing the space E of locally integrable functions $f(x)$ having a finite norm

$$(12.15)\qquad \|f\| = \int_{R^n} \{\exp [(1/(b_0 + \delta_0))|x| \log |x|]\}\ |f(x)|\, dx,$$

where δ_0 is a fixed positive number, we conclude that $\Phi \subset \Phi_1 \subset E$ if a is taken to be sufficiently large and r is taken sufficiently close to 1. We can now apply Theorem 6, Sec. 3 and get:

Theorem 19. *If $f_{jk} \in \Phi'$ where Φ is given by* (12.10), (12.11), *and if the $\tilde{f}_{jk}$ are entire functions satisfying* (12.8), *then there exists at most one generalized solution over Φ_1 of* (12.1), (12.2) *in $0 \leqslant t \leqslant T$, which is a measurable function satisfying almost everywhere*

$$|u(x, t)| \leqslant \hat{C} \exp [(1/(b_0 + \delta))|x| \log |x|] \tag{12.16}$$

for any given $\delta > 0$.

As in Sec. 3, the inequality (12.16) can also be replaced by an integral inequality.

The existence of a generalized solution of the system (12.1), (12.2) can be proved by following the method used to prove Theorem 10 in Sec. 4. Details are left to the reader.

An example of f_{jk} to which Theorem 19 applies is given by

$$f_{jk}(x) = \delta(x - x_{jk}).$$

The equations of (12.1) become difference-differential equations. Here $b_0 = \max |x_{jk}|$.

13. Difference-Differential Equations

Consider the differential system

$$\frac{\partial u(x, t)}{\partial t} = P\left(i \frac{\partial}{\partial x}\right) u(x, t), \tag{13.1}$$

$$u(x, 0) = u_0(x). \tag{13.2}$$

Let $e^k = (e_1^k, \ldots, e_n^k)$ where $e_{kj} = \delta_{kj}$. If we replace $\partial u(x, t)/\partial x_j$ by

$$\frac{\Delta_h u(x, t)}{\Delta x_j} = \frac{u(x + he^j, t) - u(x - he^j, t)}{2h} \qquad (h > 0),$$

and similarly replace any derivative $D_x^r u = \partial^{|r|}u/\partial x_1^{r_1} \cdots \partial x_n^{r_n}$ by the corresponding finite difference $\Delta_h^{|r|}u/\Delta x_1^{r_1} \cdots \Delta x_n^{r_n}$, then $P\left(i \dfrac{\partial}{\partial x}\right)$ becomes a finite-difference operator which we shall denote by $P(i\Delta_h/\Delta x)$. The system (13.1), (13.2), with u replaced by u_h, becomes

$$\frac{\partial u_h(x, t)}{\partial t} = P\left(i \frac{\Delta_h}{\Delta x}\right) u_h(x, t), \tag{13.3}$$

$$u_h(x, 0) = u_0(x). \tag{13.4}$$

We define $M(x)$, $\Omega(y)$ by

$$M_j(x_j) = (x_j + 1)\log(x_j + 1) - x_j, \qquad \Omega_j(y_j) = \frac{y_j^{(2p_0-1)/2p_0}}{(2p_0 - 1)/(2p_0)}$$

for $x_j > 0$, $y_j > 0$. We further assume that $2p_0 > 1$.

Given any $a > 0$, $b > 0$ define a', b' by

$$\frac{1}{a'} = \frac{1}{a} + b_0, \qquad \frac{1}{b'} = \frac{1}{b} - a_0$$

where a_0, b_0 are defined in (13.14) below, c_0 being some constant depending only on P, $\epsilon > 0$, and $a_0 < 1/b$ by assumption. We shall prove:

Theorem 20. *There exists at most one generalized solution of* (13.3), (13.4) (*for* $0 \leqslant t \leqslant T$) *over* $\Phi_1 = W_{M,a'}^{\Omega,b'}$ *which is a measurable function satisfying, for each t almost everywhere in x,*

$$(13.5) \qquad |u(x, t)| \leqslant A \exp[(a'' - \delta')|x| \log |x|], \qquad \frac{1}{a''} = \frac{1}{a} + 2p_0h,$$

where δ' *is any positive number.*

The proof is similar to that of Theorem 19, but parts of it will be needed later on and we therefore describe it in detail.

Proof. We wish to apply Theorem 6, Sec. 3. We therefore consider first the adjoint problem

$$(13.6) \qquad \frac{\partial \varphi_h(x, t)}{\partial t} = -P^*\left(-i\frac{\Delta_h}{\Delta x}\right)\varphi_h(x, t),$$

$$(13.7) \qquad \varphi_h(x, t_0) = \varphi(x)$$

for $0 \leqslant t \leqslant t_0$, where P^* is the transpose matrix of P and $0 < t_0 \leqslant T$.

Taking, formally, the Fourier transforms of (13.6), (13.7) and noting that

$$\mathfrak{F}\left[\frac{g(x + he^j) - g(x - he^j)}{2h}\right] = \frac{\exp[-ih\sigma\cdot e_j] - \exp[ih\sigma\cdot e_j]}{2h}(\mathfrak{F}g)(\sigma)$$
$$= -i\frac{\sin \sigma_j h}{h}\tilde{g}(\sigma),$$

we obtain the system

$$(13.8) \qquad \frac{\partial \psi_h(\sigma, t)}{\partial t} = -P^*\left(-\frac{\sin \sigma h}{h}\right)\psi_h(\sigma, t),$$

$$(13.9) \qquad \psi_h(\sigma, t_0) = \psi(\sigma),$$

where $\dfrac{\sin \sigma h}{h} = \left(\dfrac{\sin \sigma_1 h}{h}, \ldots, \dfrac{\sin \sigma_n h}{h}\right)$. The solution of (13.8), (13.9) is given, formally, by

$$\psi_h(\sigma, t) = Q_h^*(\sigma, t - t_0)\psi(\sigma), \tag{13.10}$$

where

$$Q_h^*(s, t) = \exp\,[-tP^*[(-\sin sh)/h]]. \tag{13.11}$$

Using (2.20) and the elementary inequality

$$\left|\frac{\sin s_j h}{h}\right| = |s_j|\left|\frac{\sin s_j h}{s_j h}\right| \leqslant C_1|s_j| \exp\,[h|\tau_j|] \qquad (C_1 \text{ a constant}), \tag{13.12}$$

we get

$$\|Q_h^*(s, t)\| \leqslant C_2(1 + |s| \exp\,[h|\tau|])^{(N-1)p} \exp\,[ct|s|^{p_0}e^{p_0 h|\tau|}].$$

Using the inequality $\alpha\beta \leqslant \frac{1}{2}\alpha^2 + \frac{1}{2}\beta^2$ and taking ϵ to be a fixed positive number we obtain

$$\|Q_h^*(s, t)\| \leqslant C_3\,\{\exp\,[c_0 T|\sigma|^{2p_0}]\}\,\{\exp\,[Te^{(2p_0 h+\epsilon)|\tau|}]\}, \tag{13.13}$$

where c_0 is a constant independent of t, and C_3 depends on T, ϵ and tends to ∞ if $T \to 0$.

We introduce the functions Ω^0, M^0 which are the conjugates of M and Ω, respectively, and are given by

$$\Omega_j^0(y_j) = \exp\,[y_j] - y_j - 1, \qquad M_j^0(x_j) = \frac{x_j^{2p_0}}{2p_0} \qquad \text{if } y_j > 0,\ x_j > 0.$$

From the inequality (13.13) it follows that $Q_h^*(s, t)$ is a multiplier from the space $\tilde{\Phi} = W_{M^0,1/b}^{\Omega^0,1/a}$ into $\tilde{\Phi}_1 = W_{M^0,(1/b)-a_0}^{\Omega^0,(1/a)+b_0}$, where

$$a_0 = 2p_0(c_0T)^{1/(2p_0)}, \qquad b_0 = 2p_0h + \epsilon. \tag{13.14}$$

We can now proceed as in Sec. 3 and show that $\psi_h(\sigma, t)$, as defined in (13.10), is a solution of (13.8), (13.9) in the sense of $\tilde{\Phi}_1$. We prove this statement by showing that

(α) the finite difference $\Delta\psi_h/\Delta t$ and ψ_h tend to the right sides of (13.8) and (13.9), respectively, uniformly with respect to s in bounded sets of C^n, and

(β) $\Delta\psi_h/\Delta t$ and ψ_h are bounded by a constant times $\exp\,[-M^0(\bar{a}\sigma) + \Omega^0(\bar{b}\tau)]$ for any $\bar{b} > (1/a) + b_0$, $\bar{a} < (1/b) - a_0$.

We next conclude, as in Sec. 3, that $\varphi_h = \mathfrak{F}^{-1}\psi_h$ is a solution of (13.3), (13.4) in $\Phi_1 = W_{M,a'}^{\Omega,b'}$. We finally define a space $E \supset \Phi_1$ to be the space of the locally integrable functions $g(x)$ having a finite norm

$$\|g\| = \int_{R^n} \{\exp\,[(a' - \delta)|x| \log |x|]\}\,|g(x)|\,dx.$$

E' consists of all the measurable functions $f(x)$ satisfying almost everywhere

$$|f(x)| \leqslant C \exp [(a' - \delta)|x| \log |x|]. \tag{13.15}$$

We now apply Theorem 6, Sec. 3, and conclude that if $u_0(x) \equiv 0$ and u is a generalized solution over Φ_1 satisfying (13.15), then $u(x, t) = 0$ for $0 \leqslant t \leqslant T$, provided T is sufficiently small so that a_0 in (13.14) is $< 1/b$. If T is not sufficiently small, then we proceed step by step to show that $u(x, t) \equiv 0$, taking in each step a t-interval sufficiently small, so that the corresponding a_0 is $< 1/b$. This completes the proof.

We next prove a uniqueness theorem for generalized functions over a $W_{M,a''}$ space, where $\dfrac{1}{a''} = \dfrac{1}{a} + 2p_0h$.

Theorem 21. *There exists at most one generalized solution of* (13.3) (13.4) *(for* $0 \leqslant t \leqslant T$*) over* $W_{M,a''}$ *which is a measurable function satisfying, for each* t *almost everywhere in* x,

$$|u(x, t)| \leqslant A \exp [(a'' - \delta')|x| \log |x|], \tag{13.16}$$

where δ' *is any positive number.*

Proof. The proof is similar to that of Theorem 20. The main difference is that instead of using the inequality (13.12), we use the inequality

$$\left| \frac{\sin s_j h}{h} \right| \leqslant \frac{\exp [h|\tau_j|]}{h}. \tag{13.17}$$

We then obtain, instead of (13.13),

$$\|Q_h^*(s, t)\| \leqslant C_4 \exp [c_1 T h^{-p_0} e^{(2p_0h+\epsilon)|\tau|}] \qquad (C_4 \text{ depends on } h), \tag{13.18}$$

and we proceed to prove uniqueness. The proof now consists of one step only.

As in Sec. 3 one can show that classical solutions satisfying (13.5) are generalized solutions. Hence Theorems 20, 21 imply the uniqueness of classical solutions satisfying

$$|u(x, t)| \leqslant A \exp [a_1|x| \log |x|] \tag{13.19}$$

for some $a_1 > 0$.

We shall now consider the question of existence of a generalized solution $u_h(x, t)$ of (13.3), (13.4). Taking the Fourier transforms we obtain the system

$$\frac{\partial v_h(\sigma,t)}{\partial t} = P\left(\frac{\sin \sigma h}{h}\right) v_h(\sigma, t), \tag{13.20}$$

$$v_h(\sigma, 0) = v_0(\sigma) \qquad (v_0 = \tilde{u}_0). \tag{13.21}$$

We shall prove that a solution of this system, in $\tilde{\Phi}'$, is given by

$$v_h(\sigma, t) = Q_h(\sigma, t)v_0(\sigma), \tag{13.22}$$

where

$$Q_h(\sigma, t) = \exp [tP((\sin \sigma h)/h)]. \tag{13.23}$$

That means that we have to prove the following two statements:

(γ) for any $\psi \in \tilde{\Phi}$,

$$\left(\frac{1}{\Delta t}[Q_h(\sigma, t + \Delta t) - Q_h(\sigma, t)]v_0(\sigma), \psi(\sigma)\right) \rightarrow \left(P\left(\frac{\sin \sigma h}{h}\right) Q_h(\sigma, t)v_0(\sigma), \psi(\sigma)\right) \tag{13.24}$$

as $\Delta t \rightarrow 0$, and uniformly so when ψ varies in bounded sets of $\tilde{\Phi}$, and

(δ) for any $\psi \in \tilde{\Phi}$,

$$(Q_h(\sigma, t)v_0(\sigma), \psi(\sigma)) \rightarrow (v_0(\sigma), \psi(\sigma)) \tag{13.25}$$

as $t \rightarrow 0$, and uniformly so with respect to ψ in bounded sets of $\tilde{\Phi}$.

The proof of (γ), (δ) follows by transferring the multipliers which multiply v_0 in (13.24), (13.25) into the second term so that they will multiply ψ (recall that $(\lambda v_0, \psi) = (v_0, \lambda\psi)$ if λ is a multiplier) and then using statements similar to (α), (β) of the proof of Theorem 20.

Setting

$$G_h(x, t) = \mathfrak{F}^{-1}[Q_h(\sigma, t)] \tag{13.26}$$

we can now conclude, as in the proof of Theorem 10, Sec. 4, that

$$u_h(x, t) = G_h(x, t) * u_0(x) \tag{13.27}$$

is a solution in Φ' of (13.3), (13.4) provided $u_0(x)$ belongs to Φ_1' and provided a_0 (defined in (13.14)) is $< 1/b$.

We next turn to the question of whether $u_h \rightarrow u$ in Φ', as $h \rightarrow 0$. Before proving it we have to observe that the spaces Φ_1, $\tilde{\Phi}_1$ depend on h and as h decreases they become smaller. Hence, Φ_1', $\tilde{\Phi}_1'$ increase as h decreases, and the initial values $u_0(x)$ will henceforth be taken to belong to some Φ_1' with $h = h^*$ and we shall consider the u_h only for $h < h^*$. We also recall that existence of u (i.e., u_h with $h = 0$) was proved in Theorem 10, Sec. 4, for different spaces Φ, Φ_1. The proof for the present spaces Φ, Φ_1 is quite similar. u has a representation similar to that of u_h, namely,

$$u(x) = G(x, t) * u_0(x), \tag{13.28}$$

where

$$G(x, t) = \mathfrak{F}^{-1}[Q(\sigma, t)], \qquad Q(\sigma, t) = e^{tP(\sigma)}. \tag{13.29}$$

Instead of proving that $u_h - u \to 0$ in Φ', it is sufficient to prove that $v_h - v \to 0$ in $\tilde{\Phi}'$, i.e., for any $\psi(\sigma)$ in $\tilde{\Phi}$,

$$([Q_h(\sigma, t) - Q(\sigma, t)]v_0(\sigma), \psi(\sigma)) \to 0 \qquad \text{as } h \to 0.$$

This is equivalent to proving that

$$(v_0(\sigma), [Q_h(\sigma, t) - Q(\sigma, t)]^*\psi(\sigma)) \to 0 \qquad \text{as } h \to 0,$$

or to proving that

$$\chi_h(\sigma) \equiv \left[\exp\left[tP^*\left(\frac{\sin \sigma h}{h}\right)\right] - \exp\,[tP^*(\sigma)]\right]\psi(\sigma) \to 0 \quad \text{as } h \to 0 \tag{13.30}$$

in the topology of the $\tilde{\Phi}_1$ which corresponds to $h = h^*$. The proof of (13.30) follows by verifying that the family $\{\chi_h(\sigma)\}$ is bounded in $\tilde{\Phi}_1$ and is properly convergent to zero as $h \to 0$. In proving boundedness we make use of the inequality (13.13).

We sum up our results.

Theorem 22. (i) *If u_0 belongs to Φ_1' then* (13.27) *is a generalized solution in Φ' of* (13.3), (13.4) *for* $0 \leqslant t \leqslant T$, *provided T is such that* $a_0 < 1/b$ (*see* (13.14)). (ii) *If u_0 belongs to a Φ_1' corresponding to a fixed h, say $h = h^*$, then the solution u_h* $(0 < h \leqslant h^*)$ *of* (i) *converges in Φ' to the solution u of* (13.1), (13.2) *given by* (13.28).

The restriction $a_0 < 1/b$ is actually superfluous. Indeed, if we use the inequality $\alpha\beta \leqslant \frac{1}{2}\delta\alpha^2 + \frac{1}{2}\beta^2/\delta$ for $\delta > 0$ sufficiently small, then we obtain for Q^* an estimate different from (13.13) (the expression in the first exponent is now $\delta'|\sigma|^{2p_0}$, $\delta' = \delta \cdot \text{const.}$) and all the previous considerations remain true with the same b_0 but with $a_0 = \delta^{1/(2p_0)} \cdot \text{const.}$

Both parts of Theorem 22 can be strengthened if some assumptions are made on $\Lambda(\sigma)$. We shall consider here in detail the case where the system (13.1) is a correctly posed system, that is,

$$\Lambda(\sigma) \leqslant C_1 \qquad (C_1 \text{ a constant}). \tag{13.31}$$

We shall prove that the difference-differential problem (13.3), (13.4) has a classical solution under assumptions on the initial values similar to the assumptions made in the case $h = 0$, and that $u_h \to u$ uniformly in bounded sets (in fact, in a somewhat stronger sense) as $h \to 0$.

Using (2.14), it follows from (13.31) that

$$\|Q(\sigma, t)\| \leqslant C_2(1 + |\sigma|)^k \qquad (k = (N - 1)p). \tag{13.32}$$

Assume that $p_0 > 1$. Applying Theorem 2″, Sec. 1, Chap. 5, it follows that

$$\|Q(\sigma + i\tau, t)\| \leqslant C_3(1 + |\sigma|)^k \qquad \text{if } |\tau_j| \leqslant K(1 + |\sigma_j|)^\mu \qquad (j = 1, \ldots, n) \tag{13.33}$$

for some $\mu \geqslant 1 - p_0$. If (13.33) is satisfied for some μ, then it is clearly satisfied also for any $\mu' < \mu$ provided K is appropriately modified. We may therefore take $\mu \leqslant 0$ in (13.33).

Lemma 8. *If s belongs to a domain $H_\mu(K_1)$ defined by*

$$|\tau_j| \leqslant K_1(1 + |\sigma_j|)^\mu \qquad (j = 1, \ldots, n)$$

and if $\mu < 0$ then $(\sin sh)/h$ $(0 < h \leqslant h^*)$ belongs to some domain $H_\mu(K_2)$, where K_2 is independent of h.

Proof. It is obviously sufficient to consider the case $n = 1$. Write

$$\frac{\sin sh}{h} = \frac{e^{\tau h} + e^{-\tau h}}{2}\,\frac{\sin \sigma h}{h} + i\,\frac{e^{\tau h} - e^{-\tau h}}{2h} \cos \sigma h.$$

Since $|\tau| \to 0$ as $|\sigma| \to \infty$ in $H_\mu(K_1)$,

$$\frac{e^{\tau h} + e^{-\tau h}}{2} \sim 1, \qquad \frac{e^{\tau h} - e^{-\tau h}}{2h} \sim \tau$$

for large $|s|$. Hence we only have to prove that for some K_3 independent of h,

$$|\tau \cos \sigma h| \leqslant K_3\left(1 + |\sigma|\left|\frac{\sin \sigma h}{\sigma h}\right|\right)^\mu. \tag{13.34}$$

Since $\mu < 0$ and $|(\sin \sigma h)/(\sigma h)| \leqslant 1$, (13.34) is a consequence of

$$|\tau| \leqslant K_3(1 + |\sigma|)^\mu,$$

which is true if $s \in H_\mu(K_1)$ and $K_3 = K_1$.

Note that K_2 can be made arbitrarily small if K_1 is taken to be sufficiently small. Hence, combining Lemma 8 and (13.33) we find that, for some K_1 sufficiently small,

$$\|Q_h(s, t)\| \equiv \left\| \exp\left[tP\left(\frac{\sin sh}{h}\right)\right] \right\| \leqslant C_4(1 + |\sigma|)^k \tag{13.35}$$

if $s = \sigma + i\tau$ belongs to $H_\mu(K_1)$. We can now prove the following:

Theorem 23. *Assume that $p_0 > 1$. Let the initial values $u_0(x)$ belong to the differentiability class $C^{Np+\nu}$ and satisfy the inequalities*

$$|D^q u_0(x)| \leqslant C_5(1 + |x|)^\gamma \qquad (0 \leqslant |q| \leqslant Np + \nu), \tag{13.36}$$

where $\nu \geqslant n + 2$, $\nu - n \geqslant |\mu|(\gamma + n + 1)$, $\gamma \geqslant 0$. Then (i) *there exists* *a unique classical solution of the system* (13.3), (13.4) *for* $0 \leqslant t \leqslant T$, *satisfying*

$$|D^q u_h(x, t)| \leqslant C_6(1 + |x|)^\gamma \qquad (0 \leqslant |q| \leqslant p), \tag{13.37}$$

and (ii) *for any* $\gamma' > \gamma$,

$$|D^q[u_h(x, t) - u(x, t)]| \leqslant C_5\epsilon(h)(1 + |x|)^{\gamma'} \qquad (0 \leqslant |q| \leqslant p), \tag{13.38}$$

where $\epsilon(h)$ *is a function of* h *only, and* $\epsilon(h) \to 0$ *if* $h \to 0$.

Proof. The proof of (i) follows by the proof of Theorem 15, Sec. 8, upon using the inequality (13.35). To prove (ii) note that $Q_h(s, t)$ is a continuous function of (s, t, h) provided $Q_0(s, t)$ is defined to be $Q(s, t)$. Since the inequality (13.35) holds, we can apply Theorem 7, Sec. 4, Chap. 6, and conclude that

$$G_h(x, t) = (H - \Delta)^{(k+\nu)/2} f(x, t, h) \qquad (\nu \geqslant n + 2),$$

where $f(x, t, h)$ is a continuous function in (x, t, h) $(x \in R^n,\ 0 \leqslant t \leqslant T,\ 0 \leqslant h \leqslant h^*)$ and

$$|f(x, t, h)| \leqslant C_7 \frac{1}{(1 + |x|)^{(\nu-n)/|\mu|}} \tag{13.39}$$

(compare (8.5′), (8.6′)). We have, if $|q| \leqslant p$,

$$D_x^q u_h(x, t) - D_x^q u(x, t) = \int_{R^n} [f(x - \xi, t, h) - f(x - \xi, t, 0)](H - \Delta)^{(k+\nu)/2} D^q u_0(\xi)\, d\xi, \tag{13.40}$$

and it remains to estimate the right side of (13.40).

Given any $\delta > 0$, we take $R = 1/\delta$. If $|x| > R$ then the integral in (13.40) is bounded by

$$C_8(1 + |x|)^\gamma \leqslant C_8 \delta^{\gamma' - \gamma}(1 + |x|)^{\gamma'}.$$

On the other hand, if $|x| < R$ then we divide the integral into two parts, I_1 with $|\xi| < 2R$ and I_2 with $|\xi| \geqslant 2R$. I_2 is bounded by

$$C_9/R = C_9\delta.$$

As for I_1, if h is sufficiently small then I_1 is smaller than δ. Hence,

$$|D_x^q u_h(x, t) - D_x^q u(x, t)| \leqslant C_{10}\delta^{\gamma'-\gamma}(1 + |x|)^{\gamma'}$$

if h is sufficiently small and $x \in R^n$, $0 \leqslant t \leqslant T$. The assertion (13.38) thereby follows.

Remark 1. If $p_0 \leqslant 1$, then, considering $Q(s, t)$ as being of order $\leqslant 1 + \epsilon$ (for any $\epsilon > 0$) we obtain (13.33) with $\mu \geqslant -\epsilon$. Hence, *if* $p_0 \leqslant 1$ *then Theorem 23 holds for* $\nu = n + 2$ *and for arbitrary positive number* γ.

Remark 2. If instead of assuming (13.31) one assumes that

$$\Lambda(\sigma) \leqslant -C|\sigma|^h + C_1 \qquad (C > 0, h > 0),$$

then one cannot prove an inequality for $Q_h(s, t)$ which is analogous to the inequality for $Q(s, t)$ in the parabolic case. Hence, there is no parabolic theory for the system (13.3), (13.4). One can establish, however, an analogue of Theorem 23 for mildly incorrectly posed systems. Details are omitted.

14. Inverse Theorems

The Cauchy problem

$$\frac{\partial u(x, t)}{\partial t} = P\left(i\frac{\partial}{\partial x}\right)u(x, t) \qquad (0 < t \leqslant T), \tag{14.1}$$

$$u(x, 0) = u_0(x) \tag{14.2}$$

is said to be *correctly posed* with respect to classes M_0, M of functions, if for any $u_0 \in M_0$ there exists one and only one solution $u(x, t)$ which for every fixed t belongs to M, and if, for every fixed t, $u(x, t)$ depends continuously on u_0, where the continuity of u_0, u is taken in the sense of the topology (or the norm) of M_0 and M, respectively. The reader will easily verify that in all the existence theorems of Secs. 6–10 we have implicitly also proved that the Cauchy problem is correctly posed with respect to appropriate classes M_0, M, depending on the type of P.

In this section we shall prove that the conditions imposed on u_0 in the existence theorems cannot be weakened. In fact, we shall prove that if (14.1), (14.2) is a correctly posed problem with respect to some classes M_0, M which appear in the existence theorems of Secs. 6–10, then the $\lambda_j(s)$ must have the asymptotic behavior assumed in the corresponding existence theorems. We shall consider here only two cases: the hyperbolic case and the correctly posed case. For the other cases the reader is referred to the literature; see also Problems 8, 10.

We shall need the following theorem.

Theorem 24 (*Seidenberg-Tarski*). *Let S be a finite system of polynomial equations and inequalities in unknowns $\sigma = (\sigma_1, \ldots, \sigma_n)$ and $a = (a_1, \ldots, a_m)$, and let the coefficients of the polynomials be real. Then there exists a finite set $T_1(a), \ldots, T_N(a)$, where each $T_j(a)$ consists of a finite system of polynomial equations and inequalities in a having real coefficients, such that, for any given real σ, the system S has at least one real solution (σ, a) if and only if, for some h, all the polynomial equations and inequalities in $T_h(a)$ are satisfied.*

The proof is rather lengthy and is given in the next section.

We shall give some applications of Theorem 24. Consider the function

$$\Lambda(r) = \max_{|\sigma|=r} \Lambda(\sigma) \equiv \max_{|\sigma|=r} \max_{1 \leqslant j \leqslant N} \operatorname{Re} \{\lambda_j(\sigma)\}.$$

We shall prove the following:

Lemma 9. *$\Lambda(r)$ is an algebraic function of r for r sufficiently large.*

Proof. For any r, if $\Lambda = \Lambda(r)$ then there exist some $\lambda_j = \mu_1 + i\mu_2$ and $\sigma = (\sigma_1, \ldots, \sigma_n)$ such that

$$\text{(14.3)} \qquad \begin{aligned} &\sum_{m=1}^{n} \sigma_m^2 - r^2 = 0, \\ &P_1(\sigma, \mu_1, \mu_2) = 0, \\ &P_2(\sigma, \mu_1, \mu_2) = 0, \\ &\Lambda = \mu_1, \end{aligned}$$

where P_1, P_2 are the real and imaginary parts of $\det(P(\sigma) - \lambda I)$. Using Theorem 24 with $a = (r, \Lambda)$ it follows that there exists a finite set $T_1(r, \Lambda), \ldots, T_m(r, \Lambda)$ where each $T_j(r, \Lambda)$ consists of a finite number of polynomial equations and inequalities in (r, Λ) having real coefficients, and all the equations and inequalities appearing in some $T_h(r, \Lambda)$ are satisfied for $\Lambda = \Lambda(r)$. At least one equation must occur in T_h. Indeed, if all the relations in T_h are inequalities for $\Lambda = \Lambda(r)$, then these inequalities are satisfied also for (r, Λ') for any $\Lambda' > \Lambda$ and sufficiently close to Λ. But then the system (14.3) has a solution $(\sigma', \mu_1', \mu_2', r, \Lambda')$, which means that $\Lambda(\sigma') = \Lambda' > \Lambda(r)$ and this is impossible since $|\sigma'| = r$.

If we take the product $\Pi(r, \Lambda)$ of all the polynomials appearing in all the $T_j(r, \Lambda)$, then by what we have proved it follows that $\Lambda = \Lambda(r)$ is a solution of $\Pi(r, \Lambda) = 0$. Hence, $\Lambda(r)$ is a piecewise algebraic curve. Since there are only a finite number of algebraic curves which satisfy $\Pi(r, \Lambda) = 0$, and since any two different algebraic curves intersect only in a finite number of points, we conclude that for r sufficiently large $\Lambda(r)$ is an algebraic curve.

Using the Puiseux series for algebraic curves it follows that

$$\text{(14.4)} \qquad \Lambda(r) = Ar^h + \text{lower terms} \qquad (A \neq 0).$$

The following statement is an immediate consequence of (14.4).

Corollary. *If $\Lambda(\sigma) \leqslant A_1 \log(1 + |\sigma|)$ then $\Lambda(\sigma) \leqslant A_2$.*

It is clear that if $A < 0$, $h > 0$, then and only then the system (14.1) is parabolic. If $A > 0$, $h > 0$, then the system (14.1) is incorrectly posed. For hyperbolic systems $h = 0$, $p_0 \leqslant 1$ and for correctly posed systems $h = 0$ or $h > 0$, $A < 0$.

By slightly extending the argument of Lemma 9 one can easily prove:

Lemma 10. *Let* μ *be a rational number and* $\epsilon_j = \pm 1$ $(j = 1, \ldots, n)$. *Then the function*

$$(14.5) \qquad \Lambda_\mu^\epsilon(r) = \max_{|\sigma+i\tau|=r, |\tau_j| \leqslant K_j(1+|\sigma_j|)^\mu, \operatorname{sgn}\sigma_j=\epsilon_j} \Lambda(\sigma + i\tau)$$

is an algebraic function of r, for r sufficiently large.

Corollary.

$$(14.6) \quad \Lambda_\mu(r) \equiv \max_\epsilon \Lambda_\mu^\epsilon(r) = A_\mu r^h + \text{lower terms} \qquad (A_\mu \neq 0).$$

From (6.5), (14.6) we see that in the parabolic case $A_\mu < 0$, $h > 0$ for some μ. If μ is a rational number, then (6.5) holds if and only if $A_\mu < 0$ for that μ (this follows using (2.14)). Let μ_0 be the l.u.b. of all the rational numbers μ for which $A_\mu < 0$. Let $\mu_0(t)$ be the l.u.b. of all the numbers μ for which (6.5) holds for fixed t. Then $\mu_0(t) = \mu_0$ and it is therefore independent of t. In order to obtain the "best" results in the existence theorems for parabolic systems we take $\mu = \mu_0 - \epsilon$ for any small $\epsilon > 0$, or possibly $\mu = \mu_0$. Thus, roughly speaking, the "best" choice of $\mu(t)$, for any fixed t, is independent of t.

These considerations immediately extend also to the other cases.

We shall now prove an inverse of Theorem 14, Sec. 7, concerning hyperbolic systems.

Theorem 25. *Assume that* (a) *if* u_0 *is of differentiability class* C^M, *for some* $M > 0$, $u(x, t) = G(x, t) * u_0(x)$ *(taken in the sense of* (4.10)), *is a continuous function, and* (b) *if a series* $\sum_{m=1}^{\infty} u_0^{(m)}(x)$ *is uniformly convergent in bounded sets of* R^n, *together with its first M derivatives, then the corresponding series of functions* $u^{(m)}(x, t) = G(x, t) * u_0^{(m)}(x)$ *is convergent at* $x = 0$. *Then* (14.1) *is hyperbolic.*

Proof. Let $\varphi(x)$ be a nonzero function of differentiability class C^M which vanishes for $|x| > 1$. Let $\{x^j\}$ be an arbitrary sequence of points such that $|x^j - x^k| > 1$ if $j \neq k$, and let $\{p_j\}$ be an arbitrary sequence of complex numbers. If

$$u_0(x) = \Sigma\, p_j \varphi(x + x_j),$$

then the series

$$\Sigma\, p_j[G(x, t) * \varphi(x + x_j)] = \Sigma\, p_j \tau_{x_j} v(x, t).$$

where $v(x, t) = G(x, t) * \varphi(x)$, is convergent for $x = 0$ and we thus conclude that

$$\Sigma\, p_j v(x_j, t)$$

is convergent. Since the p_j, x_j are arbitrary, it easily follows that the function $v(x, t)$ must have a compact support.

We shall need the following lemma:

Lemma 11. *Let Φ satisfy (Φ_3), (Φ_4) of Chap. 4, Sec. 1, and assume that $\tilde{\Phi}$ is sufficiently rich. If $f \in L^2(R^n)$, $f \in \Phi'$, and if $\mathfrak{F}f = g_1$ in $L^2(R^n)$, $\mathfrak{F}f = g_2$ in $\tilde{\Phi}'$, where g_2 is of function type $g_2(\sigma)$, then $g_1(\sigma) = g_2(\sigma)$ almost everywhere.*

Proof. By the definition of $\mathfrak{F}f$ in $\tilde{\Phi}'$ and by Plancherel's equality we obtain

$$(g_1, \psi) = (g_2, \psi) \qquad \text{for all } \psi \in \tilde{\Phi},$$

i.e.,

$$\int [g_1(\sigma) - g_2(\sigma)]\psi(\sigma)\, d\sigma = 0.$$

Since $\tilde{\Phi}$ is sufficiently rich, $g_1(\sigma) = g_2(\sigma)$ almost everywhere.

Since $v(x, t) = G(x, t) * \varphi(x)$ is a continuous function having a compact support, by Lemma 11 its Fourier transform (over an appropriate W space) $Q(\sigma, t)\psi(\sigma)$ can be extended into an entire function of exponential type which must then coincide with $Q(s, t)\psi(s)$. Using this fact we shall prove:

(A) $Q(s, t)$ is of exponential type $\leqslant B_1$.

To prove it we take all the components of ψ to be zero except an ath component ψ_a which we take to be

$$\psi_a(s) = \prod_{j=1}^{n} \left(\frac{\sin ks_j}{ks_j}\right)^{M+2} \qquad \left(nk(M + 2) < \frac{1}{n}\right). \tag{14.7}$$

Then $\varphi = \mathfrak{F}^{-1}\psi$ is of differentiability class C^M and it vanishes if $|x| > 1$ since $\psi_a(s)$ is of exponential type $< 1/n$ (use is being made of the Paley-Wiener theorem). Hence, $Q(s, t)\psi(s)$ is of exponential type. Since, for $\epsilon > 0$,

$$|\psi_a(s)| \geqslant B_2|s|^{-(M+2)n}$$

if $s_j = r_j \exp[i\theta_j]$, $\epsilon < \theta_j < \pi - \epsilon$ or $\pi + \epsilon < \theta_j < 2\pi - \epsilon$ $(j = 1, \ldots, N)$, we find that $Q(s, t)$ is bounded by $B_3 \exp[B_4|s|]$ if $\epsilon < \theta_j < \pi - \epsilon$ or $\pi + \epsilon < \theta_j < 2\pi - \epsilon$ $(j = 1, \ldots, N)$. Since $Q(s, t)$ is of finite order, by Problem 2, Chap. 5 it follows that $Q(s, t)$ is bounded by $B_3 \exp[B_4|s|]$ for all s. This completes the proof of (A).

We shall next prove:

(B) $Q(\sigma, t)$ is $0(|\sigma|^h)$ as $|\sigma| \to \infty$, for some $h > 0$.

Since $G(x, t) * \varphi(x)$ is a continuous function having a compact support, its Fourier transform $Q(\sigma, t)\psi(\sigma)$ is a bounded function. Taking

$$\psi_a(\sigma) = \sum_{(\sigma_1, \ldots, \sigma_n)} \prod_{j=1}^{n} \left(\frac{\sin k\hat{\sigma}_j}{k\hat{\sigma}_j}\right)^{M+\nu} \qquad \left(nk(M + \nu) < \frac{1}{n}\right), \tag{14.8}$$

where $\nu \geqslant 2$ and is such that $M + \nu$ is an even number and where each $\hat{\sigma}_j$ varies on the set $\{\sigma_j, \sigma_j - \pi/2k\}$, we see, as in the proof of (A), that $\mathfrak{F}\psi_a = \varphi_a$ is of differentiability class C^M and it vanishes for $|x| > 1$. Since

$$(14.9) \qquad |\psi_a(\sigma)| \geqslant \frac{B_5}{(1 + |\sigma|)^{(M+\nu)n}},$$

it follows that

$$\|Q(\sigma, t)\| \leqslant B_6(1 + |\sigma|)^{(M+\nu)n}$$

and the proof of (B) is completed. Using (2.14) we get the inequalities

$$\Lambda(\sigma) \leqslant B_7 \log (1 + |\sigma|), \qquad |\Lambda(s)| \leqslant B_8(1 + |s|).$$

By the corollary to Lemma 9, the first inequality implies $\Lambda(\sigma) \leqslant B_9$. Hence, (14.1) is a hyperbolic system.

We now turn to correctly posed systems.

Theorem 26. *Assume that for any initial values $u_0(x)$ of some differentiability class C^M which satisfy*

$$(14.10) \qquad |D^q u_0(x)| \leqslant A \qquad (0 \leqslant |q| \leqslant M),$$

$u(x, t) = G(x, t) * u_0(x)$ (taken in the sense of (4.10)), is a continuous function satisfying

$$(14.11) \qquad |u(x, t)| \leqslant A_1 A(1 + |x|)^\gamma$$

for some positive constants A_1, γ independent of u_0. Then (14.1) is a correctly posed system.

Proof. Consider first the case $n = 1$ and choose $u_0(x)$ in such a way that all the components of $\psi = \mathfrak{F}u_0$ are zero except for an ath component ψ_a which satisfies

$$|\psi_a(\sigma)| \geqslant A_2(1 + |\sigma|)^{-m} \qquad (A_2 > 0, m > 0).$$

We can define ψ_a by (14.8) (with $n = 1$), noting that $\mathfrak{F}^{-1}\psi_a$ satisfies (14.10), for some A. Let $Q = (Q_{ba})$. Then the bth component of $\psi(\sigma, t) = [\mathfrak{F}u(x, t)](\sigma)$ can be written in the form

$$\psi_b(\sigma, t) = Q_{ba}(\sigma, t)\psi_a(\sigma).$$

If we prove that

$$(14.12) \qquad |\psi_b(\sigma, t) \leqslant A_3(1 + |\sigma|)^{m'} \qquad (m' > 0),$$

then it follows that

$$(14.13) \qquad |Q_{ba}(\sigma, t)| \leqslant A_4(1 + |\sigma|)^{m+m'}.$$

Using the inequality (2.14) we conclude that $\Lambda(\sigma) \leqslant A_5 \log (1 + |\sigma|)$, and the proof of the theorem is completed by the corollary to Lemma 9. It thus remains to prove (14.12).

We shall need the following obvious fact which we state as a lemma:

Lemma 12. *Let $\Phi_2 \subset \Phi_1$ and assume that Φ_1, Φ_2 satisfy (Φ_3), (Φ_4) of Chap. 4, Sec. 1, and that the topology of Φ_2 is stronger than the topology induced by Φ_1. If $f \in \Phi_1'$ (hence also $f \in \Phi_2'$), and if $\mathfrak{F}f = g_1$ in $\tilde{\Phi}_1'$ and $\mathfrak{F}f = g_2$ in $\tilde{\Phi}_2'$, then $g_1 = g_2$ on $\tilde{\Phi}_2$.*

To prove (14.12) we use the assumption (14.11), i.e., for any b,

$$|u_b(x, t)| \leqslant A_1 A(1 + |x|)^\gamma,$$

where $\mathfrak{F}u_b = \psi_b$ over some W space.

The family of functions $\{u_b(x, t)\}$ (t is a parameter) may be considered as a bounded family of tempered distributions. Therefore, the Fourier transforms $\{\tilde{u}_b(\sigma, t)\}$ also form a bounded family of tempered distributions. By Lemma 12 (with $\Phi_1 = S$ and Φ_2 being some $W_{q,c}^{q,d}$ space), $\tilde{u}_b(\sigma, t) = \psi_b(\sigma, t)$ on $\tilde{\Phi}_2$. Using Theorem 13, Sec. 5, Chap. 2, we have

$$\psi_a(\sigma, t) = D^r[(1 + \sigma^2)^h F(\sigma, t)], \tag{14.14}$$

where r, h are independent of t, and $F(\sigma, t)$ is a continuous function of σ, satisfying

$$|F(\sigma, t)| \leqslant A_5. \tag{14.15}$$

Consider the function $g(s, t)$ which is obtained by integrating $\psi_a(s, t)$ r times with respect to s, from 0 to s. $g(s, t)$ is an entire function of order $\leqslant \max(1, p_0)$ and, on $\tilde{\Phi}_2$,

$$D^r[(1 + \sigma^2)^h F(\sigma, t) - g(\sigma, t)] = 0. \tag{14.16}$$

By Problem 8, Chap. 5 it follows that

$$g(\sigma, t) = (1 + \sigma^2)^h F(\sigma, t) + \sum_{m=0}^{r-1} a_m(t)\sigma^m \tag{14.17}$$

for some coefficients $a_m(t)$. The coefficients $a_m(t)$ can be calculated by substituting $\sigma = 1, \ldots, r$ in (14.17). We conclude that the $a_m(t)$ are bounded functions of t. Hence,

$$|g(\sigma, t)| \leqslant A_6(1 + \sigma^2)^{h_0} \qquad \left(h_0 = \max\left(h, \frac{r}{2}\right)\right). \tag{14.18}$$

Using Theorems 2′, 6′, Sec. 1, Chap. 5, we get

$$|\psi_b(\sigma, t)| = |D^r g(\sigma, t)| \leqslant A_7(1 + |\sigma|)^{2h_0 - \mu r} \quad (\mu = \min(0, 1 - p_0)), \tag{14.19}$$

which completes the proof of (14.12).

Note that actually it would have been sufficient to derive (14.12) for just one value of t. The reason for establishing (14.12) uniformly with respect to t is that the method of proof will be used in proving the theorem for $n > 1$.

Let ω vary on the unit sphere of R^n and let R_ω^{n-1} be the hyperplane through the origin of R^n which is orthogonal to ω. Choosing an orthonormal coordinate system in R_ω^{n-1} we have the decomposition

$$x = \alpha\omega + x', \qquad \sigma = \beta\omega + \sigma',$$

where $\alpha = x \cdot \omega$, $\beta = \sigma \cdot \omega$, $x' = (x_1', \ldots, x_{n-1}')$, $\sigma' = (\sigma_1', \ldots, \sigma_{n-1}')$ and x_j', σ_j' are the components of x' and σ', respectively, in R_ω^{n-1}. The transformation

$$x = (x_1, \ldots, x_n) \to (\alpha, x') = (\alpha, x_1', \ldots, x_n')$$

is an orthogonal transformation. Hence $dx = d\alpha\, dx'$.

Consider the function

$$u_0(x) \equiv u_0(x \cdot \omega) = u_0(\alpha)$$

for ω fixed, where $u_0(\alpha)$ is the function used in the proof for $n = 1$. We have

$$\begin{aligned} \psi_0(\sigma) = \tilde{u}_0(\sigma) &= \int e^{ix\cdot\sigma}\, u_0(x \cdot \omega)\, dx \\ &= \int [\exp[i\alpha\beta + ix'\cdot\sigma']]\, u_0(\alpha)\, d\alpha\, dx' \\ &= (2\pi)^{n-1}\, \psi(\beta) \otimes \delta(\sigma'), \end{aligned} \tag{14.20}$$

where $\psi(\beta)$ is the function used in the proof for $n = 1$, and $\otimes$ denotes the tensor product. By Lemma 12 we have

$$[\mathfrak{F}u(x, t)](\sigma) = \psi(\sigma, t) = Q(\sigma, t)\psi_0(\sigma) \qquad \text{on } \tilde{\Phi}_2, \tag{14.21}$$

where Φ_2 is some $W_{q,c}^{q,d}$ space.

Let $v_2(x_2, \ldots, x_n)$ be a fixed function in Ψ_{n-1} satisfying $v_2(0) = (2\pi)^{1-n}$, and let $v_1(x_1)$ be a variable function in Ψ_1, where Ψ_k denotes the space $\tilde{\Phi}_2$ when the variable is k-dimensional. Applying (14.21) to

$$v(\sigma) \equiv v_1(\beta)v_2(\sigma')$$

and using (14.20) we get

$$\begin{aligned} ([\mathfrak{F}u(x, t)](\sigma), v(\sigma)) &= (2\pi)^{n-1}(Q(\beta\omega+\sigma', t)[\psi(\beta) \otimes \delta(\sigma')], v_1(\beta)v_2(\sigma')) \\ &= (2\pi)^{n-1}\, (\psi(\beta), [\delta(\sigma'), Q^*(\beta\omega+\sigma', t)v_1(\beta)v_2(\sigma')]) \\ &= (\psi(\beta),\, Q^*(\beta\omega,\, t)v_1(\beta)) \\ &= (Q(\beta\omega,\, t)\psi(\beta),\, v_1(\beta)). \end{aligned} \tag{14.22}$$

If $v_1(x_1)$ varies in a bounded set of S_1 (S_1 is the space S for $n = 1$) then $v(\sigma)$ varies in a bounded set of S and this set is bounded uniformly with respect to ω since $\sigma \to (\beta, \sigma')$ is an orthogonal transformation. Recalling the assumption (14.11) we conclude that the left side of (14.22) remains bounded independently of ω. Writing this side in the form

$$(\hat{\psi}(\beta; t, \omega), v_1(\beta))$$

we can say that the family $\{\hat{\psi}(\beta; t, \omega)\}$, with (t, ω) as parameters, is a bounded family of tempered distributions in S_1'.

From (14.22) we have

$$\hat{\psi}(\beta; t, \omega) = Q(\beta\omega, t)\psi(\beta) \qquad \text{on } \Psi_1. \tag{14.23}$$

We can now proceed similarly to the case $n = 1$, replacing t by (t, ω) everywhere, since the constants A_j appearing in the analogues of (14.15), (14.18), (14.19) are independent of (t, ω). We thus obtain (compare (14.13))

$$|Q_{ba}(\beta\omega, t)| \leqslant A_4(1 + |\beta|)^{m+m'} = A_4(1 + |\beta\omega|)^{m+m'},$$

where A_4 is independent of (ω, t). Using (2.14) and the corollary to Lemma 9, the proof is completed.

15. Proof of the Seidenberg-Tarski Theorem

The proof is given in several steps.

15.1. Reduction to one equation

A polynomial inequality $R(\sigma, a) > 0$ can be written in the form $\xi^2 R(\sigma, a) - 1 = 0$, where ξ is a new real parameter. An inequality $R(\sigma, a) \geqslant 0$ can be written in the form $R(\sigma, a) - \xi^2 = 0$. A finite number of real equalities $R_j(\sigma, a) = 0$ can be written as one equality

$$\Sigma\, (R_j(\sigma, a))^2 = 0.$$

It thus follows that it suffices to prove Theorem 24 in the case of one equation, i.e.,

$$P(\sigma, a) = 0 \qquad (\sigma = (\sigma_1, \ldots, \sigma_n),\ a = (a_1, \ldots, a_m)). \tag{15.1}$$

15.2. Proof for $n = 1$

Let $\sigma = \sigma_1$, and take a to have any fixed value a^0. Set $P(\sigma) = P(\sigma, a^0)$. If $D(\sigma)$ is the highest common divisor of $P(\sigma)$ and its derivative $P'(\sigma)$, then

$$P_0(\sigma) = \frac{P(\sigma)}{D(\sigma)} \tag{15.2}$$

has the same roots as $P(\sigma)$ but without multiplicities. We construct for $P_0(\sigma)$ a Sturm's series

$$P_0(\sigma),\ P_1(\sigma),\ \ldots,\ P_k(\sigma). \tag{15.3}$$

This can be done in the following way: $P_1(\sigma)$ is taken to be $dP_0(\sigma)/d\sigma$. $-P_j(\sigma)$ (for $j \geqslant 2$) is taken to be the remainder of the division of $P_{j-1}(\sigma)$ by $P_{j-2}(\sigma)$. $P_k(\sigma)$ is the first polynomial which has no real roots. Sturm's theorem asserts that the number $v(\sigma)$ of variations of signs in the sequence (15.3) (zeroes are omitted) is a monotone-decreasing function and for any $a < b$ such that $P_0(a) \neq 0$, $P_0(b) \neq 0$, the number $v(a) - v(b)$ is equal to the number of roots of $P_0(\sigma)$ in the interval $a < \sigma < b$.

Let

$$P_0(\sigma) = c_0\sigma^p + c_1\sigma^{p-1} + \cdots + c_p \qquad (c_0 \neq 0). \tag{15.4}$$

If $P_0(\sigma_0) = 0$ and $|\sigma_0| \geqslant 1$ then

$$|c_0|\,|\sigma_0| \leqslant |c_1| + \cdots + |c_p| < (1 + c_1^2) + \cdots + (1 + c_p^2).$$

Hence the real roots of (15.4) lie in the interval

$$|\sigma| < 1 + \frac{\operatorname{sgn} c_0}{c_0}\left(p + \sum_{j=1}^{p} c_j^2\right) \equiv \gamma. \tag{15.5}$$

We conclude that $P(\sigma) = 0$ has real roots if and only if $v(-\gamma) > v(\gamma)$.

Consider now the case when a varies. Let $D(\sigma, a)$ be the highest common divisor of $P(\sigma, a)$ and $\partial P(\sigma, a)/\partial\sigma$. $D(\sigma, a)$ can be constructed by means of Euclid's algorithm, performing a finite number of divisions, i.e.,

$$\begin{aligned}
P(\sigma, a) &= Q_0(\sigma, a)P'_\sigma(\sigma, a) + R_1(\sigma, a),\\
P'_\sigma(\sigma, a) &= Q_1(\sigma, a)R_1(\sigma, a) + R_2(\sigma, a),\\
&\ \,\vdots\\
R_{\rho-2}(\sigma, a) &= Q_{\rho-1}(\sigma, a)R_{\rho-1}(\sigma, a) + R_\rho(\sigma, a)\\
R_{\rho-1}(\sigma, a) &= D(\sigma, a)R_\rho(\sigma, a).
\end{aligned} \tag{15.6}$$

The coefficients of the polynomial (in σ) $R_1(\sigma, a)$ are rational functions of a. For some values of a some of the denominators may vanish. To see exactly what happens we first take a close look at the leading term in a division of any two polynomials:

$$(A_0(a)\sigma^\alpha + \cdots) = (B_0(a)\sigma^\beta + \cdots)\left(\frac{A_0(a)}{B_0(a)}\sigma^{\alpha-\beta} + \cdots\right) + \text{the remainder.}$$

For all a for which $A_0(a) \neq 0$, $B_0(a) \neq 0$,

$$\frac{A_0(a)}{B_0(a)}\,\sigma^{\alpha-\beta}$$

is the leading term of the quotient. If a is such that $B_0(a) = 0$, then we have

$$(A_0(a)\sigma^\alpha + \cdots) = (B_1(a)\sigma^{\beta-1} + \cdots)\left(\frac{A_0(a)}{B_1(a)}\,\sigma^{\alpha-\beta+1} + \cdots\right) + \text{the remainder},$$

and

$$\frac{A_0(a)}{B_1(a)}\,\sigma^{\alpha-\beta+1}$$

is the leading term, provided $A_0(a) \neq 0$, $B_1(a) \neq 0$. If $B_1(a) = 0$ we proceed to $B_2(a)$, etc. Since at least one of the $B_j(a)$ is not equal to zero, we conclude that there is a smallest index j for which $B_j(a) \neq 0$. The leading coefficient of the quotient is then

$$\frac{A_0(a)}{B_j(a)}\,\sigma^{\alpha-\beta+j}$$

provided $A_0(a) \neq 0$. If $A_0(a) = 0$, the whole argument can be repeated with the first $A_i(a)$ which is $\neq 0$.

Having constructed the leading term (under certain equalities and inequalities on a), one can now proceed to construct each of the terms in both the quotient and the remainder. Similarly we construct all the other quotients and remainders in (15.6). We conclude that there exists a finite number of finite sets of real polynomials $W_j(a)$, $\tilde{W}_j(a)$ such that (i) for every a there exists at least one j such that all the polynomials in $W_j(a)$ are $\neq 0$ and in $\tilde{W}_j(a)$ are $= 0$, and (ii) for all a for which all the polynomials in $W_j(a)$ are $\neq 0$ and in $\tilde{W}_j(a)$ are $= 0$, $D(\sigma, a) \equiv D_j(\sigma, a)$ has the same coefficients, and all the denominators of its coefficients are $\neq 0$.

A similar statement is true also for Sturm's functions

$$P_0^{(j)}(\sigma, a),\ P_1^{(j)}(\sigma, a),\ \ldots,\ P_k^{(j)}(\sigma, a) \tag{15.7}$$

of $P_0(\sigma, a) \equiv P_0^{(j)}(\sigma, a)$. Since the denominators do not vanish if a satisfies all the inequalities in any one of the finite sets of the real polynomials (we shall denote these sets again by $W_j(a)$ (and the sets of equalities by $\tilde{W}_j(a)$), and since $v(\sigma)$ does not change if any polynomial in (15.7) is multiplied by a positive number, we may assume that the coefficients of the polynomials (15.7) are real polynomials in a.

We next observe that an inequality $v(\delta_1) > v(\delta_2)$ can be described in terms of all the possible arrangements of signs of the series (15.7) at $\sigma = \delta_1$ and at $\sigma = \delta_2$. Hence, for any j we can construct a finite number of sets $S_i(a)$, each set consisting of a finite number of equalities and inequalities of real polynomials in a, such that if $P_0^{(j)}(\delta_1, a) \neq 0$, $P_0^{(j)}(\delta_2, a) \neq 0$ then $P(\sigma, a)$ has real roots in $\delta_1 < \sigma < \delta_2$ if and only if at least for one i all the relations in $S_i(a)$ are satisfied. The relations of the $S_i(a)$ depend, of course, on δ_1, δ_2.

Writing

$$P_0^{(j)}(\sigma, a) = c_0^{(j)}\sigma^{p_j} + \cdots + c_{p_j}^{(j)}(a) \qquad (c_0^{(j)}(a) \neq 0)$$

and comparing with (15.4), (15.5) we see that all the real roots of $P_0^{(j)}$ lie in the interval

$$|\sigma| < 1 + \frac{\operatorname{sgn} c_0^{(j)}(a)}{c_0^{(j)}(a)} [p_j + \sum_k (c_k^{(j)}(a))^2] \equiv C_j(a).$$

Taking $\delta_2 = -\delta_1 = C_j(a)$ and observing that the resulting relations in each S_i are equalities and inequalities of real rational functions of a having nonvanishing denominators (since $c_0^{(j)}(a) \neq 0$), we can replace each S_i by an equivalent set of real polynomial equalities and inequalities. Denoting this set again by S_i we conclude that $P_0^{(j)}(\sigma, a)$ has at least one real root if and only if for at least one i all the relations in S_i are satisfied. The proof of the theorem for $n = 1$ is completed by taking the sets T_k to be all the sets $\{S_i \equiv S_{i(j)}, W_j, \tilde{W}_j\}$.

Note, incidentally, that any number of equalities $H_1 = 0, \ldots, H_h = 0$ (the H_i are real) can be written in the equivalent form

$$H_1^2 + \cdots + H_h^2 = 0,$$

and any number of relations $G_1 \neq 0, \ldots, G_\rho \neq 0$ can be written in the equivalent form $G_1G_2 \cdots G_\rho \neq 0$.

15.3. An auxiliary lemma for $n = 2$

The considerations for $n = 2$ are crucial in the proof of the theorem. We set $\sigma_1 = \xi$, $\sigma_2 = \eta$ and first prove the following elementary lemma.

Lemma 13. *Let $P(\xi, \eta, a)$ be a real polynomial in two variables ξ, η and a parameter $a = (a_1, \ldots, a_m)$. Then there exists a finite number of polynomials $Q_1(a), \ldots, Q_N(a)$, $\tilde{Q}_1(a), \ldots, \tilde{Q}_N(a)$ in a and polynomials $P_1(\xi, \eta, a), \ldots, P_N(\xi, \eta, a)$ in (ξ, η, a) such that* (i) *for any a, at least for one i, $Q_i(a) \neq 0$, $\tilde{Q}_i(a) = 0$ and* (ii) *for a satisfying $Q_j(a) \neq 0$, $\tilde{Q}_j(a) = 0$, the polynomial $P_j(\xi, \eta, a)$ has the same irreducible factors (up to multiples) as $P(\xi, \eta, a)$ (over the field of real numbers) but with no repeated factors, and*

$$P_j(\xi, \eta, a), \qquad \frac{\partial}{\partial \eta} P_j(\xi, \eta, a)$$

have no factors involving η.

Proof. Let $a = a^0$ and set $P(\xi, \eta) = P(\xi, \eta, a^0)$. Writing

$$P(\xi, \eta) = \sum_{k=0}^{p} \eta^k c_k(\xi),$$

we observe that the highest common divisor of $P(\xi, \eta)$ which involves only ξ is the highest common divisor $c(\xi)$ of $c_0(\xi), \ldots, c_p(\xi)$. Let

$$P(\xi, \eta) = c(\xi)\bar{P}(\xi, \eta). \tag{15.8}$$

From $\partial P/\partial\eta = c(\partial\bar{P}/\partial\eta)$ we see that P and $\partial P/\partial\eta$ have a common divisor involving η if and only if $\bar{P}$ and $\partial P/\partial\eta$ have a common divisor involving η. Consider $\bar{P}$, $\partial P/\partial\eta$ as polynomials in η with coefficients in the field $R[\xi]$ of the real rational functions of ξ. Euclid's algorithm can be used to find the highest common divisor φ of $\bar{P}$ and $\partial P/\partial\eta$. We set

$$\bar{P} = \varphi\tilde{P}. \tag{15.9}$$

$\tilde{P}$ and φ are polynomials in η having coefficients which are rational functions in ξ. Multiplying both sides of (15.9) by the least common multiples of the denominators of φ, $\tilde{P}$ and then factoring out all the factors which depend only on ξ, we get

$$d_1(\xi)\bar{P} = d_2(\xi)\varphi_1(\xi, \eta)\tilde{P}_1(\xi, \eta), \tag{15.10}$$

where φ_1, $\tilde{P}_1$ are polynomials in ξ, η and they do not have factors which are polynomials of ξ only. The same is true also for $\bar{P}$ and, by a well known theorem, also for $\varphi_1\tilde{P}_1$. It therefore follows from (15.10) that

$$d_1(\xi) = Ad_2(\xi),$$

where A is a constant. For simplicity we may assume that $\varphi_1 = \varphi$, $\tilde{P}_1 = \tilde{P}$, so that φ and $\tilde{P}$ are polynomials in (ξ, η) and have no factors depending only on ξ.

Considering $\tilde{P}$ as a polynomial over $R[\xi]$, it follows by the definition of $\tilde{P}$ that each irreducible factor of P which involves η occurs only once in $\tilde{P}$, and $\tilde{P}$ is a product of all these factors, up to a multiple μ which is a rational function in ξ. We now claim that if a polynomial T in (ξ, η) is an irreducible factor of P such that $\tilde{P} = TW$ for some polynomial W over $R[\xi]$, and if there are no nontrivial divisors of T which depend only on ξ, then W is a polynomial in (ξ, η). The proof is given by the argument given in connection with (15.10). It follows that $\mu = \text{const}$. If we now take $\tilde{c}(\xi)$ to be the polynomial which contains all the irreducible factors of $c(\xi)$, each factor repeated once only, then it follows that all the irreducible factors of P occur once only in $\tilde{c}(\xi)\tilde{P}(\xi, \eta)$ and

$$\tilde{c}\tilde{P}, \qquad \frac{\partial}{\partial\eta}(\tilde{c}\tilde{P})$$

have no common factor involving η.

Consider now the case where a varies. The construction of $\tilde{P}$, $\tilde{c}$ can be accomplished as before, provided a satisfies one set of a finite number of finite sets of polynomial inequalities, say, $W_j(a)$, and equalities, say, $\tilde{W}_j(a)$. These inequalities (equalities) state that certain coefficients of some powers $\xi^\alpha \eta^\beta$ are different from (equal to) zero. For every a, at least one set $W_i(a)$, $\tilde{W}_i(a)$ is satisfied. Furthermore, for all a satisfying $W_j(a)$, $\tilde{W}_j(a)$ the constructed polynomial $\tilde{c}\tilde{P}$, which will be denoted by $P_j(\xi, \eta, a)$, has rational coefficients in a, but the denominators do not vanish. Hence we may multiply P_j by the least common multiple of the denominators and thus conclude that the new polynomial in (ξ, η), which again will be denoted by $P_j(\xi, \eta, a)$, is a polynomial in (ξ, η, a). Noting finally, by the remark made at the end of 15.2, that the set of inequalities (equalities) in each W_j ($\tilde{W}_j$) can be replaced by one inequality $Q_j \neq 0$ (equality $\tilde{Q}_j = 0$), the proof of Lemma 13 is completed.

15.4. The main lemma for $n = 2$

In this subsection we shall prove the following lemma.

Lemma 14. *Let $P(\xi, \eta, a)$ be a real polynominal in ξ, η and $a = (a_1, \cdots, a_m)$. Then there exisits a finite set of real polynomials in a, $Q_1(a), \cdots, Q_N(a)$, $\tilde{Q}_1(a), \ldots, \tilde{Q}_N(a)$ and real polynomials in (ξ, a), $q_1(\xi, a), \ldots, q_N(\xi, a)$ such that the equation $P(\xi, \eta, a) = 0$ has at least one real solution if and only if for at least one j, $Q_j(a) \neq 0$, $\tilde{Q}_j(a) = 0$ and $q_j(\xi, a) = 0$ has at least one real solution.*

Proof. By Lemma 13 it follows that it is enough to prove Lemma 14 in the case where $P(\xi, \eta, a)$ has no multiple factors and P, $\partial P/\partial \eta$ have no common factors involving η. Take first $a = a^0$ and set

$$P(\xi, \eta) = P(\xi, \eta, a^0).$$

Suppose that there exists a real solution of

$$P(\xi, \eta) = 0, \tag{15.11}$$

and let $(\lambda, 0)$ be any real point. Let (α, β) be the closest point on (15.11) to $(\lambda, 0)$. If $(\lambda, 0) \neq (\alpha, \beta)$ and if (α, β) is not a singular point of (15.11), then the curve (15.11) exists in some neighborhood of (α, β) and is orthogonal to the straight line connecting $(\lambda, 0)$ to (α, β), i.e.,

$$(\alpha - \lambda) \frac{\partial P(\alpha, \beta)}{\partial \beta} - \beta \frac{\partial P(\alpha, \beta)}{\partial \alpha} = 0.$$

Hence, (α, β) is a real solution of the system of equations

$$(15.12) \qquad Q_\lambda(\xi) \equiv (\xi - \lambda)\frac{\partial P(\xi, \eta)}{\partial \eta} - \eta \frac{\partial P(\xi, \eta)}{\partial \xi} = 0,$$

$$P(\xi, \eta) = 0.$$

If $(\alpha, \beta) = (\lambda, 0)$ or if (α, β) is a singular point of (15.11) (i.e.,

$$\partial P/\partial \xi = \partial P/\partial \eta = 0$$

for $(\xi, \eta) = (\alpha, \beta)$), then the system (15.12) is also satisfied, We conclude that, given any real λ, there exists a real solution of (15.11) if and only if there exists a real solution of (15.12).

The polynomials P and Q_λ cannot have nontrivial common divisors which are polynomials in ξ only. Indeed, such divisors should divide also $\partial P/\partial \eta$ and, therefore, $\partial P/\partial \xi$, which is against our assumptions on P. Thus, the common divisors of P, Q_λ are polynomials in (ξ, η) of positive degree in η and they are products of some of the irreducible factors $\gamma_1, \ldots, \gamma_q$ of P, each γ_j being of positive degree in η.

We claim that if $\lambda \neq \lambda'$, then Q_λ and $Q_{\lambda'}$ do not have a common factor γ_i. Indeed, such a factor would divide $Q_\lambda - Q_{\lambda'}$, i.e., $\partial P/\partial \eta$, which is impossible since it is also a factor of P. We conclude that there exists a $\lambda = \lambda_0$ for which Q_{λ_0} and P do not have any common divisors. For simplicity we take $\lambda_0 = 0$.

We shall now prove that P and Q_0, as polynomials in η with coefficients in $R[\xi]$, do not have a nontrivial common divisor. Indeed, let D be a common divisor. Then

$$D\delta_1 = P, \qquad D\delta_2 = Q_0.$$

We multiply D by a rational function δ_3 in ξ only such that $D\delta_3$ becomes a polynomial in (ξ, η) and such that it does not have a nonconstant factor depending on ξ only. We then get

$$(D\delta_3)\hat{\delta}_1 = P, \qquad (D\delta_3)\hat{\delta}_2 = Q_0,$$

where $\hat{\delta}_1 = \delta_1/\delta_3$, $\hat{\delta}_2 = \delta_2/\delta_3$, We next use the argument given in connection with (15.10) and conclude that $\hat{\delta}_1$ and $\hat{\delta}_2$ are polynomials in (ξ, η). But then $D\delta_3$ is a constant and D is a trivial common divisor.

From what we just proved it follows that the resultant $R_1(\xi)$ of P, Q_0 as polynomials in η, over $R[\xi]$, is a nonzero element of $R[\xi]$, i.e., it is a polynomial which does not vanish identically, provided the highest coefficient of P with respect to η is assumed to be $\not\equiv 0$; this assumption we can always make (since if $P(\xi, \eta) \equiv P(\xi)$ the lemma is trivial). The same considerations apply to P, Q_0 as polynomials in ξ over $R[\eta]$ and we conclude that the resultant $R_2(\eta)$ of P, Q_0 as polynomials in ξ does not vanish identically.

Since $R_1(\xi) \not\equiv 0$, $R_2(\eta) \not\equiv 0$, there exists only a finite number of complex solutions of $R_1 = 0$ and of $R_2 = 0$. Let $\alpha_1, \ldots, \alpha_{N_1}$ and $\beta_1, \ldots, \beta_{N_2}$ be all the complex roots of $R_1(\xi) = 0$ and $R_2(\eta) = 0$, respectively. Since two polynomials have a common complex solution if and only if their resultant vanishes (the leading term of at least one of them is assumed to be $\neq 0$), it follows that the only possible solutions of

$$P(\xi, \eta) = 0, \qquad Q_0(\xi, \eta) = 0 \tag{15.13}$$

are the points (α_i, β_j) and the problem is to determine whether at least one of these points is a real solution of (15.13).

We shall assume in the sequel that the degree of P is equal to its degree with respect to η (otherwise a suitable linear transformation can be used).

We make a transformation

$$\xi = \xi' + \eta', \qquad \eta = m\eta' \qquad (m \neq 0) \tag{15.14}$$

which takes (15.13) into

$$P'(\xi', \eta') = 0, \qquad Q_0'(\xi', \eta') = 0 \tag{15.15}$$

and a solution (α, β) of (15.13) into a solution (α', β') of (15.15) given by

$$\alpha' = \alpha - \frac{1}{m}\beta, \qquad \beta' = \frac{1}{m}\beta. \tag{15.16}$$

If (α, β) is real then α' is real. Conversely, if α' is real then (α, β) is real provided m is not given by

$$m = \frac{\operatorname{Im}\{\beta\}}{\operatorname{Im}\{\alpha\}}. \tag{15.17}$$

We shall construct a number $m = m_0$ such that

$$m \neq \frac{\operatorname{Im}\beta_i}{\operatorname{Im}\alpha_j} \tag{15.18}$$

for all i, j. It then follows that if α' is real then (α, β) is real; in other words, (15.13) *has a real solution if and only if* (15.15) *has a solution with α' real.*

To construct m_0, we may assume that $R_1(\xi)$, $R_2(\eta)$ have no multiple roots (otherwise we can achieve this situation by Euclid's algorithm), and we introduce the polynomial in m

$$p_0(m) = \prod_{h,k}\prod_{i \neq j}\left(m - \frac{\beta_k - \beta_h}{\alpha_j - \alpha_i}\right) = \frac{\Pi\,\Pi[(\alpha_j - \alpha_i)\,m - (\beta_k - \beta_h)]}{\Pi\,\Pi(\alpha_j - \alpha_i)}.$$

If α_i (or β_j) is a complex root of $R_1(\xi)$ (or $R_2(\eta)$) then its conjugate is also a root and therefore all the numbers $(\operatorname{Im}\beta_i)/(\operatorname{Im}\alpha_j)$ are roots of $p_0(m)$. The coefficients of $p_0(m)$ are fractions. Each numerator is a symmetric polynomial in the α's and, separately, in the β's. From the proof of the fundamental theorem of symmetric polynomials one sees that one can express each numerator as a polynomial in the elementary symmetric polynomials of the α's with coefficients which are symmetric polynomials in the β's. Hence, each numerator can be expressed as a polynomial in the coefficients of $P_1(\xi)$ and $P_2(\eta)$. The denominator can be expressed as a polynomial in the coefficients of $P_1(\xi)$.

We next introduce

$$p(m) = p_0(m)P_h(1, m)$$

where $P_h(\xi, \eta)$ (of degree h) is the principal part of the polynomial $P(\xi, \eta)$. Since we assumed that the degree of $P(\xi, \eta)$ is equal to its degree in η, $P_h(1, m) \not\equiv 0$ and therefore $p(m) \not\equiv 0$. Writing

$$p(m) = c_0 m^k + c_1 m^{k-1} + \cdots + c_{k-1} m + c_k$$

we find that the number

$$m_0 = 1 + \frac{\operatorname{sgn} c_0}{c_0}\left(k + \sum_{j=1}^{k} c_j^2\right)$$

is not a root of $p(m)$ (compare (15.5)). Since $P'_h(0, 1) = P_h(1, m_0) \neq 0$ (as $p(m_0) \neq 0$) the leading part of $P'(\xi', \eta')$ (which is a polynomial of degree h) contains a term $A(\eta')^h$ $(A \neq 0)$. Hence, the resultant $R'(\xi')$ of the polynomials P', Q'_0 considered as polynomials in η' is not identically zero. By the above italicized statement it follows that (15.13) has a real solution if and only if $R'(\xi')$ has a real root.

We now consider the general case where a varies. All the previous arguments can be carried out without change provided we restrict a by a finite set of polynomial inequalities and equalities of the form $W(a) \neq 0$, $\tilde{W}(a) = 0$. There is a finite number of such sets and each a satisfies all the inequalities and equalities of at least one set. Each set of inequalities (equalities) can be replaced by just one inequality (equality). This completes the proof of the lemma.

15.5. Extension of Lemma 14

We shall now extend the previous lemma.

Lemma 15. *Let $P(\xi, \eta, a)$ be a real polynomial in ξ, η and $a = (a_1, \cdots, a_m)$ and let $P_0(\xi, a)$ be a real polynomial in ξ, a. Then there exists a finite set of real polynomials in a, $Q_1(a), \ldots, Q_N(a)$, $\tilde{Q}_1(a), \ldots, \tilde{Q}_N(a)$ and real polynomials in (ξ, a), $q_1(\xi, a), \ldots, q_N(\xi, a)$ such that $P(\xi, \eta, a) = 0$ has a* *real solution satisfying* $P_0(\xi, a) \neq 0$ *if and only if for at least one* j, $Q_j(a) \neq 0$, $\tilde{Q}_j(a) = 0$ *and* $q_j(\xi, a) = 0$ *has at least one real solution.*

Proof. By Euclid's algorithm we find the highest common divisor $D(\xi, a)$ of $P(\xi, \eta, a)$ and $P_0(\xi, a)$ and then proceed with

$$P_1(\xi, \eta, a) \equiv P(\xi, \eta, a)/D(\xi, a)$$

and $P_0(\xi, a)$. If we prove the lemma for P_1 and P_0, then the proof for P and P_0 follows (with possibly different polynomials Q_j, $\tilde{Q}_j$, q_j). Hence without loss of generality we may assume that $P(\xi, \eta, a)$ and $P_0(\xi, a)$ have no common divisors. Their resultant $R(\eta, a)$ is therefore not identically zero. Writing

$$R(\eta, a) = c_0\eta^k + c_1\eta^{k-1} + \cdots + c_k$$

it is clear (compare (15.5)) that (if $c_0 \neq 0$) for

$$\eta_0 = 1 + \frac{\operatorname{sgn} c_0}{c_0}\left(k + \sum_{j=1}^{k} c_j^2\right)$$

$R(\eta_0, a) \neq 0$. Hence $P(\xi, \eta_0, a)$ and $P_0(\xi, a)$ have no common factors. It follows that if the polynomial

$$\hat{P}(\xi, \eta', a) \equiv P(\xi, \eta' P_0(\xi, a) + \eta_0, a)$$

has a real root, then $P_0(\xi, a) \neq 0$. We can now apply Lemma 14 to the polynomial $\hat{P}(\xi, \eta', a)$. If $c_0 = 0, \ldots, c_h = 0$, $c_{h+1} \neq 0$, we proceed similarly.

15.6. Completion of the Proof

We next extend Lemma 14 to any n.

Lemma 16. *Let* $P(\sigma, a)$ *be a real polynomial in* $\sigma = (\sigma_1, \ldots, \sigma_n)$ *and* $a = (a_1, \ldots, a_m)$. *Then there exists a finite set of real polynomials in* a $Q_1(a), \ldots, Q_N(a)$, $\tilde{Q}_1(a), \ldots, \tilde{Q}_N(a)$ *and real polynomials in* (ξ, a), $q_1(\xi, a). \ldots, q_N(\xi, a)$ (ξ *is one variable*) *such that for any real* a, *the equation* $P(\sigma, a) = 0$ *has at least one real solution if and only if, for at least one* j, $Q_j(a) \neq 0$, $\tilde{Q}_j(a) = 0$ *and the equation* $q_j(\xi, a) = 0$ *has at least one real solution.*

Proof. The proof is by induction on n. The cases $n = 0$, $n = 1$ are trivial and the case $n = 2$ coincides with the statement of Lemma 14. Assume the truth of the lemma for $n \leqslant k$ and consider

$$P(\sigma_1, \ldots, \sigma_k, \sigma_{k+1}, a) = 0. \tag{15.19}$$

We take σ_{k+1} as a parameter and, by the inductive assumption, there exist polynomials $Q_1^*(\sigma_{k+1}, a), \ldots, Q_M^*(\sigma_{k+1}, a)$, $\tilde{Q}_1^*(\sigma_{k+1}, a), \ldots, \tilde{Q}_M^*(\sigma_{k+1}, a)$ and

$q_1^*(\xi, \sigma_{k+1}, a), \ldots, q_M^*(\xi, \sigma_{k+1}, a)$ such that (15.19) has a real solution if and only if, for at least one j, $Q_j^*(\sigma_{k+1}, a) \neq 0$, $\tilde{Q}_j^*(\sigma_{k+1}, a) = 0$ and the equation $q_j^*(\xi, \sigma_{k+1}, a) = 0$ has at least one real solution, i.e., if and only if $Q_j^*(\sigma_{k+1}, a) \neq 0$ and

$$\tilde{q}_j^*(\xi, \sigma_{k+1}, a) \equiv [\tilde{Q}_j^*(\sigma_{k+1}, a)]^2 + [q_j(\xi, \sigma_{k+1}, a)]^2 = 0$$

has at least one real solution. Applying Lemma 15, the proof is thereby completed.

We can now immediately complete the proof of Theorem 24. Indeed, for each j we apply the result of Sec. 15.2 (i.e., Theorem 24 for $n = 1$) to $q_j(\xi, a)$ and add to each of the finite sets of equalities and inequalities the inequality $Q_j(a) \neq 0$ and the equality $\tilde{Q}_j(a) = 0$.

PROBLEMS

1. Prove that for systems parabolic in the sense of Petrowski, $\mu = 1$ (and consequently, by (6.11),

$$|D_x^q G(x, t)| \leqslant \frac{A_q'}{t^{(n+q)/p}} \exp\left[-\beta_0 t^{-1/(p-1)}|x|^{p/(p-1)}\right]).$$

[*Hint:* If $|\sigma + i\tau| = 1$ and $|\tau_j| \leqslant \epsilon_0$ $(j = 1, \ldots, n)$ for $\epsilon_0 > 0$ sufficiently small then $\operatorname{Re}\{\lambda_j^0(\sigma + i\tau)\} < -\delta_0$ for some $\delta_0 > 0$. Setting $\tilde{s} = s/|s|$, $\tilde{\lambda}(\tilde{s}) = \lambda(s)/|s|^p$, one can write

$$\det(P(s) - \lambda(s)I) = |s|^{Np} \det(P^0(\tilde{s}) + \epsilon(s) - \tilde{\lambda}(\tilde{s})I),$$

where $\epsilon(s) \to 0$ if $|s| \to \infty$. Hence, $Re\{\tilde{\lambda}_j(\tilde{\sigma} + i\tilde{\tau})\} < -\delta_0/2$ if $|s| \geqslant A$ and $|\tilde{\tau}_j| \leqslant \epsilon_0$ for $j = 1, \ldots, n$.]

2. (a) Derive (6.10) for $n = 1$ in the following way:

   (i) Using Theorem 6, Sec. 1, Chap. 5, prove that

$$|[\sigma^k Q(\sigma, t)]^{(q)}| \leqslant C_1^q q^{q(1-\mu/h)} t^{(q\mu-\gamma)/h} |\sigma|^k \exp\left[-C_2 t\, |\sigma|^h\right].$$

   (ii) Using

$$(ix)^q G^{(k)}(x, t) = \frac{1}{(2\pi)^n} \int_{R^n} [(-i\sigma)^k Q(\sigma, t)]^{(q)} e^{-i\sigma\cdot x}\, d\sigma$$

   and (i), estimate $G^{(k)}$ by making a good choice of q.

   (b) Prove (6.11) for $n = 1$ in a similar way.

3. Find the possible types of the equation $\dfrac{\partial^q u}{\partial t^q} = a\dfrac{\partial^p u}{\partial x^p} + b\dfrac{\partial^r u}{\partial x^r}$ $\quad (p > r)$.

4. Construct explicitly Green's matrix $G(x, t)$ for the heat equation $u_t = a^2 u_{xx}$ and for the wave equation $u_{tt} = a^2 u_{xx}$ (a is real, $a \neq 0$).

5. Prove that the solution $u(x, t)$ of Theorem 17 can be extended into an entire function $u(z, t)$ in z of order $\leqslant p_0'$ and type $\leqslant d$.

6. (a) If $P(x)$ is a polynomial and $P(x) > 0$ in $0 < |x| \leqslant 1$, $P(0) = 0$, then $P(x) > A|x|^\alpha$ for some $A > 0$, $\alpha > 0$.
   (b) If $P(x)$ is a polynomial and $P(x) > 0$ in $R \leqslant |x| < \infty$, for some $R > 0$, then $P(x) > B|x|^\beta$ for some $B > 0$. [*Hint:* Use Theorem 24.]

7. Prove that if (14.1) is a correctly posed system then for any $u_0 \in S$ the solution $u = G * u_0$ also belongs to S, for any fixed t.

8. Prove the following statement (which is stronger than the converse of Problem 7): If for any $u_0(x) \in C_c^\infty$, $u = G * u_0$ is a classical solution of (14.1), (14.2) satisfying

$$|u(x, t)| \leqslant \frac{A}{(1 + |x|)^{n+1}},$$

then (14.1) is a correctly posed system.

9. Inequalities of the form

$$(*) \qquad \left\| A\left(i\frac{\partial}{\partial x}\right) u(x, t)\right\| \leqslant C \left\| B\left(i\frac{\partial}{\partial x}\right) u_0(x)\right\| \qquad (0 \leqslant t \leqslant T)$$

can be established for some differential systems, and converse theorems can also be proved. Taking $\|\cdot\|$ to be the $L^2(R^n)$ norm prove that if (14.1) is a correctly posed system then for any A there exists a B such that $(*)$ holds whenever $u_0 \in S$. [*Hint:* Rewrite $(*)$ in the equivalent form

$$\|A(\sigma)Q(\sigma, t)v_0(\sigma)\| \leqslant C\|B(\sigma)v_0(\sigma)\|.]$$

10. Prove that if $A(\sigma) \neq 0$ for any $|\sigma| > R$ (for some $R > 0$) and if $(*)$ holds for any $u_0 \in S$ (u and all the derivatives of u appearing in $(*)$ are assumed to belong to $L^2(R^n)$), then (14.1) is a correctly posed system. [*Hint:* Use the previous hint and Problem 6(b).]

CHAPTER 8

THE CAUCHY PROBLEM IN SEVERAL TIME VARIABLES

In this chapter we shall solve the problems

$$\begin{aligned} \frac{\partial u_j(x, t)}{\partial t_j} &= \sum_{k=1}^{\nu} P_{jk}\left(t, i\frac{\partial}{\partial x}\right) u_k(x, t), \\ u_j(x, t)\Big|_{t_j=0} &= u_{0j}(x) \qquad (j = 1, \ldots, \nu), \end{aligned} \tag{0.1}$$

and

$$\begin{aligned} \frac{\partial^{\nu} u(x, t)}{\partial t_1 \cdots \partial t_{\nu}} &= P\left(i\frac{\partial}{\partial x}\right) u(x, t), \\ u(x, t)\Big|_{t_j=0} &= u_{0j}(x, t_1, \ldots, t_{j-1}, t_{j+1}, \ldots, t_{\nu}) \qquad (j = 1, \ldots, \nu), \end{aligned} \tag{0.2}$$

where the u_j and u are vectors. We shall refer to (0.1) as a Cauchy problem in several time variables. The problem (0.2) is called the *Goursat problem.* It will be shown that many, but not all, of the results of Chap. 7 can be extended to the present problems.

1. Uniqueness and Existence of Generalized Solutions

We shall prove an analogue of Theorem 6, Sec. 3, Chap. 7. Let Φ_j, Φ_{0j}, E_j $(j = 1, \ldots, \nu)$ be linear topological spaces such that

$$\Phi_j \subset \Phi_{0j} \subset E_j$$

and assume that the topology of Φ_j is stronger than the topology of Φ_{0j} and that the topology of Φ_{0j} is stronger than the topology of E_j. Assume further that Φ_j is dense in E_j. It is clear that

$$E_j' \subset \Phi_{0j}' \subset \Phi_j'.$$

Set $\Phi = \Phi_1 \times \cdots \times \Phi_{\nu}$, $\Phi_0 = \Phi_{01} \times \cdots \times \Phi_{0\nu}$, $E = E_1 \times \cdots \times E_{\nu}$.

Let $A_{ij}(t)$ be continuous linear operators from Φ_{0j} into Φ_{0i}, for any value of $t = (t_1, \ldots, t_{\nu})$ in the interval $0 \leqslant t \leqslant T$ (i.e., $0 \leqslant t_m \leqslant T_m$, $m = 1, \ldots, \nu$) and denote by $A_{jk}^*(t)$ the adjoint of $A_{jk}(t)$. Assume that, if

$t \to \tau$, $A_{jk}(t)\varphi \to A_{jk}(\tau)\varphi$ in Φ_{0j} uniformly with respect to φ in bounded sets of Φ_{0k}.

Theorem 1. *Assume that for any $t^0 = (t_1^0, \ldots, t_\nu^0)$ in the interval $0 < t^0 < T$ and for any $\varphi_0 = (\varphi_{01}, \ldots, \varphi_{0\nu})$ in Φ there exists a solution $\varphi(t) = (\varphi_1(t), \ldots, \varphi_\nu(t))$ for $0 \leqslant t \leqslant t^0$ of the system*

$$\frac{\partial \varphi_j(t)}{\partial t_j} = \sum_{k=1}^{\nu} A_{jk}(t)\varphi_k \qquad (j = 1, \ldots, \nu), \tag{1.1}$$

$$\varphi_j(t)\Big|_{t_j = t_j{}^0} = \varphi_{0j} \qquad (j = 1, \ldots, \nu) \tag{1.2}$$

which belongs to Φ_0. Then there exists at most one solution $u(t)$ in Φ_0', for $0 \leqslant t \leqslant T$, of the system

$$\frac{\partial u_j(t)}{\partial t_j} = -\sum_{k=1}^{\nu} A_{kj}^*(t)u_k(t) \qquad (j = 1, \ldots, \nu), \tag{1.3}$$

$$\lim_{t_j \to 0} u_j(t) = u_{0j} \qquad (j = 1, \ldots, \nu) \tag{1.4}$$

which belongs to E' for each t, where $u_0 = (u_{01}, \ldots, u_{0\nu})$ is any given element of Φ_0'.

The equations (1.1), (1.3) are taken in the sense of the topologies of Φ_0 and Φ_0', respectively, (as in the remark following Theorem 6, Sec. 3, Chap. 7) and, in addition, $\varphi(t)$ $(0 \leqslant t \leqslant t^0)$ and $u(t)$ $(0 < t \leqslant T)$ are assumed to be continuous in t in the topologies of Φ_0 and Φ_0' respectively. (The continuity in Φ_0' is taken in the following sense: if $t \to \tau$ along any sequence then $u(t) \to u(\tau)$ in the strong topology of Φ_0'.)

Proof. For any solutions u, φ,

$$\begin{aligned}\sum_j \frac{\partial}{\partial t_j}(u_j, \varphi_j) &= \sum_j \left[\left(\frac{\partial u_j}{\partial t_j}, \varphi_j\right) + \left(u_j, \frac{\partial \varphi_j}{\partial t_j}\right)\right] \\ &= -\sum_{j,k}(A_{kj}^* u_k, \varphi_j) + \sum_{j,k}(u_j, A_{jk}\varphi_k) = 0.\end{aligned}$$

Hence, integrating with respect to $t_1, \ldots, t_\nu$ and taking $u_0 = 0$, we get

$$\sum_j \int_0^{t_1} \cdots \int_0^{t_{j-1}} \int_0^{t_{j+1}} \cdots \int_0^{t_\nu} (u_j(\tau_1, \ldots, \tau_{j-1}, t_j, \tau_{j+1}, \ldots, \tau_\nu), \varphi_j) \, d\tau_1 \cdots d\tau_{j-1}\, d\tau_{j+1} \cdots d\tau_\nu = 0. \tag{1.5}$$

We want to prove that for any $t^0 = (t_1^0, \ldots, t_\nu^0)$ and for any h, $u_h(t^0) = 0$. Taking a solution of (1.1), (1.2) with $\varphi_{0j} = 0$ if $j \neq h$ and $\varphi_{0h} = \psi \in \Phi_h$, we obtain from (1.5), with $t = t^0$,

$$\int_0^{t_1{}^0} \cdots \int_0^{t_{h-1}{}^0} \int_0^{t_{h+1}{}^0} \cdots \int_0^{t_\nu{}^0} (u_h(\tau_1, \ldots, \tau_{h-1}, t_h^0, \tau_{h+1}, \ldots, \tau_\nu), \psi) \, d\tau_1 \cdots d\tau_{h-1}\, d\tau_{h+1} \cdots d\tau_\nu = 0.$$

Differentiating with respect to $t_1^0, \ldots, t_{h-1}^0, t_{h+1}^0, \ldots, t_\nu^0$ we get

$$(u_h(t^0), \psi) = 0.$$

Since this is true for any $\psi \in \Phi_h$, it follows that $u_h(t^0) = 0$, and the proof is completed.

Let $P_{jk}(t, s)$ be $N_j \times N_k$ matrices whose elements are polynomials in s of degrees $\leqslant p$; at least one of the polynomials is assumed to be of degree p. Consider the differential system

$$\frac{\partial u_j(x, t)}{\partial t_j} = \sum_{k=1}^{\nu} P_{jk}\left(t, i\frac{\partial}{\partial x}\right) u_k(x, t) \qquad (j = 1, \ldots, \nu), \tag{1.6}$$

$$u_j(x, t)\Big|_{t_j=0} = u_{0j}(x) \qquad (j = 1, \ldots, \nu), \tag{1.7}$$

where u_j is a vector of dimension N_j. The concepts of a classical solution and of a generalized solution of (1.6), (1.7) for $0 \leqslant t \leqslant T$ are similar to those introduced in Sec. 1, Chap. 7. The generalized solution is required to be continuous in $t = (t_1, \ldots, t_\nu)$ for $0 < t \leqslant T$, and each component u_j of a classical solution is continuous in (x, t) for $x \in R^n$, $0 \leqslant t_j \leqslant T_j$, $0 < t_k \leqslant T_k$ if $k \neq j$, and all its derivatives which occur in (1.6) are continuous functions in (x, t) for $x \in R^n$, $0 < t \leqslant T$.

In order to prove uniqueness of the solutions of (1.6), (1.7) we consider the adjoint problem

$$\frac{\partial \varphi_j(x, t)}{\partial t_j} = -\sum_{k=1}^{\nu} P_{kj}^*\left(t, -i\frac{\partial}{\partial x}\right) \varphi_k(x, t), \tag{1.8}$$

$$\varphi_j(x, t)\Big|_{t_j=t_j^0} = \varphi_{0j}(x), \tag{1.9}$$

where P_{jk}^* is the transpose of P_{jk}. In order to apply Theorem 1, we have to solve the system (1.8), (1.9) for $0 \leqslant t \leqslant t^0$. Taking, formally, the Fourier transforms of (1.8), (1.9) we obtain

$$\frac{\partial \psi_j(\sigma, t)}{\partial t_j} = -\sum_{k=1}^{\nu} P_{kj}^*(t, -\sigma)\psi_k(\sigma, t), \tag{1.10}$$

$$\psi_j(\sigma, t)\Big|_{t_j=t_j^0} = \psi_{0j}(\sigma). \tag{1.11}$$

To solve (1.10), (1.11) we first consider the system

$$\frac{\partial Q_{jk}(s, t)}{\partial t_j} = \sum_{h=1}^{\nu} P_{jh}(t, s) Q_{hk}(s, t) \qquad (j, k = 1, \ldots, \nu), \tag{1.12}$$

$$Q_{jk}(s, t)\Big|_{t_j=0} = \delta_{jk} I_{jk} \qquad (j, k = 1, \ldots, \nu), \tag{1.13}$$

where I_{jk} is an $N_j \times N_k$ matrix whose elements (m, m) are equal to 1 and whose elements (m, h) are equal to 0 if $m \neq h$. We shall prove the following:

Theorem 2. *If the coefficients of $P(t, s)$ have $[\nu/2] + 1$ continuous derivatives with respect to t, then there exists a unique classical solution $\{Q_{jk}\}$ of* (1.12), (1.13). *The solution is an entire function in s, a continuous function in (s, t), and*

$$\|Q_{jk}(s, t)\| \leqslant B_0 \exp [B|t|\,|s|^p] \qquad (|t| = t_1 + \cdots + t_\nu) \tag{1.14}$$

for some constants B_0, B.

The proof is given in Sec. 3 below.

Now let $\hat{Q}_{jk}$ be the solution of the system

$$\frac{\partial \hat{Q}_{jk}(s, t)}{\partial t_j} = \sum_{h=1}^{\nu} P^*_{hj}(t, -s)\hat{Q}_{hk}(s, t),$$

$$\hat{Q}_{jk}(s, t)\Big|_{t_j=0} = \delta_{jk} I_{jk}.$$

By Theorem 2, $\hat{Q}_{jk}$ is an entire function of s, continuous in (s, t), and

$$\|\hat{Q}_{jk}(s, t)\| \leqslant B_0 \exp [B|t|\,|s|^p], \tag{1.15}$$

where for simplicity we take B_0, B to be the same constants as in (1.14). It is clear that the functions

$$\psi_j(\sigma, t) = \sum_{k=1}^{\nu} \hat{Q}_{jk}(\sigma, t_0 - t)\psi_{0k}(\sigma) \tag{1.16}$$

form, formally, a solution of (1.10), (1.11).

Defining

$$\tilde{\Phi}_j = W^{p,b}_{p,a}, \qquad \tilde{\Phi}_{0j} = W^{p,b+c_0}_{p,a-c_0},$$

where

$$c_0 = (pB|T|)^{1/p}, \tag{1.17}$$

and assuming that $c_0 < a$, one can verify analogously to Sec. 3, Chap. 7, that the $\hat{Q}_{jk}$ are multipliers from $\tilde{\Phi}_k$ into $\tilde{\Phi}_{0j}$, and that the $\psi_j(\sigma, t)$ form a solution of (1.10), (1.11) in the sense of $\tilde{\Phi}_0 = \tilde{\Phi}_{01} \times \cdots \times \tilde{\Phi}_{0\nu}$ provided the ψ_{0j} belong to $\tilde{\Phi}_j$. Details are left to the reader.

Taking the inverse Fourier transforms we obtain a solution in Φ_0 of (1.8), (1.9). Defining E_j to be the space E of Sec. 3, Chap. 7, we can then apply Theorem 1 and thus arrive at the following result:

Theorem 3. *Assume that $p > 1$ and let β be any positive number. Then there exists at most one measurable function $u(x, t) = (u_1(x, t), \ldots, u_\nu(x, t))$ (for each fixed t) which is a generalized solution of* (1.6), (1.7), *for $0 \leqslant t \leqslant T$, over $\Phi_0 = \Phi_{01} \times \cdots \times \Phi_{0\nu}$ ($\Phi_{0j} = W^{p',1/(b+c_0)}_{p',1/(a-c_0)}$) and which satisfies*

$$\int_{R^n} [\exp[-\beta r|x|^{p'}]] \, |u(x, t)|^r \, dx < A \qquad \text{for some } 1 < r < \infty, \tag{1.18}$$

or, for each t,

$$|u(x, t)| \leqslant A \exp[\beta|x|^{p'}] \qquad \text{almost everywhere,} \tag{1.19}$$

where A *is a constant independent of* t.

Note that c_0 can be taken to be any positive number, but that then the proof consists of several steps in each of which the t-interval is sufficiently small.

By an argument of Chap. 7, Sec. 3, one easily shows that classical solutions of (1.6), (1.7) satisfying (1.18) or (1.19) are also generalized solutions. We thus have the following theorem:

Theorem 4. *Let* $p > 1$, $\beta > 0$. *Then there exists at most one classical solution of* (1.6), (1.7) *in* $0 \leqslant t \leqslant T$, *satisfying* (1.18) *or* (1.19) *with a constant* A *independent of* t.

The existence of a generalized solution of (1.6), (1.7) can also be proved by obvious modifications of the proof of Theorem 10, Sec. 4, Chap. 7. We obtain the following result:

Theorem 5. *Let* $p > 1$ *and assume that* $c_0 < a$ (c_0 *is defined in* (1.17)). *For any* $u_0 \in \Phi_0'$ *there exists a generalized solution* $u(x, t)$ *in* Φ' *of* (1.6), (1.7) *for* $0 \leqslant t \leqslant T$. $u(x, t)$ *is given by*

$$u_j(x, t) = \sum_{k=1}^{\nu} G_{jk}(x, t) * u_{0k}(x), \tag{1.20}$$

where $G_{jk} = \mathfrak{F}^{-1}[Q_{jk}]$.

For later reference we write the Fourier transforms of (1.20):

$$v_j(\sigma, t) = \sum_{k=1}^{\nu} Q_{jk}(\sigma, t) v_{0k}(\sigma). \tag{1.21}$$

Summing up, we see that the results of Chap. 7, for the Cauchy problem, concerning the uniqueness of generalized and of classical solutions and the existence theorem of generalized solutions remain true for the case of several time variables, provided we replace p_0 by p and provided we establish Theorem 2. In the next section we prove a useful inequality that will be used in the proof of Theorem 2, in Sec. 3.

2. Sobolev's Lemma

Theorem 6 (*Sobolev's lemma*). *If* $f(x)$ *is a function defined in an open bounded set* D *of* R^n *and if its first* m *derivatives exist and are continuous in*

D, where $m > n/2$, then for any $y \in D$,

$$|f(y)|^2 \leqslant C \sum_{|\alpha| \leqslant m} R^{2|\alpha|-n} \int_D |D^\alpha f(x)|^2 \, dx, \tag{2.1}$$

where R is the distance from y to the boundary of D and C is a constant depending only on m, n.

Proof. Let $g(t)$ be a C^∞ function in one real variable $t \geqslant 0$ such that $g(t) = 1$ if $0 \leqslant t \leqslant \frac{1}{2}$ and $g(t) = 0$ if $t \geqslant 1$. The function

$$\zeta(t) = g\left(\frac{t}{R}\right)$$

satisfies the inequalities

$$\left|\frac{d^k \zeta(t)}{dt^k}\right| \leqslant \frac{B_k}{R^k}, \tag{2.2}$$

where the B_k are constants depending only on $g(t)$ and k. Setting $r = |x - y|$ and noting that $\zeta(0) = 1$, $\zeta(R) = 0$, we have

$$f(y) = -\int_0^R \frac{\partial}{\partial r} [\zeta(r) f(x)] \, dr.$$

Integrating over the $(n - 1)$-dimensional unit sphere Ω with center y, we obtain

$$\Omega_n f(y) = -\int_\Omega \int_0^R \frac{\partial}{\partial r} (\zeta f) \, dr \, d\Omega \qquad \left(\Omega_n = \int_\Omega d\Omega\right).$$

Integrating by parts $(m - 1)$ times with respect to r we get

$$\Omega_n f(y) = \frac{(-1)^m}{(m-1)!} \int_\Omega \int_0^R r^{m-1} \frac{\partial^m}{\partial r^m} (\zeta f) \, dr \, d\Omega.$$

Writing $r^{m-1} = r^{m-n} r^{n-1}$ and using Schwarz' inequality we find that

$$|f(y)|^2 \leqslant C_1 \left\{\int\int \left|\frac{\partial^m}{\partial r^m} (\zeta f)\right|^2 dV\right\} \left\{\int\int r^{2(m-n)} \, dV\right\}, \tag{2.3}$$

where $dV = r^{n-1} \, dr \, d\Omega$ is the volume element and C_1 is a constant depending only on m, n. Since $2m > n$, the second integral on the right side of (2.3) is bounded by $C_2 R^{2m-n}$, where C_2 depends only on m, n. Using (2.2) and Leibniz' rule we can estimate also the first integral on the right side of (2.3) and thus obtain

$$|f(y)|^2 \leqslant C_3 \sum_{k=0}^{m} R^{2k-n} \int_D \left|\frac{\partial^k f}{\partial r^k}\right|^2 dV, \tag{2.4}$$

where C_3 depends only on m, n. Noting that

$$\left|\frac{\partial^k f(x)}{\partial r^k}\right| \leqslant C_4 \sum_{|\alpha| = k} |D^\alpha f(x)|,$$

where C_4 depends only on k and on the direction cosines of the radial direction r and hence is bounded by a constant depending only on m and n, the inequality (2.1) follows from (2.4).

The inequality (2.1) is not suitable in estimating $f(y)$ near the boundary of D since R becomes arbitrarily small. In order to obtain an appropriate extension of (2.1) for y near the boundary, D has to satisfy some additional property.

Definition. We say that D satisfies the *cone condition* if every y in $\bar{D}$ is the vertex of a finite cone $\Gamma_y(\rho)$, i.e., the intersection of a cone with a sphere of radius ρ about the vertex y, such that $\Gamma_y(\rho)$ lies in $\bar{D}$ and its volume $\geqslant \gamma\rho^n$, where ρ, γ are positive numbers independent of y.

Assuming that D satisfies the cone condition, we can modify the proof of Theorem 6, replacing R by ρ and integrating only in the solid angle of the cone $\Gamma_y(\rho)$ (instead of Ω). We obtain the following result:

Theorem 6′. *Assume that D satisfies the cone condition and that D and $f(x)$ satisfy the assumptions of Theorem 6. If $m > n/2$ then, for any $y \in D$,*

$$|f(y)|^2 \leqslant C' \sum_{|\alpha| \leqslant m} \int_D |D^\alpha f(x)|^2 \, dx,$$

where C' is a constant depending only on m, n, ρ, and γ.

Note that the cone condition is satisfied if D is a convex set; also if the boundary of D can be represented in local parameters by continuously differentiable functions.

3. Proof of Theorem 2

Consider first the special case where the P_{jk} do not depend on t, i.e., $P_{jk}(s, t) \equiv P_{jk}(s)$. We try to solve (1.12), (1.13) by setting

$$Q_{jk}(s, t) = \sum Q^{jk}_{h_1 \cdots h_\nu}(s) t_1^{h_1} \cdots t_\nu^{h_\nu}. \tag{3.1}$$

Equation (1.13) is formally equivalent to

$$Q^{jk}_{h_1 \cdots h_{j-1} 0 h_{j+1} \cdots h_\nu} = \begin{cases} 0 & \text{if } j \neq k \text{ of if } j = k \text{ and } h_i \neq 0 \text{ for at least one } i \\ I_{jj} & \text{if } j = k \text{ and } h_i = 0 \text{ for all } i, \end{cases} \tag{3.2}$$

and (1.12) is formally equivalent to

$$(h_j + 1) Q^{jk}_{h_1 \cdots h_{j-1}(h_j+1)h_{j+1} \cdots h_\nu} = \sum_{m=1}^{\nu} P_{jm} Q^{mk}_{h_1 \cdots h_\nu}. \tag{3.3}$$

The formulas (3.2), (3.3) clearly define the $Q^{ij}_{h_1 \cdots h_\nu}$ recursively in a unique way. We proceed to estimate them. Use will be made of the inequalities

$$\|P_{jk}(s)\| \leqslant B_1(|s|^p + 1) \qquad (B_1 \text{ a constant}). \tag{3.4}$$

We shall prove that

$$\|Q^{jk}_{h_1 \cdots h_\nu}(s)\| \leqslant N_0 \frac{B^{|h|}(|s| + 1)^{|h|p}}{h!} \qquad (N_0 = \max_j N_j), \tag{3.5}$$

where $h = (h_1, \ldots, h_\nu)$, $|h| = h_1 + \cdots + h_\nu$, $h! = h_1! \cdots h_\nu!$ and B is a constant. The proof given below uses the classical Cauchy method of majorants.

From (3.3) it follows that

$$(h_j + 1)\|Q^{jk}_{h_1 \cdots h_{j-1}(h_j+1)h_{j+1} \cdots h_\nu}\| \leqslant \sum_{m=1}^{\nu} \|P_{jm}\| \, \|Q^{mk}_{h_1 \cdots h_\nu}\|.$$

Hence, if there exist numbers $g^{jk}_{h_1 \cdots h_\nu}$, π_{jm} such that

$$\|Q^{jk}_{h_1 \cdots h_{j-1} 0 h_{j+1} \cdots h_\nu}\| \leqslant g^{jk}_{h_1 \cdots h_{j-1} 0 h_{j+1} \cdots h_\nu}, \qquad \|P_{jm}\| \leqslant \pi_{jm},$$

and such that (3.3) holds with the Q's and the P's replaced by the g's and the π's respectively, then

$$\|Q^{jk}_{h_1 \cdots h_\nu}\| \leqslant g^{jk}_{h_1 \cdots h_\nu}.$$

Denoting the right side of (3.4) by α, we take $\pi_{jm} = \alpha$ and try to solve the system

$$\frac{\partial g_{mk}}{\partial t_m} = \alpha \sum_{q=1}^{\nu} g_{qk}, \tag{3.6}$$

$g_{mk}\Big|_{t_m=0}$ = a power series with non-negative terms, which reduces to N_0 when $t = 0$.

A solution is given by

$$g_{mk} = N_0 \exp[\nu\alpha(t_1 + \cdots + t_\nu)].$$

Writing $g_{mk} = \Sigma\, g^{mk}_{h_1 \cdots h_\nu} t_1^{h_1} \cdots t_\nu^{h_\nu}$, we have

$$g^{mk}_{h_1 \cdots h_\nu} = N_0 \frac{\nu^{|h|}\alpha^{|h|}}{h!},$$

which completes the proof of (3.5). From (3.5) it also follows that

$$\|Q_{jk}(s, t)\| \leqslant B_2 \exp[B|t|\,|s|^p] \qquad (|t| = t_1 + \cdots + t_\nu), \tag{3.7}$$

where B_2 is a constant depending on T, and $0 \leqslant t \leqslant T$. One can also easily verify, using (3.5), that the Q_{jk} form a solution, in the classical sense, of (1.12), (1.13).

The previous proof can be extended to the case where the P_{jk} are analytic functions of t. Let each element in $P_{jk}(s, t)$ be majorized by a series in t of the form $B_3\alpha[1 - B_4(t_1 + \ldots + t_\nu)]^{-1}$. We take B_3, B_4

to be the same constant for all the elements of all the matrices P_{jk} and to be independent of s. Instead of (3.6) we now have

$$\frac{\partial g_{mk}}{\partial |t|} = \frac{B_5\alpha}{1 - B_4|t|} \sum_{q=1}^{\nu} g_{qk},$$

provided $g_{mk}(t_1, \ldots, t_\nu) = g_{mk}(t_1 + \cdots + t_\nu) \equiv g_{mk}(|t|)$. The coefficients $g^{jk}_{h_1 \cdots h_\nu}$ majorize $\|Q^{jk}_{h_1 \cdots h_\nu}\|$.

A solution is given by

$$g_{mk}(t) = B_6(1 - B_4|t|)^{-\nu B_5\alpha/B_4},$$

where B_6 is independent of s and is the same for all the m, k. We conclude that the series in (3.1) is convergent and the Q_{jk} form a solution of (1.11), (1.12) provided $|t| < 1/B_4$. Note that B_4 can be taken arbitrarily small if the radii of convergence of the series expansions of the P_{jk} are arbitrarily large. This is the case, for instance, if the $P_{jk}(s, t)$ are polynomials in t. An inequality similar to (3.7) follows easily by using the estimates on the $Q^{jk}_{h_1 \cdots h_\nu}$, but we shall not need to use this inequality in the future.

We now turn to the general case. We make the transformation

$$\hat{Q}_{jk} = Q_{jk} - \delta_{jk} I_{jk}$$

which transfers the system (1.12), (1.13) into the system

$$\frac{\partial \hat{Q}_{jk}(s, t)}{\partial t_j} = \sum_{m=1}^{\nu} P_{jm}(t, s)\hat{Q}_{mk}(s, t) + P_{jk}(s, t), \tag{3.8}$$

$$\hat{Q}_{jk}(s, t)\Big|_{t_j=0} = 0.$$

In order to prove the theorem we first observe that it suffices to consider the system (3.8) for each k fixed. Next it suffices to consider it for each fixed column-vector in $\hat{Q}_{jk}$. We shall consider only the case where each column vector is replaced by a scalar function, since the proof for vectors is then obtained by minor modifications. All we have to prove then is the following lemma.

Lemma 1. *Let $f_i(t)$, $p_{ij}(t)$ have $[\nu/2] + 1$ continuous derivatives in the interval $0 \leqslant t \leqslant T$, and let*

$$|D^q f_i(t)| \leqslant \beta, \qquad |D^q p_{ij}(t)| \leqslant \alpha \qquad (i, j = 1, \ldots, \nu; |q| \leqslant [\nu/2] + 1), \tag{3.9}$$

where α, β are constants and $\alpha > 1$. Then there exists one and only one classical solution of the system

$$\frac{\partial w_i}{\partial t_i} = \sum_{j=1}^{\nu} p_{ij}(t) w_j + f_i(t), \tag{3.10}$$

$$w_i\Big|_{t_i=0} = 0 \qquad (i = 1, \ldots, \nu) \tag{3.11}$$

in $0 \leqslant t \leqslant T$, *and it satisfies the inequality*

$$|w_i(t)| \leqslant C\beta \exp [A|t|\alpha] \qquad (i = 1, \ldots, \nu), \tag{3.12}$$

where A depends only on ν *and C depends only on* ν *and* $|T|$.

Proof. Suppose that $\{w_i\}$ is a solution of (3.10), (3.11), and that the first $[\nu/2] + 1$ derivatives of the w_i are continuous functions. We shall derive estimates on the w_i and on their derivatives. Substituting

$$w_i = v_i \exp [\lambda(t_1 + \cdots + t_\nu)]$$

into (3.10), (3.11) we obtain

$$\frac{\partial v_i}{\partial t_i} = \sum_{j=1}^{\nu} (p_{ij} - \lambda\delta_{ij})v_j + f_i \exp [-\lambda|t|], \tag{3.13}$$

$$v_i \Big|_{t_i=0} = 0. \tag{3.14}$$

We multiply (3.13) by $\bar{v}_i$ and add to it the product of v_i by the complex conjugate of (3.13). Summing over i and integrating the result with respect to $t_1, \ldots, t_\nu$ we find, using (3.14),

$$2 \int_0^T \sum_j |f_j|\,|v_j|\,dt \geqslant \frac{\lambda}{2} \int_0^T \sum_j |v_j|^2\,dt \qquad (dt = dt_1 \cdots dt_\nu),$$

provided $\lambda = A_1\alpha$ where A_1 is a sufficiently large constant depending only on ν. Take $A_1 > 4$ so that $\lambda > 4$.

Using Schwarz' inequality and then replacing the v_j in terms of the w_j, we obtain the inequality

$$\int_0^T \sum_j |w_j|^2\,dt \leqslant \{\exp [2\lambda|T|]\} \int_0^T \sum_j |f_j|^2\,dt. \tag{3.15}$$

We next differentiate (3.10) with respect to any fixed t_k, and consider the system obtained for $\partial w_i/\partial t_k \equiv \hat{w}_i$. The initial conditions for the $\hat{w}_i$ are

$$\hat{w}_i \Big|_{t_i=0} = 0 \quad \text{if } i \neq k, \qquad \hat{w}_k \Big|_{t_k=0} = [\sum_j p_{kj}w_j + f_k]_{t_k=0} = \hat{f}_k.$$

Setting $\tilde{w}_i = \hat{w}_i$ for $i \neq k$, $\tilde{w}_k = \hat{w}_k - \hat{f}_k$, we obtain a differential system for the $\tilde{w}_j$, similar to (3.10), (3.11). Before applying (3.15), for this system, we have to estimate $\hat{f}_k$. Writing (3.15) with T replaced by $(T_1, \ldots, T_{k-1}, t_k, T_{k+1}, \ldots, T_\nu)$, dividing by t_k and taking $t_k \to 0$ we obtain

$$\int_0^{T_k} \sum_j |w_j(\bar{t}_k)|^2\,d\bar{t}_k \leqslant \{\exp [2\lambda|T|]\} \int_0^{T_k} \sum_j |f_j(\bar{t}_k)|^2\,d\bar{t}_k, \tag{3.16}$$

where $\bar{T}_k = (T_1, \ldots, T_{k-1}, T_{k+1}, \ldots, T_\nu)$, $d\bar{t}_k = dt_1 \cdots dt_{k-1}\, dt_{k+1} \cdots dt_\nu$, and $\bar{t}_k = t \Big|_{t_k=0}$.

We can now apply (3.15) to the $\tilde{w}_j$ and thus obtain an L^2 estimate also on $\hat{w}_j$:

$$(3.17) \qquad \int_0^T \sum_j \left| \frac{\partial}{\partial t_k} w_j \right|^2 dt \leqslant C_1 \{\exp [A_2 \alpha |T|]\} \left\{ \int_0^T \sum_j |f_j(\bar{t}_k)|^2\, dt + \int_0^T \sum_j |f_j|^2\, dt + \int_0^T \sum_{i,j} \left| \frac{\partial}{\partial t_i} f_j \right|^2 dt \right\},$$

where the A_m are used to denote constants depending only on ν and the C_m are used to denote constants depending only ν, T, which remain bounded if some components of $T \to 0$. Using the proof of Sobolev's lemma for $n = 1$ one sees that the first integral on the right side of (3.17) is bounded by a constant (which remains bounded if $T_k \to 0$) times the sum of the other two integrals. Hence, we have

$$(3.18) \qquad \int_0^T \sum_j \left| \frac{\partial}{\partial t_k} w_j \right|^2 dt \leqslant C_2 \{\exp [A_2 \alpha |T|]\} \left\{ \int_0^T \sum_j |f_j|^2\, dt + \int_0^T \sum_{i,j} \left| \frac{\partial}{\partial t_i} f_j \right|^2 dt \right\}.$$

We can now proceed as before: differentiate (3.10) twice with respect to two components of t. Derive, for the nonzero boundary conditions, bounds analogous to (3.16) (by using (3.15) and (3.18)), and then apply (3.15). Proceeding step by step we obtain the inequalities

$$(3.19) \qquad \int_0^T \sum_j |D_t^q w_j|^2\, dt \leqslant C_3 \{\exp [A_3 \alpha |T|]\} \int_0^T \sum_j \sum_{|r| \leqslant |q|} |D_t^r f_j|^2\, dt$$

for all q, $|q| \leqslant [\nu/2] + 1$.

Using Theorem 6′, (3.12) follows for $t = T$. The proof for any t follows by noting that in the above proof T is an arbitrary positive vector.

From the inequality (3.15) follows the uniqueness of the solution of (3.10), (3.11). To prove existence, we approximate f_i, p_{ij} and their first $[\nu/2] + 1$ derivatives by sequences of analytic functions $f_i^{(m)}$, $p_{ij}^{(m)}$ which can be taken to be polynomials in t. Denote by $w_j^{(m)}$ the corresponding solutions. Then, $w_j^{(m)} - w_j^{(k)} \equiv W_j^{mk}$ satisfies

$$\frac{\partial W_i^{mk}}{\partial t_i} = \sum_j p_{ij}^{(m)}(t) W_j^{mk} + \{\sum_j [p_{ij}^{(m)}(t) - p_{ij}^{(k)}(t)] w_j^{(k)} + [f_i^{(m)}(t) - f_i^{(k)}(t)]\},$$

$$W_i^{mk} \Big|_{t_i=0} = 0.$$

To this system we apply the a priori inequalities (3.19) for $0 \leqslant |q| \leqslant [\nu/2] + 1$.

We find that the $w_j^{(m)}$ together with their first $[\nu/2] + 1$ derivatives form a Cauchy sequence in the L^2 sense. Let $\lim w_j^{(m)} = w_j$. Writing (3.10) in an integrated form and noting (by Theorem 6′) that $w_j^{(m)} \to w_j$ uniformly, we conclude that the w_j form a solution of (3.10), (3.11).

Finally, since (3.12) holds for the $w_j^{(m)}$, it also holds for the w_j. This completes the proof.

Remark. If the $P_{jk}(t, s)$ are square symmetric matrices, then

$$\|P_{jk}(t, s)\| = \max_k |\lambda_{jk,m}(t, s)|,$$

where the $\lambda_{jk,m}(t, s)$ are the eigenvalues of $P_{jk}(t, s)$. By Theorem 4, Sec. 2, Chap. 7,

$$|\lambda_{jk,m}(t, s)| \leqslant C'(|s|^{p_{0,jk}} + 1),$$

where $p_{0,jk}$ is the reduced order of $P_{jk}(t, s)$ $(p_{0,jk} \leqslant p)$. It follows that

$$\|P_{jk}(t, s)\| \leqslant C''(|s|^{p_0} + 1) \qquad (p_0 = \max_{j,k} p_{0,jk}). \tag{3.20}$$

Theorem 2 is therefore valid with p replaced by p_0. Hence also (1.15) and Theorems 3–5 are valid with p replaced by p_0.

4. Existence of Classical Solutions

In this section we shall prove that under some conditions on the $Q_{jk}(\sigma, \mathrm{t})$, the generalized solution of Theorem 5, Sec. 1, is a classical solution provided the boundary values u_{0j} belong to an appropriate class of functions. The results are similar to those derived in the case of one time variable, and we shall therefore state results only for systems analogous to correctly posed systems.

Condition (A). The system (1.6) is said to satisfy the condition (A) if the solution $\{Q_{jk}\}$ of (1.12), (1.13) satisfies the inequalities

$$\|Q_{jk}(\sigma, \mathrm{t})\| \leqslant A(1 + |\sigma|)^h, \tag{4.1}$$

where A, h are positive constants and h is an integer.

By Theorem 2″, Sec. 1, Chap. 5 it follows that, if $p > 1$,

$$\|Q_{jk}(\sigma + i\tau, t)\| \leqslant A_1(1 + |\sigma|)^h \qquad \text{in some region} \tag{4.2}$$

$$|\tau_m| \leqslant K(1 + |\sigma_m|)^\mu \qquad (m = 1, \ldots, n)$$

for some $\mu \geqslant 1 - p$. μ depends on j, k, t. Observing, however, that if (4.2) is valid for a given μ, then it is also valid for any smaller value of μ (with a different K), we can take μ to be independent of j, k, t. We can now proceed as in Sec. 8, Chap. 7, and obtain the following result:

Theorem 7. *Assume that* $p > 1$, $\mu < 0$. *If the* $u_{0j}(x)$ *are continuous functions together with their first* $h + p + \kappa$ *derivatives, for any* $\kappa \geqslant n + 2$, *and if*

$$|D^q u_{0j}(x)| \leqslant B_1(1 + |x|)^\gamma \quad \text{for } 0 \leqslant |q| \leqslant h + p + \kappa,\ 1 \leqslant j \leqslant \nu, \tag{4.3}$$

where $\gamma + n + 1 \leqslant (\kappa - n)/|\mu|$, $\gamma \geqslant 0$, *then there exists a unique classical solution* $u(x, t) = \{u_j(x, t)\}$ *of* (1.6), (1.7) *in* $0 \leqslant t \leqslant T$ *satisfying*

$$|D_x^q u_j(x, t)| \leqslant B_1 B^*(1 + |x|)^\gamma \qquad \text{for } 0 \leqslant |q| \leqslant p, 0 \leqslant t \leqslant T, \tag{4.4}$$

where B^* *is a constant independent of the* u_{0m}.

In the case that $\mu \geqslant 0$, we can establish an analogue of Theorem 7′, Sec. 8, Chap. 7.

As for inverse theorems, we have the following analogue of Theorem 26, Sec. 14, Chap. 7:

Theorem 8. *Assume that for any initial values* $u_{0j}(x)$ *satisfying*

$$|D^q u_{0j}(x)| \leqslant B_2 \qquad \text{for } 0 \leqslant |q| \leqslant M;\ 1 \leqslant j \leqslant \nu, \tag{4.5}$$

where all the derivatives in (4.5) *are continuous, the generalized solution*

$$u_j(x, t) = \sum_{k=1}^{\nu} G_{jk}(x, t) * u_{0k}(x)$$

is a continuous function satisfying

$$|u_j(x, t)| \leqslant B_2 \bar{B}(1 + |x|)^\delta \qquad (1 \leqslant j \leqslant \nu), \tag{4.6}$$

where $\bar{B}$, δ *are positive constants independent of* u_0, *then the* Q_{jk} *satisfy the condition* (*A*).

Proof. Taking $u_{0j} = 0$ if $j \neq m$ and all the components of u_{0m} to be 0 except one component, say an ath component, and comparing the bth components of

$$v_k(\sigma, t) = Q_{km}(\sigma, t) v_{0m}(\sigma) \qquad (v_k = \mathfrak{F} u_k)$$

(see (1.21)), we get

$$v_k^b(\sigma, t) = Q_{km}^{ba}(\sigma, t) v_{0m}^a(\sigma).$$

We can now proceed as in the proof of Theorem 26.

In order to get an existence theorem analogous to that for hyperbolic systems we have to assume that the condition (A) holds and, in addition,

$$\|Q_{jk}(s, t)\| \leqslant A' \exp [a'|s|].$$

Existence theorems analogous to those for parabolic systems and for incorrectly posed problems can also be obtained if appropriate assumptions are made on the growth of the $Q_{jk}(\sigma, t)$.

5. The Goursat Problem

In this section we consider the special system

$$\frac{\partial u_j(x, t)}{\partial t_j} = P_j\left(i\frac{\partial}{\partial x}\right) u_{j+1}(x, t) \qquad (j = 1, \ldots, \nu; u_{\nu+1} = u_1), \tag{5.1}$$

$$u_j(x, t)\Big|_{t_j=0} = u_{0j}(x, \hat{t}_j), \tag{5.2}$$

where the $P_j(s)$ are $N_j \times N_{j+1}$ matrices, $N_{\nu+1} = N_1$, and

$$\hat{t}_j = (t_1, \ldots, t_{j-1}, t_{j+1}, \ldots, t_\nu).$$

We first prove that *for sufficiently smooth solutions, the system* (5.1), (5.2) *can be reduced to the system*

$$\frac{\partial^\nu w(x, t)}{\partial t_1\, \partial t_2 \cdots \partial t_\nu} = P\left(i\frac{\partial}{\partial x}\right) w(x, t), \tag{5.3}$$

$$w(x, t)\Big|_{t_j=0} = w_{0j}(x, \hat{t}_j) \qquad (j = 1, \ldots, \nu), \tag{5.4}$$

where

$$P = P_1P_2 \cdots P_\nu. \tag{5.5}$$

More precisely: Given u_{0j} we can find w_{0j} such that upon solving (5.3), (5.4) we can deduce from w the solution of (5.1), (5.2). The problem of solving (5.3), (5.4) is called the *Goursat problem.*

We first observe that if we apply $\partial/\partial t_2$ to the first equation in (5.1) and use the second equation in (5.1), then apply $\partial/\partial t_3$ to the resulting equation and use the third equation in (5.1), and so on, we arrive at (5.3) with $w = u_1$. This leads to the idea that the w_{0j} should be chosen as the values of u_1 on $t_j = 0$. Thus we take $w_{01} = u_{01}$. To define w_{02}, we substitute in $\partial u_1/\partial t_1 = P_1u_2$, $t_2 = 0$ and thus find $\partial u_1/\partial t_1$. Since u_1 is given on $t_1 = 0$, we obtain u_1 on $t_2 = 0$ by integration. We then take $w_{02} = u_1$ on $t_2 = 0$. To find w_{03} (that is, u_1 on $t_3 = 0$) we first use $\partial u_2/\partial t_2 = P_2u_3$ on $t_3 = 0$. Since u_2 is given on $t_2 = 0$, u_2 on $t_3 = 0$ is obtained by integrating $\partial u_2/\partial t_2$. We then use $\partial u_1/\partial t_1 = P_1u_2$ as before. Proceeding in the same manner, we construct, in a unique way, the w_{0j}.

Suppose now that we have solved (5.3) with the w_{0j} just constructed. We then can construct a solution of (5.1), (5.2) as follows. We define u_ν by $\partial u_\nu/\partial t_\nu = P_\nu w$, $u_\nu = u_{0\nu}$ on $t_\nu = 0$. Next we define $u_{\nu-1}$ by

$$\partial u_{\nu-1}/\partial t_{\nu-1} = P_{\nu-1}u_\nu,$$

$u_{\nu-1} = u_{0,\nu-1}$ on $t_{\nu-1} = 0$. Proceeding in the same way, we define $u_{\nu-2}, \ldots, u_2$. Since $w = u_{01}$ on $t_1 = 0$, all we need to show is that $\partial w/\partial t_1 = P_1u_2$ and then $(w, u_2, \ldots, u_\nu)$ is a solution of (5.1), (5.2).

Defining $H = \partial w/\partial t_1 - P_1 u_2$ and using (5.3), we get

$$\frac{\partial}{\partial t_2}\frac{\partial}{\partial t_3}\cdots\frac{\partial}{\partial t_\nu} H = 0.$$

On $t_2 = 0$, $H = 0$ by the definition of w on $t_2 = 0$. Hence,

$$\frac{\partial}{\partial t_3}\cdots\frac{\partial}{\partial t_\nu} H = 0.$$

Again, on $t_3 = 0$, $H = 0$ by the definition of w on $t_3 = 0$. Proceeding step by step we thus arrive at $H = 0$ and the proof is completed.

By a similar procedure we can also prove that the system (5.3), (5.4) (for solutions for which all the derivatives occurring in (5.3) are continuous for $0 \leqslant t \leqslant T$, $x \in R^n$) can be reduced to a system (5.1), (5.2) in the sense discussed above. Note that the P_j in (5.1) can be chosen in different ways.

It is somewhat simpler to study the system (5.3), (5.4) than to study the system (5.1), (5.2). Note that boundary conditions in (5.2) and (5.4) depend on $(x, \hat{t}_j)$ (and not only on x) so that the results of Secs. 1, 4 cannot be applied to (5.1), (5.2) without some additional analysis. Throughout this section we shall study the system (5.3), (5.4) and we shall not rely on the results of previous sections concerning (5.1), (5.2).

We first introduce the matrix

$$Q(s, t) = \sum_{m=0}^{\infty} \frac{(t_1 t_2 \cdots t_\nu)^m}{(m!)^\nu} (P(s))^m \tag{5.6}$$

which satisfies (compare (1.12), (1.13))

$$\frac{\partial^\nu Q(s, t)}{\partial t_1 \cdots \partial t_\nu} = P(s) Q(s, t), \tag{5.7}$$

$$Q(s, t)\Big|_{t_j=0} = I_{NN} \qquad (1 \leqslant j \leqslant \nu), \tag{5.8}$$

$$\frac{\partial^k Q(s, t)}{\partial t_1 \cdots \partial t_k}\Big|_{t_{k+1}=0} = 0 \qquad (1 \leqslant k \leqslant \nu - 1),$$

where $N = N_1$. Taking the Fourier transforms of (5.3), (5.4) we get

$$\frac{\partial^\nu \chi(\sigma, t)}{\partial t_1 \cdots \partial t_\nu} = P(\sigma)\chi(\sigma, t), \tag{5.9}$$

$$\chi(\sigma, t)\Big|_{t_j=0} = \chi_{0j}(\sigma, \hat{t}_j). \tag{5.10}$$

Let $\chi^0(\sigma, t)$ be a function satisfying (5.10). Thus, if $\nu = 2$ we take

$$\chi^0(\sigma, t) = \chi_{01}(\sigma, t_2) + \chi_{02}(\sigma, t_1) - \chi_{01}(\sigma, 0). \tag{5.11}$$

(We assume that the w_{0j}, and hence their transforms χ_{0j}, satisfy the necessary consistency conditions which ensure that they are boundary values of a smooth function.) For $\nu > 2$ we can take $\chi^0(\sigma, t)$ to be also a linear combination of the χ_{0j} with some of the t_j's replaced by 0. Consequently, $\partial^\nu \chi^0/\partial t_1 \cdots \partial t_\nu = 0$.

The function

$$\hat{\chi}(\sigma, t) = \chi(\sigma, t) - \chi^0(\sigma, t)$$

satisfies the system

$$\frac{\partial^\nu \hat{\chi}(\sigma, t)}{\partial t_1 \cdots \partial t_\nu} = P(\sigma)\hat{\chi}(\sigma, t) + f(\sigma, t), \tag{5.12}$$

$$\hat{\chi}(\sigma, t)\Big|_{t_j=0} = 0, \tag{5.13}$$

where

$$f(\sigma, t) = P(\sigma)\chi^0(\sigma, t). \tag{5.14}$$

The formal solution $\hat{\chi}$ of (5.12), (5.13), or rather the formal solution χ of (5.9), (5.10), is then given by

$$\chi(\sigma, t) = \chi^0(\sigma, t) + \int_0^{t_1} \cdots \int_0^{t_\nu} Q(\sigma, t - \tau) f(\sigma, \tau)\, d\tau_1 \cdots d\tau_\nu. \tag{5.15}$$

The formal solution of (5.3), (5.4) is then

$$w(x, t) = w^0(x, t) + \int_0^{t_1} \cdots \int_0^{t_\nu} [G(x, t - \tau) * g(x, \tau)]\, d\tau_1 \cdots d\tau_\nu, \tag{5.16}$$

where

$$G(x, t) = \mathfrak{F}^{-1}[Q(\sigma, t)], \qquad g(x, t) = \mathfrak{F}^{-1}[f(\sigma, t)]. \tag{5.17}$$

Note, by (5.11), that for $\nu = 2$

$$w^0(x, t) = w_{01}(x, t_2) + w_{02}(x, t_1) - w_{01}(x, 0). \tag{5.18}$$

Formulas (5.15), (5.16) are analogous to formulas (1.21), (1.20).

We now observe that the following analogue of Theorem 1 is valid: *If for any $\varphi_0 \in \Phi$ there exists a solution in Φ_0 of*

$$\frac{\partial^\nu \varphi(t)}{\partial t_1 \cdots \partial t_\nu} = (-1)^\nu A^*(t)\varphi(t) + \varphi_0 \qquad (0 \leqslant t \leqslant t^0), \tag{5.19}$$

$$\varphi(t)\Big|_{t_j = t_j^0} = 0 \qquad (1 \leqslant j \leqslant \nu), \tag{5.20}$$

then $w \equiv 0$ *is the only solution in the sense of* Φ_0', *of*

$$\frac{\partial^\nu w(t)}{\partial t_1 \cdots \partial t_\nu} = A(t)w(t) \qquad (0 < t < T), \tag{5.21}$$

$$w(t)\Big|_{t_j=0} = 0 \qquad (1 \leqslant j \leqslant \nu), \tag{5.22}$$

which belongs to E'.

The proof is obtained by evaluating the integral of

$$\left(\frac{\partial^\nu w}{\partial t_1 \cdots \partial t_\nu}, \varphi\right) + (-1)^\nu \left(w, \frac{\partial^\nu \varphi}{\partial t_1 \cdots \partial t_\nu}\right),$$

thereby deducing the relation

$$\int_0^{t^0} (w(t), \varphi_0)\, dt = 0$$

which yields, upon differentiation, $(w(t^0), \varphi_0) = 0$, i.e., $w(t^0) = 0$.

With the aid of the matrix $Q(-s, t)$ we can solve the system (5.19), (5.20) where $A^*(t) = P^*(-i\,(\partial/\partial x))$ in some W space, and thus derive an analogue of Theorem 3. The existence of a generalized solution can also be established analogously to Theorem 5. Moreover, from the inequality (5.30) proved below we see that the analogues of Theorems 3–5 are valid with p replaced by p_0/ν, where p_0 is the reduced order of $P(s)$.

The analogue of Theorem 7 can be proved using (5.16) and assuming that

$$\|Q(\sigma, t)\| \leqslant A(1 + |\sigma|)^h, \tag{5.23}$$

where A, h are positive constants and h is an integer.

Theorem 9. *Assume that* $p_0 > \nu$ *and that* $\mu < 0$. *If*

$$|D_x^q w_{0j}(x, \hat{t}_j)| \leqslant B_1(1 + |x|)^\gamma \qquad (0 \leqslant |q| \leqslant h + 2p + \kappa; \quad \gamma \geqslant 0; \quad 1 \leqslant j \leqslant \nu) \tag{5.24}$$

where $\kappa \geqslant n + 2$ *and all the derivatives are continuous functions for* $x \in R^n$, $0 \leqslant t \leqslant T$, *and if* $\gamma + n + 1 \leqslant (\kappa - n)/|\mu|$, *then there exists a unique classical solution* $w(x, t)$ *of* (5.3), (5.4) *for* $0 \leqslant t \leqslant T$ *satisfying*

$$\sum_{k_1 < \cdots < k_m} \left|\frac{\partial^m w(x, t)}{\partial t_{k_1} \cdots \partial t_{k_m}}\right| + |D_x^q w(x, t)| \leqslant B_1 B'(1 + |x|)^\gamma \qquad (0 \leqslant |q| \leqslant p), \tag{5.25}$$

where B' *is a constant independent of the* w_{0j}.

The proof is left to the reader.

The converse of Theorem 9, i.e., the analogue of Theorem 8 for (5.3), (5.4), can also be proved directly. We leave the details to the reader.

In order to study the growth behavior of $Q(s, t)$ and, in particular, express the condition (5.23) in terms of the eigenvalues of $P(\sigma)$, we write Q in the form

$$Q(s, t) = f_\nu(t_1 t_2 \cdots t_\nu P(s)), \tag{5.26}$$

where

$$f_\nu(z) = \sum_{m=0}^{\infty} \frac{z^m}{(m!)^\nu}. \tag{5.27}$$

$f_\nu(z)$ is an entire function of order $1/\nu$ and type ν. Hence, for any $\epsilon > 0$,

$$|f_\nu(z)| \leqslant C_\epsilon \exp\,[(\nu + \epsilon)|z|^{1/\nu}]. \tag{5.28}$$

Similar inequalities hold for the derivatives of $f_\nu(z)$.

Let p_0 be the reduced order of P. By Theorem 4, Sec. 2, Chap. 7,

$$M(s) \leqslant C(|s|^{p_0} + 1). \tag{5.29}$$

Applying Theorem 2, Sec. 2, Chap. 7, and making use of (5.28) and of the similar inequalities which hold for $d^k f_\nu(z)/dz^k$ for $1 \leqslant k \leqslant N$, we get

$$\|Q(s, t)\| \leqslant H_1(1 + |s|^{p(N-1)}) \exp\,[H(t_1 \cdots t_\nu)^{p_0/\nu}|s|^{p_0/\nu}], \tag{5.30}$$

where H_1, H are constants independent of (s, t) and H is also independent of T.

We shall next show that (5.23) is satisfied under some rather general conditions on the eigenvalues of $P(\sigma)$, provided $\nu = 2$. If, however, $\nu > 2$, then $f_\nu(z)$ is of order $< \frac{1}{2}$ and therefore it cannot be bounded by $0(|z|^h)$ (for any $h > 0$) along any ray issuing from the origin (as follows by Theorem 1, Sec. 1, Chap. 5). Hence, the condition (5.23) is likely not to be satisfied, if $\nu > 2$, even in the most simple cases, for instance, if $P(\sigma)$ is a real nonconstant diagonal matrix.

In considering the case $\nu = 2$ we use the identity

$$f_2(z) = J_0(2i\sqrt{z}), \tag{5.31}$$

where $J_\lambda(z)$ is Bessel's function of order λ.

We shall need the asymptotic formula (see, for instance Reference 48, p. 22)

$$\begin{aligned} J_\lambda(z) = {} & \sqrt{\frac{2}{\pi z}}\left[\cos\left(z - \frac{\lambda\pi}{2} - \frac{\pi}{4}\right)\right]\left[1 + 0\left(\frac{1}{|z|^2}\right)\right] \\ & - \sqrt{\frac{2}{\pi z}}\left[\sin\left(z - \frac{\lambda\pi}{2} - \frac{\pi}{4}\right)\right]\left[\frac{4\lambda^2 - 1}{8z} + 0\left(\frac{1}{|z|^3}\right)\right] \end{aligned} \tag{5.32}$$

for $\lambda \geqslant 0$, $-\delta < \arg z < \delta$ (for any $0 < \delta < \pi$), and $|z|$ sufficiently large. We shall also need the well-known identity

$$J'_m(z) = \frac{m}{z} J_m(z) - J_{m+1}(z) \qquad (m = 0, 1, 2, \ldots). \tag{5.33}$$

From (5.32) we conclude that, for some $\epsilon > 0$,

$$|J_\lambda(z)| \leqslant \frac{1-\epsilon}{\sqrt{|z|}} e^{|y|} \tag{5.34}$$

if $|z|$ is sufficiently large.

Substituting (5.34) with $\lambda = m$ and $\lambda = m + 1$ in (5.33) we get

$$|J'_m(z)| \leqslant \frac{1-(\epsilon/2)}{\sqrt{|z|}} e^{|y|} \tag{5.35}$$

if $|z|$ is sufficiently large.

We next differentiate (5.33) and use (5.34), (5.35). We obtain, for sufficiently large $|z|$,

$$|J''_m(z)| \leqslant \frac{1-(\epsilon/3)}{\sqrt{|z|}} e^{|y|}.$$

Proceeding step by step we conclude that for any given k there exists an $\epsilon_0 > 0$ such that for all $|z|$ sufficiently large (depending only on k, ϵ_0, m, δ), $-\delta < \arg z < \delta$,

$$|J_m^{(q)}(z)| \leqslant \frac{1-\epsilon_0}{\sqrt{|z|}} e^{|y|} \qquad \text{for } q = 0, 1, \ldots, k. \tag{5.36}$$

Since $J_m(-z) = (-1)^m J_m(z)$, these inequalities hold without any restriction on $\arg z$.

We proceed to estimate $f_2(z)$ and its derivatives.

Setting $z = x + iy$, $\sqrt{z} = x_1 + iy_1$, we get

$$x_1^2 - y_1^2 = x, \; 2x_1y_1 = y. \tag{5.37}$$

We shall take (x_1, y_1) such that

$$y_1 \geqslant C_1 > 0, \qquad \exp\,[2|x_1|] \leqslant \sqrt{2y_1} \tag{5.38}$$

for C_1 sufficiently large. Hence, by (5.31), (5.36) (with $q = 0$),

$$|f_2(z)| \leqslant 1 - \epsilon_0. \tag{5.39}$$

Next, using

$$f'_2(z) = \frac{d}{dz} J_0(2i\sqrt{z}) = J'_0(2i\sqrt{z}) \frac{i}{\sqrt{z}} \tag{5.40}$$

and (5.36), we obtain

$$|f'_2(z)| \leqslant \frac{1-\epsilon_0}{\sqrt{|z|}}. \tag{5.41}$$

Differentiating (5.40), we obtain in a similar way

$$|f_2''(z)| \leqslant \frac{1 - (\epsilon_0/2)}{|z|}.$$

Proceeding step by step we arrive at

$$(5.42) \qquad |f_2^{(q)}(z)| \leqslant \frac{1}{|z|^{q/2}} \qquad \text{for } q = 0, 1, \ldots, k,$$

provided C_1 is sufficiently large, depending only on k. We take $k = N - 1$.

(5.42) was proved for points z for which (5.38) holds with C_1 sufficiently large. Hence (5.42) holds for the subset of points z satisfying

$$(5.43) \qquad |y| \leqslant \frac{1}{4}\sqrt{|x|}\log|x| \qquad (x < -C < 0)$$

for some C sufficiently large.

Suppose now that the eigenvalues $\lambda_j(\sigma)$ of $P(\sigma)$ lie for all σ sufficiently large in the domain (5.43). Since this domain is convex, the convex hull of the $\lambda_j(\sigma)$ also lies in the domain. Using Theorem 2, Sec. 2, Chap. 7, we get

$$(5.44) \qquad \|f_2(P(\sigma))\| \leqslant C'(1 + |\sigma|)^{(N-1)p}.$$

Recalling (5.26) we get

$$(5.45) \qquad \|Q(\sigma, t)\| \leqslant C''(1 + |\sigma|)^{(N-1)p},$$

where C'' is a constant depending on $|T|$.

It is easily seen that the inequality (5.45) remains true if the $\lambda_j(\sigma)$ vary in a region

$$(5.46) \qquad |y| \leqslant \frac{1}{4}\sqrt{|x|}\log(|x| + E_1) \qquad (x \leqslant E_2)$$

for any positive constants E_1, E_2.

We have proved the following theorem:

Theorem 10. *Assume that* $p_0 > \nu$. *Then there exists at most one generalized solution of* (5.3), (5.4) *which is a measurable function satisfying*

$$(5.47) \qquad \int_{R^n} \{\exp[-\beta r|x|^{\tilde{p}}]\}\, |u(x, t)|^r\, dx < A_0 \qquad (1 < r < \infty),$$

or satisfying almost everywhere (for each t*)*

$$(5.48) \qquad |u(x, t)| \leqslant A_0 \exp[\beta|x|^{\tilde{p}}] \qquad \left(\frac{1}{\tilde{p}} + \frac{\nu}{p_0} = 1\right),$$

where A_0 *is a constant independent of* t. *If* $\nu = 2$ *and if for any* $\sigma \in R^n$ *the eigenvalues of* $P(\sigma)$ *lie in a region of the form* (5.46), *then there exists a unique classical solution of* (5.3), (5.4) *as asserted by Theorem 9, and* $h = (N - 1)p$.

Remark 1. If $p_0 \leqslant \nu$ the assertion of uniqueness remains true provided $\tilde{p}$ is replaced by any positive number. If, in addition, the eigenvalues of $P(\sigma)$ lie in (5.46), then the existence and uniqueness of solutions can be established as for hyperbolic systems.

Remark 2. If we rewrite the equations of (5.1) in a different order, taking the mth equation to be the first one, the $(m + 1)$st equation to be the second one, and so on, then we obtain instead of P the matrix $P_m P_{m+1} \cdots P_{m-1}$ whose eigenvalues may have a different asymptotic behavior than the eigenvalues of P.

Remark 3. If (5.46) is replaced by

$$|y| \leqslant E_3 \sqrt{|x|} \log (|x| + E_1) \qquad (x \leqslant E_2), \tag{5.49}$$

then we can still derive the condition (5.23) but h will depend on E_3.

Remark 4. One can also find conditions on the eigenvalues of $P(\sigma)$ which will ensure that

$$\|Q(\sigma, t)\| \leqslant A' \exp [A''|\sigma|^h] \qquad \left(0 < h < \frac{p_0}{\nu}\right).$$

PROBLEMS

1. Let P be an elliptic operator (for definition see p. 288) of order $2m$ and set $P = P_1 + P_2$ where P_1 consists of all the terms in P of orders $> m$. Prove that if the coefficients of P_1 are real then (5.46) is satisfied for either P or $-P$.

CHAPTER 9

S SPACES

In Chaps. 7 and 8 we made an extensive use of generalized functions over W spaces. For one of the applications given in Chap. 10 we shall need to use generalized functions over other fundamental spaces, namely, S spaces. The basic properties of these spaces are given in the present chapter.

1. Definition of S Spaces

We shall introduce three types of S spaces: $S_{\alpha,A}$ (and their union spaces S_α), $S^{\beta,B}$ (and their union spaces S^β), and $S_{\alpha,A}^{\beta,B}$ (and their union spaces S_α^β). We shall refer to all these spaces as S spaces.

The space $S_{\alpha,A}$ where $\alpha = (\alpha_1, \ldots, \alpha_n) \geqslant 0$, $A = (A_1, \ldots, A_n) > 0$, consists of all the C^∞ functions $\varphi(x)$ satisfying for any $0 \leqslant |k| < \infty$, $0 \leqslant |q| < \infty$,

$$|x^k D^q \varphi(x)| \leqslant C_{q\bar{A}} \bar{A}^k k^{k\alpha} \tag{1.1}$$

for any $\bar{A} > A$, where $C_{q\bar{A}}$ are constants depending on φ. If $x^k \neq 0$, then, dividing both sides of (1.1) by x^k, we obtain

$$|D^q \varphi(x)| \leqslant C_{q\bar{A}} \inf_k \frac{\bar{A}^k k^{k\alpha}}{|x^k|} \equiv C_{q\bar{A}} \mu_\alpha\left(\frac{x}{\bar{A}}\right), \tag{1.2}$$

where

$$\mu_\alpha\left(\frac{x}{\bar{A}}\right) = \prod_{j=1}^n \mu_{\alpha_j}\left(\frac{x_j}{\bar{A}_j}\right)$$

and

$$\mu_m(t) = \inf_{0<h<\infty} \frac{h^{hm}}{|t|^h}. \tag{1.3}$$

If $m = 0$, then

$$\mu_0(t) = \begin{cases} 1 & \text{if } |t| \leqslant 1, \\ 0 & \text{if } |t| > 1. \end{cases}$$

Hence $\varphi(x) = 0$ if $|x_j| > A_j$ for some j, i.e., φ belongs to $K(A)$. Since the converse is also true, we have

$$S_{0,A} = K(A). \tag{1.4}$$

Now let $m > 0$. The minimum of the function $h^{hm}/|t|^h$ (h a real variable > 0) is obtained for $h = h_0 \equiv (1/e)t^{1/m}$, and the value of the minimum is

$$\exp [-(m/e)|t|^{1/m}].$$

Since h_0 is not necessarily an integer we evaluate $h^{hm}/|t|^h$ at the integer h_1 which is the nearest integer to h_0 from the right. We find that

$$\exp [-(m/e)|t|^{1/m}] \leqslant \mu_m(t) \leqslant C_0 \exp [-(m/e)|t|^{1/m}], \tag{1.5}$$

where $C_0 = (1 + e)^m$ if $|t| \geqslant 1$. If $|t| < 1$,

$$\mu_m(t) \leqslant 1 \leqslant [\exp (m/e)] \exp [-(m/e)|t|^{1/m}]$$

and (1.5) is therefore satisfied with $C_0 = e^{m/e}$.

Using (1.5) we obtain from (1.2)

$$|D^q\varphi(x)| \leqslant C'_{q\bar{A}} \exp \left\{- \sum_{j=1}^{n} \frac{\alpha_j}{e} \left(\frac{|x_j|}{\bar{A}_j}\right)^{1/\alpha_j}\right\} \tag{1.6}$$
$$\equiv C'_{q\bar{A}} \exp [-(\alpha/e)(|x|/\bar{A})^{1/\alpha}],$$

provided $x^k \neq 0$ for all k. By continuity, (1.6) is valid also if $x^k = 0$ for some k.

Conversely, if φ satisfies (1.6), then, in view of (1.5), φ also satisfies (1.2) and, hence, (1.1).

(1.6) can be rewritten in the form:

$$|D^q\varphi(x)| \leqslant C_{a'q} \exp \{- \sum_{j=1}^{n} a'_j|x_j|^{1/\alpha_j}\} \equiv C_{a'q} \exp [-a'|x|^{1/\alpha}] \tag{1.7}$$

for any $a' < a$ where

$$a = \frac{\alpha}{eA^{1/\alpha}}. \tag{1.8}$$

We define a topology in $S_{\alpha,A}$ by considering $S_{\alpha,A}$ to be a $K\{M_p\}$ space (see Sec. 1, Chap. 2) with

$$M_p(x) = \exp [a(1 - (1/p))|x|^{1/\alpha}] \qquad (p = 1, 2, \ldots). \tag{1.9}$$

Thus, the sequence of norms $\|\varphi\|_p$ is given by

$$\|\varphi\|_p = \sup_{|q|\leqslant p} \sup_{x\in R^n} M_p(x)|D^q\varphi(x)|. \tag{1.10}$$

$S_{\alpha,A}$ is then a complete countably normed space and since the (P) condition is also satisfied (see Sec. 2, Chap. 2), $S_{\alpha,A}$ is a perfect space. Obviously, a sequence $\{\varphi_j\}$ is convergent to 0 in $S_{\alpha,A}$ if and only if $\varphi_j \to 0$ properly and, for any $a' < a$,

$$|D^q\varphi_j(x)| \leqslant C_{a'q} \exp [-a'|x|^{1/\alpha}],$$

where the $C_{a'q}$ are independent of j.

S_α is defined to be the union space of the $S_{\alpha,A}$. It is also the countable union space of any sequence of spaces $S_{\alpha,A^{(m)}}$ where the $A^{(m)}$ tend monotonically to ∞ as $m \to \infty$. Thus a sequence $\{\varphi_j\}$ is convergent to 0 in S_α if and only if it is convergent to 0 in some space $S_{\alpha,A}$.

One can define the topology of $S_{\alpha,A}$ also in a different way, by giving the family of norms

$$(1.11) \quad \|\varphi\|_{q\rho} = \sup_{x\in R^n} \sup_k \frac{|x^k\, D^q\varphi(x)|}{(A+\rho)^k k^{k\alpha}} \qquad (0 \leqslant |k| < \infty\,;\ \rho = 1, \tfrac{1}{2}, \ldots).$$

The proof follows by noting that the right side is equal to

$$\sup_{x\in R^n} \frac{|D^q\varphi(x)|}{\mu_\alpha\left(\dfrac{x}{A+\rho}\right)}.$$

It follows that $\varphi_j \to 0$ in $S_{\alpha,A}$ if and only if $\varphi_j \to 0$ properly and

$$|x^k D^q\varphi_j(x)| \leqslant C_{q\bar{A}} \bar{A}^k k^{k\alpha} \qquad \text{for any } \bar{A} > A.$$

The space $S^{\beta,B}$, where $\beta = (\beta_1, \ldots, \beta_n) \geqslant 0$, $B = (B_1, \ldots, B_n) > 0$, consists of all the C^∞ functions $\varphi(x)$ which satisfy, for $0 \leqslant |k| < \infty$, $0 \leqslant |q| < \infty$,

$$(1.12) \qquad |x^k\, D^q\varphi(x)| \leqslant C_{k\bar{B}} \bar{B}^q q^{q\beta}$$

for any $\bar{B} > B$, where the constants $C_{k\bar{B}}$ depend on φ. The topology of $S^{\beta,B}$ is given in terms of the norms (compare (1.11))

$$(1.13) \qquad \|\varphi\|_{k\rho} = \sup_{x\in R^n} \sup_q \frac{|x^k\, D^q\varphi(x)|}{(B+\rho)^q q^{q\beta}}$$

$$(0 \leqslant |k| < \infty\,;\ \rho = 1, \tfrac{1}{2}, \ldots).$$

$S^{\beta,B}$ is not a $K\{M_p\}$ space. Nevertheless we shall prove that it is a complete countably normed space and, moreover, it is a perfect space.

Theorem 1. *$S^{\beta,B}$ is a perfect space.*

Proof. The proof is given in several steps.

(a) If a sequence $\{\varphi_j\} \subset S^{\beta,B}$ is properly convergent to $\varphi(x)$ and if, for some k, ρ, $\|\varphi_j\|_{k\rho} \leqslant C$, then $\|\varphi\|_{k\rho}$ exists and is $\leqslant C$.

Indeed, for any $A > 0$,

$$\sup_{|x|\leqslant A} \sup_{|q|\leqslant p} \frac{|x^k\, D^q\varphi(x)|}{(B+\rho)^q q^{q\beta}} = \lim_{j\to\infty} \sup_{|x|\leqslant A} \sup_{|q|\leqslant p} \frac{|x^k\, D^q\varphi_j(x)|}{(B+\rho)^q q^{q\beta}}$$
$$\leqslant \limsup_{j\to\infty} \|\varphi_j\|_{k\rho} \leqslant C.$$

Taking $p \to \infty$, $A \to \infty$ we obtain $\|\varphi\|_{k\rho} \leqslant C$.

(b) If a sequence $\{\varphi_j\}$ is pointwise convergent to 0 and if it is a Cauchy sequence in the norm $\|\cdot\|_{k\rho}$, then $\|\varphi_j\|_{k\rho} \to 0$.

Since $\{\varphi_j\}$ is a Cauchy sequence in $\|\cdot\|_{k\rho}$, it is also properly convergent to some φ. But then $\varphi(x) = 0$ and $\varphi_j - \varphi_h \to \varphi_j$ properly if $h \to \infty$. By (a),

$$\|\varphi_j\|_{k\rho} \leqslant \sup_{h \geqslant j} \|\varphi_j - \varphi_h\|_{k\rho}$$

and the assertion follows.

(c) $S^{\beta,B}$ is a complete space.

Let $\{\varphi_j\}$ be a Cauchy sequence in $S^{\beta,B}$. Then it is a Cauchy sequence in each of the norms $\|\cdot\|_{k\rho}$. Consequently $\{\varphi_j\}$ is properly convergent to some φ. By (a), $\|\varphi\|_{k\rho}$ exists for any k, ρ, i.e., $\varphi \in S^{\beta,B}$. Applying (b) to $\{\varphi - \varphi_j\}$ we conclude that $\|\varphi - \varphi_j\|_{k\rho} \to 0$ for any k, ρ, i.e., $\varphi_j \to \varphi$ in $S^{\beta,B}$.

(d) The norms $\|\cdot\|_{k\rho}$ are in concordance.

Let $\{\varphi_j\}$ be a Cauchy sequence in the norms $\|\cdot\|_{k_1\rho_1}$, $\|\cdot\|_{k_2\rho_2}$ and let $\|\varphi_j\|_{k_1\rho_1} \to 0$. Then $\varphi_j(x) \to 0$ for each x and, by (b), $\|\varphi_j\|_{k_2\rho_2} \to 0$.

(e) If a sequence $\{\varphi_j\}$ is bounded in each norm $\|\cdot\|_{k\rho}$ and is properly convergent to 0, then $\|\varphi_j\|_{k\rho} \to 0$ for any k, ρ, i.e., $\varphi_j \to 0$ in $S^{\beta,B}$.

Let k, ρ be given and let η be any positive number. Take $\rho' < \rho$. The norms $\|\varphi_j\|_{k\rho'}$ are bounded by a constant $C_{k\rho'}$. For all q with $|q|$ sufficiently large, say $|q| \geqslant q_0$,

$$\frac{(B + \rho')^q}{(B + \rho)^q} < \frac{\eta}{C_{k\rho'}}.$$

But then

$$|x^k\, D^q\varphi_j(x)| \leqslant C_{k\rho'}(B + \rho')^q q^{q\beta} < \eta(B + \rho)^q q^{q\beta}.$$

If $|q| < q_0$ and $|x_m| > C_{k+1,\rho}/\eta$ (where $C_{k+1,\rho} \geqslant \|\varphi_j\|_{k+1,\rho}$) for some m $(1 \leqslant m \leqslant n)$, then

$$|x^k\, D^q\varphi_j(x)| = \frac{1}{|x_m|}\, |x^{k+1}\, D^q\varphi_j(x)| \leqslant \frac{1}{|x_m|}\|\varphi_j\|_{k+1,\rho}(B + \rho)^q q^{q\beta}$$
$$< \eta(B + \rho)^q q^{q\beta}.$$

Finally, if $|q| < q_0$ and $|x_m| \leqslant C_{k+1,\rho}/\eta$ for all $m = 1, \ldots, n$, then

$$|x^k\, D^q\varphi_j(x)| < \eta(B + \rho)^q q^{q\beta}$$

if j is sufficiently large, say if $j \geqslant j_0$. We conclude that if $j \geqslant j_0$, then

$$\|\varphi_j\|_{k\rho} = \sup_{x\in R^n} \sup_q \frac{|x^k\, D^q\varphi_j(x)|}{(B + \rho)^q q^{q\beta}} \leqslant \eta,$$

which proves that $\varphi_j \to 0$ in each norm $\|\cdot\|_{k\rho}$.

(f) If $\{\varphi_j\}$ is a bounded sequence in each norm $\|\cdot\|_{kp}$ and if $\varphi_j \to \varphi$ properly, then $\varphi \in S^{\beta,B}$ and $\varphi_j \to \varphi$ in $S^{\beta,B}$.

Indeed, by (a), $\varphi \in S^{\beta,B}$. Now apply (e) to $\{\varphi - \varphi_j\}$.

Noting that from any sequence $\{\varphi_j\}$ bounded in the norms $\|\cdot\|_{kp}$ one can extract a properly convergent subsequence, and applying (f), the proof that $S^{\beta,B}$ is a perfect space is completed.

We denote by S^β the union space of the spaces $S^{\beta,B}$. S^β is also the countable union space of any sequence $\{S^{\beta,B^{(m)}}\}$, where the $B^{(m)}$ increase monotonically to ∞. The space S^β consists of all the C^∞ functions $\varphi(x)$ satisfying

$$(1.14)\qquad |x^k D^q\varphi(x)| \leqslant C_k B^q q^{q\beta} \qquad (0 \leqslant |q| < \infty, 0 \leqslant |k| < \infty)$$

for some constants C_k, B depending on φ. If $\beta > 1$ then, by Theorem 11, Sec. 7, Chap. 6, there exist nonanalytic functions satisfying (1.14). If $\beta \leqslant 1$ then every function satisfying (1.14) is analytic since its Taylor's series

$$(1.15)\qquad \varphi(x+h) = \sum_{0\leqslant|q|<\infty} \frac{h^q}{q!} D^q\varphi(x)$$

is convergent. For $\beta = 1$ the series is convergent for $|h_j| < 1/(B_j e)$ $(j = 1, \ldots, n)$. Hence $\varphi(x)$ can be extended into a complex analytic function $\varphi(x + iy)$ for $|y_j| < 1/(B_j e)$ $(j = 1, \ldots, n)$. If $\beta < 1$ then the series (1.15) is convergent for all h and $\varphi(x + iy)$ is therefore an entire function. Furthermore,

$$(1.16)\qquad |x^k\varphi(x+iy)| = \left| \sum \frac{(iy)^q}{q!} x^k D^q\varphi(x) \right| \leqslant C_k \sum \frac{|y^q|}{q!} B^q q^{q\beta}.$$

In order to evaluate the right side we use (1.5) and thus get, for any $\eta > 0$, $0 \leqslant r < 1$,

$$(1.17)\qquad \sup_p \frac{\eta^p e^p}{p^{p(1-r)}} = \frac{1}{\inf_p \dfrac{p^{p(1-r)}}{\eta^p e^p}} \leqslant \exp\,[c\eta^{1/(1-r)}]$$

$$\left(c = \frac{1-r}{e} \exp\,[1/(1-r)]\right).$$

Noting that the inequalities

$$(1.18)\qquad \frac{\eta^p e^p}{p^{p(1-r)}} < \frac{1}{2^p} \qquad \text{if } p > p_0$$

are satisfied if $p_0 = (2e\eta)^{1/(1-r)}$, and using (1.17), (1.18), we have

$$(1.19)\qquad \sum_{p=1}^{\infty} \frac{\eta^p e^p}{p^{p(1-r)}} = \sum_{p\leqslant p_0} + \sum_{p>p_0} \leqslant (2e\eta)^{1/(1-r)} \exp\,[c\eta^{1/(1-r)}] + 1$$
$$\leqslant C' \exp\,[c'\eta^{1/(1-r)}]$$

for any $c' > c$.

Using $q! \geqslant C''q^q e^{-q}$ and (1.19) we can estimate the right side of (1.16):

$$|x^k \varphi(x + iy)| \leqslant C'_k \exp [b'|y|^{1/(1-\beta)}] \tag{1.20}$$

for any b' satisfying

$$b' > \frac{1 - \beta}{e} (Be)^{1/(1-\beta)}. \tag{1.21}$$

Conversely, if $\varphi(z)$ is an entire function satisfying (1.20), then by Theorem 5, Sec. 1, Chap. 5, it follows that $\varphi(x)$ satisfies (1.14). We have thus proved the following theorem:

Theorem 2. *If $0 \leqslant \beta < 1$, then the space $S^{\beta,B}$ consists of all the functions $\varphi(x)$ which can be extended into entire functions $\varphi(z)$ satisfying*

$$|x^k \varphi(x + iy)| \leqslant C'_k \exp [b'|y|^{1/(1-\beta)}] \tag{1.22}$$

for any b' satisfying (1.21).

We now introduce the space $S_{\alpha,A}^{\beta,B}$ ($\alpha \geqslant 0, \beta \geqslant 0, A > 0, B > 0$). It consists of all the C^∞ functions satisfying

$$|x^k D^q \varphi(x)| \leqslant \bar{C} \bar{A}^k \bar{B}^q k^{k\alpha} q^{q\beta} \qquad (0 \leqslant |k| < \infty, 0 \leqslant |q| < \infty) \tag{1.23}$$

for any $\bar{A} > A$, $\bar{B} > B$, where $\bar{C}$ depends on $\bar{A}$, $\bar{B}$, and φ. A topology is defined in terms of the norms

$$\|\varphi\|_{\delta\rho} = \sup_{x \in R^n} \sup_{k,q} \frac{|x^k D^q \varphi(x)|}{(A + \delta)^k (B + \rho)^q k^{k\alpha} q^{q\beta}} \qquad (\delta, \rho = 1, \tfrac{1}{2}, \ldots). \tag{1.24}$$

Theorem 3. *$S_{\alpha,A}^{\beta,B}$ is a perfect space.*

The proof is similar to that of Theorem 1 and the details are left to the reader.

We denote by S_α^β the union space of the spaces $S_{\alpha,A}^{\beta,B}$. S_α^β is then also the countable union space of any sequence $\{S_{\alpha,A^{(m)}}^{\beta,B^{(m)}}\}$, where the $A^{(m)}$ and $B^{(m)}$ tend monotonically to ∞, as $m \to \infty$. If A is kept fixed while B varies, the union space of the spaces $S_{\alpha,A}^{\beta,B}$ is denoted by $S_{\alpha,A}^\beta$. $S_\alpha^{\beta,B}$ is defined in a similar way.

If

$$|x^k D^q \varphi(x)| \leqslant C A^k B^q k^{k\alpha} q^{q\beta} \qquad (0 \leqslant |k| < \infty, 0 \leqslant |q| < \infty), \tag{1.25}$$

then, as proved above,

$$|D^q \varphi(x)| \leqslant C' B^q q^{q\beta} \exp [-a|x|^{1/\alpha}] \qquad \left(a = \frac{\alpha}{eA^{1/\alpha}}\right). \tag{1.26}$$

If $\beta < 1$, then

$$|x^k \varphi(x + iy)| \leqslant C'' A^k k^{k\alpha} \exp [b'|y|^{1/(1-\beta)}] \tag{1.27}$$

for any $b' > b$, where

$$b = \frac{1-\beta}{e}(Be)^{1/(1-\beta)}. \tag{1.28}$$

Applying the argument by which one deduces (1.26) from (1.25), one derives from (1.27),

$$|\varphi(x+iy)| \leqslant C_0 \exp\left[-a|x|^{1/\alpha} + b'|y|^{1/(1-\beta)}\right]. \tag{1.29}$$

The inequalities (1.6), (1.22), (1.29) establish the following connections between S spaces and W spaces:

$$S_\alpha = W_{1/\alpha} \qquad \text{if } 0 < \alpha \leqslant 1; \tag{1.30}$$

$$S^\beta = W^{1/(1-\beta)} \qquad \text{if } 0 \leqslant \beta < 1; \tag{1.31}$$

$$S_\alpha^\beta = W_{1/\alpha}^{1/(1-\beta)} \qquad \text{if } 0 < \alpha \leqslant 1,\ 0 \leqslant \beta < 1. \tag{1.32}$$

2. Operators in S Spaces

Theorem 4. *Multiplication by polynomials and differentiation are bounded operators in all the S spaces.*

The proof is straightforward and is left to the reader. We also leave it to the reader to verify that translations (i.e., $\varphi(x) \to \varphi(x+h)$) are operators bounded uniformly with respect to h (if $|h| \leqslant 1$) in the spaces $S_{\alpha,A}$ $(\alpha > 0)$, S_0, $S^{\beta,B}$ $(\beta > 0)$, S^0, $S_{\alpha,A}^{\beta,B}$ $(\alpha > 0, \beta > 0)$, $S_0^{\beta,B}$ $(\beta > 0)$, and $S_{\alpha,A}^0$ $(\alpha > 0)$. It follows that all the results of Chap. 4 are true when the test space Φ is any one of these S spaces.

Theorem 5. *Let $f(x)$ be a C^∞ function satisfying*

$$|D^q f(x)| \leqslant C_q \exp\left[a_1|x|^{1/\alpha}\right] \qquad (\alpha > 0) \tag{2.1}$$

and let $a \equiv \alpha/(eA^{1/\alpha}) > a_1$. Then $f(x)$ is a multiplier from $S_{\alpha,A}$ into $S_{\alpha,A'}$, where $a - a_1 = \alpha/(e(A')^{1/\alpha})$.

Proof. Using (1.6) and Leibniz' formula we get

$$|D^q(f(x)\varphi(x))| \leqslant C'_{qa'} \exp\left[-(a'-a_1)|x|^{1/\alpha}\right]$$

for any $a' < a$, from which the assertion follows.

Theorem 6. *Let $f(x)$ be a C^∞ function satisfying*

$$|D^q f(x)| \leqslant C_0 B_0^q q^{q\beta}(1+|x|)^h. \tag{2.2}$$

Then $f(x)$ is a multiplier from $S^{\beta,B}$ into $S^{\beta,B+B_0}$.

Theorem 7. *Let $f(x)$ be a C^∞ function satisfying*

$$|D^q f(x)| \leqslant C_0 B_0^q q^{q\beta} \exp\,[a_0|x|^{1/\alpha}] \qquad (\alpha > 0) \tag{2.3}$$

and let $a_0 < a \equiv \alpha/(eA^{1/\alpha})$. Then $f(x)$ is a multiplier from $S_{\alpha,A}^{\beta,B}$ into $S_{\alpha,A'}^{\beta,B'}$, where $a - a_0 = \alpha/(e(A')^{1/\alpha})$, $B' = B + B_0$.

The proofs are left to the reader.

We mention that the function f appearing in Theorem 6, Sec. 1, Chap. 5, belongs to $S_{1/h}^{1-(\mu/h)}$ if $a < 0$, and is a multiplier in $S_{1/h}^{1-(\mu/h)}$ if $a > 0$. The function appearing in Theorem 4, Sec. 1, Chap. 5, belongs to $S_{1/h}^{1-(1/\gamma)}$ if $a < 0$, and is a multiplier in $S_{1/h}^{1-(1/\gamma)}$ if $a > 0$.

Theorem 8. *Let $f(z)$ be an entire function of order $\leqslant 1/\beta$ and type $< \beta/(B^{1/\beta}e^2)$. Then $f(D)$ is a bounded operator from $S^{\beta,B}$ into S^{β,Be^β} and from $S_{\alpha,A}^{\beta,B}$ into $S_{\alpha,A}^{\beta,Be^\beta}$.*

Proof. If $f(z) = \Sigma\, c_m z^m$ is the Taylor series of an entire function f, then

$$|c_m| \leqslant C\left(\frac{b_0 e p}{m}\right)^{m/p} \tag{2.4}$$

provided f is of order $\leqslant p$ and type $< b_0$. These inequalities are obtained from Cauchy's inequalities by taking the radii r_j to be $\left(\dfrac{m_j}{b_{0j}p_j}\right)^{1/p_j}$ where $b_0 = (b_{01}, \ldots, b_{0n})$, $p = (p_1, \ldots, p_n)$.

Now let $\varphi \in S^{\beta,B}$. Then, for any $\rho > 0$,

$$|x^k\, D^q\varphi(x)| \leqslant C_{k\rho}(B + \rho)^q q^{q\beta}.$$

We have to prove that $\psi(x) = \Sigma\, c_m\, D^m\varphi(x)$ belongs to S^{β,Be^β}. Since

$$(m + q)^{(m+q)\beta} = (m + q)^{m\beta}(m + q)^{q\beta}, \quad (m + q)^{m\beta} \leqslant m^{m\beta}e^{q\beta},$$

$$(m + q)^{q\beta} \leqslant q^{q\beta}e^{m\beta},$$

we get

$$\begin{aligned} |x^k\, D^q\psi(x)| &\leqslant \Sigma\, |c_m x^k D^{m+q}\varphi(x)| \\ &\leqslant C_{k\rho}(B + \rho)^q\, \Sigma\, |c_m|(B + \rho)^m(m + q)^{(m+q)\beta} \\ &\leqslant C_{k\rho}(B + \rho)^q q^{q\beta} e^{q\beta}\, \Sigma\, |c_m|(B + \rho)^m e^{m\beta} m^{m\beta}. \end{aligned} \tag{2.5}$$

Using (2.4) with $p = 1/\beta$, $b_0 = \theta\beta/(B^{1/\beta}e^2)$ for some $\theta < 1$, and taking ρ such that $1 + \rho/B < 1/\theta^{1/p}$, we obtain

$$|x^k\, D^q\psi(x)| \leqslant C_{k\rho}C'_\rho(Be^\beta + \rho')^q q^{q\beta}, \tag{2.6}$$

where C'_ρ is the sum of the series on the right side of (2.5), and $\rho' = \rho e^\beta$ can be made arbitrarily small. From (2.6) it follows that $\psi \in S^{\beta,Be^\beta}$ and, furthermore, that $f(D)$ maps bounded sets in $S^{\beta,B}$ into bounded sets in S^{β,Be^β}.

The proof for $S_{\alpha,A}^{\beta,B}$ is similar and the details are left to the reader.

3. Fourier Transforms of S spaces

Let $\{M_{jh}^{(i)}\}$ be sequences of positive numbers (for $i = 1, \ldots, n$) which satisfy the following conditions:

(E_1) For any $j, h = 1, 2, \ldots,$

$$jh \frac{M_{j-1,h-1}^{(i)}}{M_{jh}^{(i)}} \leqslant \gamma_i (j + h)^{\theta_i} \qquad \text{for some } \gamma > 0, 0 < \theta \leqslant 1. \tag{3.1}$$

(E_2) For any $j, h = 0, 1, 2, \ldots,$ and for any $\epsilon > 0$,

$$\frac{M_{j+n+1,h}^{(i)}}{M_{jh}^{(i)}} \leqslant H_{jh}^{(i)}(\epsilon)(1 + \epsilon)^{j+h} \qquad (H_{jh}^{(i)}(\epsilon) \text{ a constant}). \tag{3.2}$$

We set

$$M_{kq} = M_{k_1 q_1}^{(1)} \cdots M_{k_n q_n}^{(n)}, \qquad H_{kq}(\epsilon) = H_{k_1 q_1}^{(1)}(\epsilon) \cdots \mathrm{H}_{k_n q_n}^{(n)}(\epsilon).$$

We shall now prove a general theorem on the Fourier transform of any C^∞ fast-decreasing function, which will be used later in studying the Fourier transforms of S spaces.

Theorem 9. *Let $\varphi(x)$ be a C^∞ function satisfying the inequalities*

$$|x^k D^q \varphi(x)| \leqslant C A^k B^q M_{kq} \qquad (0 \leqslant |k| < \infty, 0 \leqslant |q| < \infty), \tag{3.3}$$

where the $M_{jh}^{(i)}$ satisfy the conditions (E_1), (E_2). Then the Fourier transform $\psi(\sigma)$ of $\varphi(x)$ satisfies the inequalities

$$|\sigma^k D^q \psi(\sigma)| \leqslant C_1 C (1 + H_{qk}(\epsilon)) A_2^q B_2^k M_{qk} \tag{3.4}$$

$$(0 \leqslant |k| < \infty, 0 \leqslant |q| < \infty),$$

where $A_2 = A_1(1 + \epsilon)$, $B_2 = B_1(1 + \epsilon)$ for any $\epsilon > 0$, and

$$A_1 = A e^{\gamma/AB}, \qquad B_1 = B e^{\gamma/AB} \qquad \text{if } \theta = 1, \tag{3.5}$$

$$A_1 = A(1 + \epsilon), \quad B_1 = B(1 + \epsilon) \qquad \text{if } \theta < 1, \tag{3.6}$$

and C_1 depends only on ϵ, A, B, γ, θ.

We need, perhaps, to mention that by an equation of the form $D = E^{F/G}$, where D, E, F, G are vectors with components D_j, E_j, F_j, G_j, we mean $D_j = E_j^{F_j/G_j}$ for all j.

We shall need the following lemma:

Lemma 1. *If the $M_{jh}^{(i)}$ satisfy the condition (E_1) and if $\varphi(x)$ satisfies (3.3) then*

$$|D^q(x^k \varphi(x))| \leqslant C' C A_1^k B_1^q M_{kq} \qquad (0 \leqslant |k| < \infty, 0 \leqslant |q| < \infty), \tag{3.7}$$

where A_1, B_1 are defined by (3.5), (3.6), and C' depends only on ϵ, A, B, γ, θ.

Proof. Assume for simplicity that $n = 1$. Using Leibniz' rule and (3.1), (3.3) we get

$$\begin{aligned}|D^q(x^k\varphi(x))| &= \left|\Sigma \binom{q}{p} D^p x^k \cdot D^{q-p}\varphi(x)\right| \\ &\leqslant C\,\Sigma \frac{1}{p!} q \cdots (q-p+1)k \cdots \\ &\qquad\qquad (k-p+1)A^{k-p}B^{q-p}M_{k-p,q-p} \\ &\leqslant CA^kB^qM_{kq}\,\Sigma \frac{\gamma^p(k+q)^{p\theta}}{p!(AB)^p} \\ &= CA^kB^qM_{kq}\exp\,[(\gamma/AB)(k+q)^\theta]\end{aligned}$$

and the proof for $\theta = 1$ follows. If $\theta < 1$ the proof follows by noting that

$$\exp\,[(\gamma/AB)(k+q)^\theta] < C_\epsilon(1+\epsilon)^{k+q}$$

for any $\epsilon > 0$.

From the previous proof one also derives the inequality

$$|x^2\,D^q(x^k\varphi(x))| \leqslant C''CA_1^{k+2}B_1^qM_{k+2,q} \tag{3.8}$$

which will be needed later on.

Proof of Theorem 9. Assume first that $n = 1$. By Lemma 1 and (3.8) we get

$$\begin{aligned}|D^q(x^k\varphi(x))| &\leqslant C\min\left\{C'A_1^kB_1^qM_{kq}, \frac{C''}{x^2}\,A_1^{k+2}B_1^qM_{k+2,q}\right\} \\ &\leqslant CC_1'A_1^kB_1^qM_{kq}\min\left\{1, \frac{A_1^2}{x^2}\frac{M_{k+2,q}}{M_{kq}}\right\}.\end{aligned}$$

Hence, using (3.2) we have

$$\begin{aligned}|\sigma^q\,D^k\psi(\sigma)| &= |\mathfrak{F}[D^q(x^k\varphi(x))]| \leqslant \int_{-\infty}^{\infty}|D^q(x^k\varphi(x))|\,dx \\ &\leqslant 2C_1'CA_1^kB_1^qM_{kq}\left\{1 + \frac{A_1^2M_{k+2,q}}{M_{kq}}\int_1^\infty \frac{dx}{x^2}\right\} \\ &\leqslant C_1CA_2^kB_2^q(1 + H_{kq}(\epsilon))M_{kq}\end{aligned}$$

and the proof is completed.

If $n > 1$ then the proof is similar, provided we use the inequality

$$\left(\sum_{j=1}^n |x_j|^{n+1}\right)|D^q(x^k\varphi(x))| \leqslant C_1''CA_1^{k+n+1}B_1^qM_{k+n+1,q}$$

instead of (3.8).

In order to apply Theorem 9 to functions in S spaces we shall need the following elementary lemma.

Lemma 2. *Let* $\{a_j^{(i)}\}$, $\{b_h^{(i)}\}$ *be sequences of positive numbers, for* $i = 1, \ldots, n$, *satisfying the conditions:*

$$\frac{a_j^{(i)}}{a_{j-1}^{(i)}} \geqslant H_1 j^{1-\lambda_i}, \qquad \frac{b_h^{(i)}}{b_{h-1}^{(i)}} \geqslant H_2 h^{1-\mu_i} \qquad (H_1 > 0,\ H_2 > 0),$$

where $\lambda_i \geqslant 0$, $\mu_i \geqslant 0$, $\lambda_i + \mu_i = \theta_i \leqslant 1$. *Then the numbers* $M_{jh}^{(i)} = a_j^{(i)} b_h^{(i)}$ *satisfy the condition* (E_1) *(with the same* θ*) and* $\gamma = 1/H_1H_2$.

Proof.

$$jh \frac{M_{j-1,h-1}^{(i)}}{M_{jh}^{(i)}} = j \cdot \frac{a_{j-1}^{(i)}}{a_j^{(i)}}\, h\, \frac{b_{h-1}^{(i)}}{b_h^{(i)}} \leqslant \frac{1}{H_1H_2} j^{\lambda_i} h^{\mu_i} \leqslant \frac{1}{H_1H_2} (j + h)^{\lambda_i + \mu_i}.$$

Now let $\varphi \in S_{\alpha,A}$, $\alpha > 0$. Then, for any $\delta > 0$,

$$|x^k D^q \varphi(x)| \leqslant C_{q\delta}(A + \delta)^k k^{k\alpha},$$

so that (3.3) is satisfied with $M_{kq} = C_{q\delta} k^{k\alpha}$, $B = 1$. Since $C_{q\delta}$ can be replaced by any larger constant, we may assume without loss of generality that

$$C_{q\delta} = C_{q_1\delta}^{(1)} \cdots C_{q_n\delta}^{(n)}$$

and

$$\frac{C_{q_i\delta}^{(i)}}{C_{q_{i-1}\delta}^{(i)}} \geqslant q_i^{1-\mu_i} \qquad \text{for any } \mu_i > 0.$$

Taking $a_j^{(i)} = j^{j\alpha_i}$, $b_h^{(i)} = C_{h\delta}^{(i)}$ we find that

$$\frac{a_j^{(i)}}{a_{j-1}^{(i)}} \geqslant j^{\alpha_i} \geqslant j^{1-\lambda_i} \qquad \text{if } \lambda_i = \max\,(0, 1 - \alpha_i).$$

Choosing $\mu_i < \min\{\alpha_i, 1\}$ we see that $\mu_i + \lambda_i < 1$ and, by Lemma 2, the condition (E_1) is satisfied with $\theta < 1$.

(E_2) is also satisfied since

$$\frac{M_{j+n+1,h}^{(i)}}{M_{jh}^{(i)}} = \frac{(j + n + 1)^{(j+n+1)\alpha_i}}{j^{j\alpha_i}} \leqslant \{\exp[(n + 1)\alpha_i]\}(j + n + 1)^{(n+1)\alpha_i} \leqslant H^{(i)}(\epsilon)(1 + \epsilon)^j$$

for any $\epsilon > 0$. Applying Theorem 9 we conclude that

$$|\sigma^k D^q \psi(\sigma)| \leqslant C'_\delta(A + \delta_1)^q (1 + \delta_1)^k q^{q\alpha} C_{k\delta} \leqslant C'_{k\delta_1}(A + \delta_1)^q q^{q\alpha}$$

for any $\delta_1 > 0$, i.e., $\psi \in S^{\alpha,A}$. From the inequalities derived in this proof it also follows that the mapping $\varphi \to \psi$ is a bounded mapping from $S_{\alpha,A}$ into $S^{\alpha,A}$.

Consider next a function $\varphi(x)$ in $S^{\beta,B}$, $\beta > 0$. Then, for any $\rho > 0$

$$|x^k D^q \varphi(x)| \leqslant C_{k\rho}(B + \rho)^q q^{q\beta}.$$

Again we may assume that $C_{k\rho} = C_{k_1\rho}^{(1)} \cdots C_{k_n\rho}^{(n)}$ and that the $C_{k_i\rho}^{(i)}$ are such that $a_j^{(i)} = C_{j\rho}^{(i)}$, $b_h^{(i)} = h^{h\beta_i}$ satisfy the assumptions of Lemma 2. Hence, the $M_{k_iq_i}^{(i)} = C_{k_i\rho}q_i^{q_i\beta_i}$ satisfy the condition (E_1). (E_2) is also satisfied since

$$\frac{M_{j+n+1,h}^{(i)}}{M_{jn}^{(i)}} = \frac{C_{j+n+1,\rho}^{(i)}}{C_{j\rho}^{(i)}} \equiv H_j^{(i)}.$$

Applying Theorem 9 we conclude that

$$|\sigma^k D^q\psi(\sigma)| \leqslant C'_\rho(1 + H_q)(1 + \rho_1)^q(B + \rho_1)^k k^{k\beta} = C'_{q\rho_1}(B + \rho_1)^k k^{k\beta}$$

for any $\rho_1 > 0$, i.e., $\psi \in S_{\beta,B}$. From the inequalities derived in this proof it also follows that the mapping $\varphi \to \psi$ is a bounded mapping from $S^{\beta,B}$ into $S_{\beta,B}$.

Combining the last two results and using (1.6), Chap. 4, we arrive at the following theorem:

Theorem 10. *If* $\alpha > 0$, $\beta > 0$, *then*

$$\widetilde{S_{\alpha,A}} = S^{\alpha,A}, \qquad \widetilde{S^{\beta,B}} = S_{\beta,B} \tag{3.9}$$

and $\mathfrak{F}$ *is a linear topological mapping.*

Consider next a function $\varphi \in S_{\alpha,A}^{\beta,B}$ where $\alpha > 0$, $\beta > 0$. Then, for any $\delta > 0$, $\rho > 0$

$$|x^k D^q\varphi(x)| \leqslant C_{\delta\rho}(A + \delta)^k(B + \rho)^q k^{k\alpha}q^{q\beta}.$$

Taking $a_j^{(i)} = j^{j\alpha_i}$, $b_h^{(i)} = h^{h\beta_i}$ we see that

$$\frac{a_j^{(i)}}{a_{j-1}^{(i)}} \geqslant j^{j(1-\lambda_i)}, \qquad \frac{b_h^{(i)}}{b_{h-1}^{(i)}} \geqslant h^{h(1-\mu_i)},$$

where $\lambda_i = \max\{1 - \alpha_i, 0\}$, $\mu_i = \max\{1 - \beta_i, 0\}$. If $\alpha + \beta > 1$ then we find that $\theta = \lambda + \mu < 1$ and, by Lemma 2, the condition (E_1) is satisfied for $M_{jh}^{(i)} = a_j^{(i)}b_h^{(i)}$. (E_2) is also satisfied and the proof is the same as for the case of $S_{\alpha,A}$.

Applying Theorem 9 we conclude that

$$|\sigma^k D^q\psi(\sigma)| \leqslant C'_{\delta_1\rho_1}(A + \delta_1)^q(B + \rho_1)^k q^{q\alpha}k^{k\beta}$$

for any $\delta_1 > 0$, $\rho_1 > 0$, i.e., $\psi \in S_{\beta,B}^{\alpha,A}$. The mapping $\varphi \to \psi$ is also easily seen to be bounded. Using (1.6), Chap. 4, we get the following theorem:

Theorem 11. *If* $\alpha > 0$, $\beta > 0$ *and* $\alpha + \beta > 1$, *then*

$$\widetilde{S_{\alpha,A}^{\beta,B}} = S_{\beta,B}^{\alpha,A} \tag{3.10}$$

and $\mathfrak{F}$ *is a linear topological mapping.*

If $\alpha = 0$ then the proof of Theorem 9 remains true, provided we apply Theorem 8 with $\theta = 1$. Hence we get.

$$\widetilde{S_{0,A}} = S^{0,Ae^{1/A}}. \tag{3.11}$$

Similarly we get

$$\widetilde{S^{0,B}} = S_{0,Be^{1/B}}, \tag{3.12}$$

$$\widetilde{S_{\alpha,A}^{\beta,B}} = S_{\beta,Be^{1/AB}}^{\alpha,Ae^{1/AB}} \qquad \text{if } \alpha \geqslant 0,\ \beta \geqslant 0,\ \alpha+\beta = 1. \tag{3.13}$$

The case $\alpha + \beta < 1$ need not be considered since the spaces S_α^β are trivial spaces if $\alpha + \beta < 1$ (see Sec. 4).

From Theorems 9, 10 and from (3.11)–(3.13) it follows that

$$\widetilde{S_\alpha} = S^\alpha, \quad \widetilde{S^\beta} = S_\beta, \quad \widetilde{S_\alpha^\beta} = S_\beta^\alpha \tag{3.14}$$

$$\text{whenever } \alpha \geqslant 0,\ \beta \geqslant 0,\ \alpha + \beta \geqslant 1.$$

Note that $S_0 = K = (D)$, $S^0 = Z$. In Theorem 1, Sec. 1, Chap. 4, it was already proved that $\widetilde{S_0} = S^0$.

4. Nontriviality and Richness of S Spaces

The spaces $S_{\alpha,A}$ contain $K(A)$ and are therefore nontrivial spaces for and $\alpha \geqslant 0$. The spaces $S^{\alpha,A}$, which are equal to the spaces $\widetilde{S_{\alpha,A}}$, are therefore also nontrivial spaces. It remains to consider the nontriviality of the spaces $S_{\alpha,A}^{\beta,B}$. By Theorem 18, Sec. 5, Chap. 5, S_α^β is a trivial space if $\alpha + \beta < 1$. It thus remains to consider the case where $\alpha + \beta \geqslant 1$.

From Theorem 11, Sec. 7, Chap. 6, it follows that S_0^β is nontrivial if $\beta > 1$. If $\beta \leqslant 1$ then S_0^β is a trivial space, since its functions are analytic functions having a compact support. Using (3.13) it also follows that S_α^0 is nontrivial if and only if $\alpha > 1$. From Theorem 19, Sec. 20, Chap. 5, and from (1.32) it follows that S_α^β ($\alpha > 0$, $\beta > 0$) is nontrivial if $\alpha + \beta = 1$. Next, since

$$S_{\alpha',A}^{\beta,B} \subset S_{\alpha,A'}^{\beta,B}, \qquad S_{\alpha,A}^{\beta,B} \subset S_{\alpha,A}^{\beta',B'} \tag{4.1}$$

for $\alpha' > \alpha$, $\beta' > \beta$ and for arbitrary A', B', it follows that $S_{\alpha,A'}^{\beta,B}$, is nontrivial if $\alpha + \beta > 1$. If $\alpha + \beta = 1$ then for some pair (A, B) $S_{\alpha,A}^{\beta,B}$ is nontrivial. Now if $\varphi(x) \in S_{\alpha,A_0}^{\beta,B_0}$, then $\varphi(\lambda x) \subset S_{\alpha,A_0/\lambda}^{\beta,\lambda B_0}$ (see Problem 1). Hence, there exists a $\gamma > 0$ (not uniquely determined when $n > 1$) such that $S_{\alpha,A}^{\beta,B}$ is nontrivial if $AB > \gamma$ and it is trivial if $AB < \gamma$.

If $\beta > 1$ and $1 < \beta_0 < \beta$ then, by (4.1),

$$S_{0,A_0}^{\beta_0,B_0} \subset S_{0,A_0}^{\beta,B}$$

for any B. Taking (A_0, B_0) such that $S_{0,A_0}^{\beta_0,B_0}$ is nontrivial, we conclude, by decreasing B, that γ can be taken arbitrarily small. Hence $S_{0,A}^{\beta,B}$ is nontrivial for any A, B. The same is true for $S_{\alpha,A}^{0,B}$, by (3.13). We sum up all these results in the following theorem:

Theorem 12. *The following spaces are nontrivial:*

$S_{\alpha,A}$, $S^{\beta,B}$ *for any* α, β, A, B;
$S_{\alpha,A}^{0,B}$, $S_{0,A}^{\beta,B}$ *for* $\alpha > 1$, $\beta > 1$ *and any* A, B;
$S_{\alpha,A}^{\beta,B}$ *for* $\alpha > 0$, $\beta > 0$, $\alpha + \beta > 1$ *and any* A, B, *and*
$S_{\alpha,A}^{\beta,B}$ *for* $\alpha > 0$, $\beta > 0$, $\alpha + \beta = 1$ *and* $AB > \gamma$.

From Theorems 5, 6, 7 of Sec. 7 it follows that $e^{i\sigma\cdot x}$ is a multiplier in $S_{\alpha,A}$ $(\alpha > 0)$, in $S^{\beta,B}$ $(\beta > 0)$ and in $S_{\alpha,A}^{\beta,B}$ $(\alpha > 0, \beta > 0)$. One can also easily verify that $e^{i\sigma\cdot x}$ is a multiplier in $S_0^{\beta,B}$ $(\beta > 0)$ and in $S_{\alpha,A}^0$ $(\alpha > 0)$. Finally, $e^{i\sigma\cdot x}$ is obviously a multiplier in $S_0 = K$ and in $S^0 = Z$. Using Theorem 21, Sec. 5, Chap. 5 we obtain the following theorem:

Theorem 13. *The following spaces are sufficiently rich:*

$S_{\alpha,A}$ $S^{\beta,B}$ *for* $\alpha > 0$, $\beta > 0$ *and any* A, B;
S_0, S^0;
$S_{\alpha,A}^{\beta,B}$ *for* $\alpha > 0$, $\beta > 0$, $\alpha + \beta > 1$ *and any* A, B;
$S_{\alpha,A}^{\beta,B}$ *for* $\alpha > 0$, $\beta > 0$, $\alpha + \beta = 1$ *and* $AB > \gamma$, *and*
$S_{\alpha,A}^0$, $S_0^{\beta,B}$ *for* $\alpha > 1$, $\beta > 1$ *and any* A, B.

We conclude this chapter with a remark concerning more general S spaces. If we replace in the definition of S_α, S^β, S_α^β the sequences $\{k^{k\alpha}\}$, $\{q^{q\beta}\}$ by $\{a_k\}$ and $\{b_q\}$ respectively, then we obtain spaces which we denote by S_{a_k}, S^{b_q} and $S_{a_k}^{b_q}$ respectively. If the a_k, b_q are assumed to satisfy some growth conditions, then theorems similar to those proved for S spaces can be derived. For example, if

$$\frac{a_{k+1}}{a_k} \leqslant Ch^k \qquad (C > 0, h > 0),$$

then polynomials are multipliers in S_{a_k}. The spaces S_{a_k} always admit differentiation. If the a_k, b_q satisfy the conditions of Lemma 2 and if $a_{k+n+1} < a_k A_0^k$, then

$$\widetilde{S_{a_k}^{b_q}} = S_{b_k}^{a_q}.$$

Since these more general S spaces will not be used in the future, we shall not give here any further details.

PROBLEMS

1. Prove that the mapping $\varphi(x) \to \varphi(\lambda x)$ is a bounded mapping from

$$S_{\alpha,A} \to S_{\alpha,A/\lambda}, \qquad S^{\beta,B} \to S^{\beta,\lambda B} \quad \text{and} \quad S_{\alpha,A}^{\beta,B} \to S_{\alpha,A/\lambda}^{\beta,\lambda B}.$$

2. Prove that if $f(z)$ is an entire function satisfying

$$|f(x + iy)| \leq C(1 + |x|)^h \exp [b|y|^\gamma]$$

then it is a multiplier from any $S^{\beta,B}$ into $S^{\beta,B'}$, where

$$(B'e)^\gamma = (b_0 + b)e\gamma \quad \text{and} \quad (Be)^\gamma = b_0 e\gamma.$$

3. Prove Theorem 7, Sec. 3, Chap. 7 for $\Phi = S_{\alpha,A}^{\beta,B}$, $\Phi_1 = S_{\alpha,A}^{\beta,B'}$, where $\alpha = 1 - \beta$, $\beta = 1/p_0$, by solving the system (3.6), (3.7) directly, taking $\varphi(x, t) = Q^*(-i\,(\partial/\partial x), t_0 - t)\varphi_0(x)$ and using Theorem 8.

CHAPTER 10

FURTHER APPLICATIONS TO PARTIAL DIFFERENTIAL EQUATIONS

In this chapter we give further applications to partial differential equations. The results in Sec. 4 are used in Sec. 5, but otherwise each section is essentially independent of the others.

1. A Phragmén-Lindelöf Type Theorem

A classical theorem of Phragmén-Lindelöf for $n = 1$ (which can be deduced from Theorem 1′, Sec. 1, Chap. 5), can be stated as follows: if $u(x, t)$ is a complex-valued function, defined for $-\infty < x < \infty, -\infty < t < \infty$ and satisfying the Cauchy-Riemann equations, i.e.,

$$\frac{\partial u}{\partial t} = i \frac{\partial u}{\partial x}, \tag{1.1}$$

and the boundedness conditions

$$\begin{aligned} |u(x, t)| &\leqslant C \exp [a_0|t|^\gamma + b_0|x|^\gamma] \qquad (\gamma < 1), \\ |u(x, 0)| &\leqslant C(1 + |x|)^h, \end{aligned} \tag{1.2}$$

then $u(x, t)$ is a polynomial in (x, t). This result is false if (1.1) is replaced by the equation

$$\frac{\partial u}{\partial t} = \frac{\partial u}{\partial x},$$

whose general solution is $f(x + t)$, f being an arbitrary function. We shall generalize the result of Phragmén and Lindelöf to solutions of a system

$$\frac{\partial u(x, t)}{\partial t} = P\left(i \frac{\partial}{\partial x}\right) u(x, t), \tag{1.3}$$

where $P(s)$ is an $N \times N$ matrix whose reduced order is p_0.

Theorem 1. *Let* $2p_0 > 1$. *Assume that the eigenvalues of* $P(\sigma)$ *are real for any real vector* σ. *Then if* u *is a classical solution of* (1.3) *for* $x \in R^n$,

$-\infty < t < \infty$, *satisfying the inequalities*

$$(1.4)\qquad |u(x, t)| \leqslant C \exp [a|t|^\gamma + b|x|^\delta] \qquad (0 \leqslant \gamma < 1, 0 \leqslant \delta < (2p_0)'),$$

where $x \in R^n$, $-\infty < t < \infty$ *and* $1/(2p_0)' + 1/(2p_0) = 1$, *and if*

$$(1.5)\qquad |u(x, 0)| \leqslant C(1 + |x|)^h \qquad (h \geqslant 0),$$

where $x \in R^n$, *then*

$$(1.6)\qquad u(x, t) = \sum_{k=0}^{r} U_k(x)t^k \qquad \left(r = 2\left[\frac{h+n}{2}\right] + N + 1\right),$$

where $U_k(x)$ *satisfy the systems*

$$(1.7)\qquad \begin{aligned} P\left(i\frac{\partial}{\partial x}\right) U_{k-1}(x) &= kU_k(x) \qquad (1 \leqslant k < r), \\ P\left(i\frac{\partial}{\partial x}\right) U_r(x) &= 0. \end{aligned}$$

If $2p_0 \leqslant 1$ then we can replace p_0 by $\hat{p} = \frac{1}{2} + \epsilon$, for any $\epsilon > 0$, in the proof given below and thus obtain the assertion of the theorem without making any assumptions on the magnitude of δ in (1.4).

Proof. The outline of the proof is quite simple: we consider u as a generalized function over some W space and prove, for any $\varphi \in W$, that $(u(x, t), \varphi(x))$ is a polynomial in t (use is being made, in this proof, of the classical Phragmén-Lindelöf theorem mentioned above). Taking $\varphi = \varphi_\nu$, where $\{\varphi_\nu\}$ converges to $\delta_{x^0}(x)$, we find that $u(x, t)$ is also a polynomial in t, for any fixed $x = x^0$.

We proceed with the details. Setting

$$(1.8)\qquad u(x, 0) = u_0(x),$$

and using (1.4), it follows from Theorem 8, Sec. 3, Chap. 7, that $u(x, t)$ is a generalized solution (for $-\infty < t < \infty$) over W_q^q of (1.3), (1.8) provided $\delta < q$. We can take $\delta < q < (2p_0)'$. Taking the Fourier transforms of both sides of (1.3), (1.8) and using the continuity of $\mathfrak{F}$ we obtain

$$(1.9)\qquad \begin{aligned} \frac{\partial v(\sigma, t)}{\partial t} &= P(\sigma)v(\sigma, t) \qquad (v = \mathfrak{F}u), \\ v(\sigma, 0) &= v_0(\sigma), \end{aligned}$$

for $-\infty < t < \infty$, $\sigma \in R^n$. $v(\sigma, t)$ belongs to $(W_{q'}^{q'})'$, for each fixed t, where $1/q + 1/q' = 1$. (1.9) is satisfied in the sense of $(W_{q'}^{q'})'$.

Now the elements of the matrix

$$Q(s, t) = \exp [tP(s)]$$

are entire functions of s of order $\leqslant p_0$ and finite type. Since $q < (2p_0)'$,

$q' > 2p_0 > p_0$, and it follows that $Q(s, t)$ is a multiplier in $W_{q'}^{q'}$. By an argument used in proving Theorem 7, Sec. 3, Chap. 7, one easily shows that the adjoint problem to (1.9) has a solution in $W_{q'}^{q'}$ for any initial values in $W_{q'}^{q'}$. Applying Theorem 6, Sec. 3, Chap. 7, with

$$\Phi = \Phi_1 = E = W_{q'}^{q'},$$

it follows that there exists at most one solution in $(W_{q'}^{q'})'$ of (1.9). Similarly to the proof of Theorem 10, Sec. 4, Chap. 7, one shows that $Q(\sigma, t)v_0(\sigma)$ is a solution in $(W_{q'}^{q'})'$ of (1.9). Since $\mathfrak{F}u$ is also a solution of (1.9) over $W_{q'}^{q'}$, we conclude that

$$(\mathfrak{F}u)(\sigma) = Q(\sigma, t)v_0(\sigma).$$

Using finally the definition of Fourier transform, we arrive at the formula

$$(e^{tP(\sigma)}v_0(\sigma), \psi(\sigma)) = (u(x, t), \varphi(x)) \qquad \text{for any } \varphi \in W_q^q, \tag{1.10}$$

where $\varphi = \tilde{\psi}$. This formula will be used very substantially in the sequel.

We introduce the function of t

$$f_\varphi(t) \equiv (u(x, t), \varphi(x)),$$

for any fixed $\varphi \in W_q^q$. Defining

$$U(x) = \frac{u(x, 0)}{(1 + |x|^2)^{m/2}} \qquad \left(\frac{m}{2} = \left[\frac{h + n + 2}{2}\right]\right),$$

we get

$$v_0(\sigma) = \mathfrak{F}[u(x, 0)](\sigma) = (1 - \Delta)^{m/2}V(\sigma),$$

where $V = \mathfrak{F}U$ is a continuous bounded function. We then have

$$\begin{aligned} f_\varphi(t) &= ([\exp[tP(\sigma)]]v_0(\sigma), \psi(\sigma)) = (v_0(\sigma), [\exp[tP^*(\sigma)]]\psi(\sigma)) \\ &= ((1 - \Delta_\sigma)^{m/2}V(\sigma), [\exp[tP^*(\sigma)]]\psi(\sigma)) \\ &= (V(\sigma), (1 - \Delta_\sigma)^{m/2}[[\exp[tP^*(\sigma)]]\psi(\sigma)]) \\ &= \int_{R^n} V(\sigma)\left\{\sum_{k=0}^{m} t^k[\exp[tP^*(\sigma)]]P_k(\sigma)R_k(D_\sigma)\psi(\sigma)\right\}d\sigma, \end{aligned} \tag{1.11}$$

where P_k, R_k are polynomials. Using the right side of (1.11) to extend $f_\varphi(t)$ into the whole complex t-plane we shall prove that $f_\varphi(t)$ is an entire function in t of order < 2.

Use will be made of the obvious inequalities

$$|P_k(\sigma)R_k(D_\sigma)\psi(\sigma)| \leqslant C_1 \exp[-2a_1|\sigma|^{q'}] \qquad (k = 0, 1, \ldots, m) \tag{1.12}$$

and of the inequality

$$\|\exp[tP^*(\sigma)]\| \leqslant C_2(1 + |\sigma|)^{(N-1)p} \exp[c|t|(1 + |\sigma|^{p_0})] \tag{1.13}$$

for complex t, where C_2, c are independent of t. The proof of (1.13) follows by writing

$$\exp[tP^*(s)] = [\exp[t_1P^*(s)]][\exp[it_2P^*(s)]] \qquad \text{where } t = t_1 + it_2,$$

and applying Theorems 3, 4, of Sec. 2, Chap. 7 to each factor.

Using (1.12), (1.13) we obtain, for t complex, $|t| > 1$.

$$|f_\varphi(t)| \leqslant C_3 \sum_{k=0}^{m} |t|^k \int_{R^n} [\exp[c|t|(1 + |\sigma|^{p_0})]][\exp[-a_1|\sigma|^{q'}]]\, d\sigma$$

$$\leqslant C_4 e^{c|t|}|t|^m \int_{R^n} [\exp[c|t|\,|\sigma|^{p_0}]][\exp[-a_1|\sigma|^{q'}]]\, d\sigma.$$

Using the inequality

$$|\alpha\beta| \leqslant \frac{\alpha^\lambda}{\lambda} + \frac{\beta^{\lambda'}}{\lambda'} \qquad \left(\frac{1}{\lambda} + \frac{1}{\lambda'} = 1\right)$$

with $\alpha = c|t|$, $\beta = |\sigma|^{p_0}$ we find that the last integrand is bounded by

$$[\exp[c_1|t|^\lambda + c_2|\sigma|^{p_0\lambda'}]][\exp[-a_1|\sigma|^{q'}]].$$

Taking λ such that, for any given $\epsilon > 0$,

$$q' - \epsilon < p_0\lambda' < q', \tag{1.14}$$

we conclude that

$$|f_\varphi(t)| \leqslant C_5\,|t|^m \exp[c|t| + c_1|t|^\lambda]. \tag{1.15}$$

If $\epsilon < q' - 2p_0$, then (1.14) implies

$$\lambda' > \frac{q' - \epsilon}{p_0} > \frac{2p_0}{p_0} = 2$$

and, consequently, $\lambda < 2$ and $f_\varphi(t)$ is an entire function of order < 2.

Now let t be real. Then

$$|f_\varphi(t)| \leqslant \int |u(x, t)\varphi(x)|\, dx \leqslant C_6 \exp[a|t|^\gamma]. \tag{1.16}$$

If t is pure imaginary, i.e., $t = i\rho$ where ρ is real, then using Theorem 2, Sec. 2, Chap. 7, with $A = iP(\sigma)$, $f(z) = e^{\rho z}$ we obtain

$$\|e^{i\rho P(\sigma)}\| \leqslant [\exp[|\rho|\Lambda'(\sigma)]] \sum_{j=0}^{N-1} 2^j|\rho|^j\|P(\sigma)\|^j,$$

where $\Lambda'(\sigma) = \max|\text{Re}\,[\lambda_k'(\sigma)]|$, $\lambda_k'(\sigma)$ being the eigenvalues of $iP(\sigma)$. Since by assumption of the theorem, the $\lambda_k'(\sigma)$ are pure imaginary numbers, we conclude that

$$\|\exp[i\rho P(\sigma)]\| \leqslant C_7(1 + |\rho|)^{N-1}(1 + |\sigma|^{(N-1)p}).$$

The same inequality holds when $P(\sigma)$ is replaced by $P^*(\sigma)$ since the eigenvalues of P and P^* are the same. Using this inequality in (1.11) we find that

$$|f_\varphi(t)| \leqslant C_8(1 + |t|^{m+N-1}) \qquad \text{if } \operatorname{Re}\{t\} = 0. \tag{1.17}$$

In view of (1.16), (1.17) and the fact that $f_\varphi(t)$ is an entire function of order < 2, we can apply the result of Problem 2, Chap. 5, (for $n = 1$) and thus conclude that

$$|f_\varphi(t)| \leqslant C_9 \exp [a_2|t|^\gamma] \tag{1.18}$$

for all complex values of t, where a_2, C_9 are constants. From (1.18), (1.17) it follows, recalling that $\gamma < 1$ and using the classical result stated in the first paragraph of this section, that $f_\varphi(t)$ is a polynomial in t of degree $\leqslant m + N - 1$.

Observe that all the previous considerations remain true if $u(x, t)$ is replaced by $u(x - x^0, t)$. Indeed, since

$$b|x - x^0|^\gamma \leqslant b_1|x|^\gamma + b_2 \tag{1.19}$$

for some constants b_1, b_2, $u(x - x^0, t)$ satisfies the same assumptions as $u(x, t)$ except that the constants are different. We conclude that

$$f_\varphi(t; x^0) \equiv \int u(x - x^0, t)\varphi(x)\, dx$$

is also a polynomial in t of degree $\leqslant m + N - 1$.

We shall now make a special choice of φ, by taking $\varphi = \varphi_k(x; x^0)$ where the φ_k are to be defined later on. It will be shown that

$$f_{\varphi_k}(t; x^0) \to u(x^0, t) \qquad \text{as } k \to \infty, \tag{1.20}$$

for any fixed t. Since

$$f_{\varphi_k}(t; x^0) = \sum_{j=0}^{m+N-1} a_{kj}t^j,$$

by substituting $t = 0, 1, \ldots, m + N - 1$ and solving for the coefficients a_{kj}, we find, using (1.20), that $a_{kj} \to U_j(x^0)$ as $k \to \infty$ and

$$u(x^0, t) = \sum_{j=0}^{m+N-1} U_j(x^0)t^j.$$

Since this equality is true for any x^0, the proof of (1.6) is completed.

Substituting in (1.6) $t = 0, 1, \ldots, m + N - 1$, we can then solve the $U_j(x)$ in terms of the $u(x, q)$ $(q = 0, 1, \ldots, m + N - 1)$ and find that they belong to the same differentiability class to which the $u(x, q)$ belong. Substituting, finally, (1.6) into (1.3), and comparing the coefficients of t^q on both sides, (1.7) follows. It thus remains to define the $\varphi_k(x; x^0)$ and to prove (1.20).

From Theorem 7, Sec. 1, Chap. 5 it follows that there exists an even real-valued function $g(\sigma) \not\equiv 0$ in $W_{q'}^{q'}$ provided $1 < q' < 2$. The existence of such a function $g(\sigma)$ for $2 < q' < \infty$ follows from (4.17), Chap. 5, since the Fourier transform of an even real function is again an even real function. For $q' = 2$ we can take $g(\sigma) = \exp[-\sigma_1^2 - \cdots - \sigma_n^2]$. The function $\psi_0(\sigma) = g^2(\sigma)$ again belongs to $W_{q'}^{q'}$ and is an even non-negative function. Define

$$\psi_1(\sigma) = \int_{R^n} \psi_0(\sigma - \xi)\psi_0(\xi)\, d\xi.$$

By Problem 6, Chap. 5, $\psi_1(\sigma)$ belongs to $W_{q'}^{q'}$. Set

$$\varphi_0 = \mathfrak{F}\psi_0, \qquad \varphi_1 = \mathfrak{F}\psi_1.$$

Then φ_0 is real-valued and by the classical convolution theorem,

$$\varphi_1(x) = \varphi_0^2(x) \geqslant 0, \qquad \varphi_1(0) = \varphi_0^2(0) = \left[\int \psi_0(\sigma)\, d\sigma\right]^2 > 0.$$

We now define

$$\varphi_k(x; x^0) = \frac{\varphi_1(k(x - x^0))}{\int_{R^n} \varphi_1(k(x - x^0))\, dx} \qquad (k = 2, 3, \ldots).$$

Then, $\varphi_k \in W_q^q$ and $\varphi_k(x; x^0) \geqslant 0$, $\int_{R^n} \varphi_k(x; x^0)\, dx = 1$. Since

$$\int_{R^n} \varphi_1(k(x - x^0))\, dx = \frac{1}{k^n} \int_{R^n} \varphi_1(y)\, dy = \frac{B_1}{k^n},$$

we also have

$$|\varphi_k(x; 0)| \leqslant \frac{k^n}{B_1} \varphi_1(kx) \leqslant B_2 k^n \exp[-a'|kx|^q], \tag{1.21}$$

where a' and the B_j are constants independent of k.

Writing

$$\begin{aligned} f_{\varphi_k}(t; x^0) - u(x^0, t) &= \int_{R^n} [u(x, t) - u(x^0, t)]\varphi_k(x; x^0)\, dx \\ &= \int_{R^n} [u(x, t) - u(x^0, t)]\varphi_k(x - x^0; 0)\, dx \\ &= \int_{R^n} [u(\xi + x^0, t) - u(x^0, t)]\varphi_k(\xi; 0)\, d\xi, \end{aligned}$$

we break the last integral into two integrals, I_1 and I_2. For I_1, $|\xi| \leqslant \beta$ where β depends on any given $\epsilon > 0$ and is such that $|u(\xi + x^0, t) - u(x^0, t)| < \epsilon$ if $|\xi| < \beta$. Then

$$|I_1| \leqslant \epsilon \int_{|\xi| < \beta} \varphi_k(\xi; 0)\, d\xi \leqslant \epsilon.$$

As for I_2, we use (1.21) and (1.4), (1.19) and obtain

$$|I_2| \leqslant \int_{|\xi|>\beta} [|u(\xi + x^0, t)| + |u(x^0, t)|]\varphi_k(\xi; 0)\, d\xi$$
$$\leqslant B_3 \int_{|\xi|>\beta} \{\exp [b_1|\xi|^\delta]\}\, k^n \{\exp [-a'|k\xi|^q]\}\, d\xi$$
$$\leqslant B_4 \int_{|\xi|>k\beta} \{\exp [b_1(\xi/k)^\delta - a'|\xi|^q]\}\, d\xi < \epsilon$$

if k is sufficiently large, since $\delta < q$. Combining the estimates for I_1 and I_2, the proof of (1.20) is completed.

2. A Liouville Type Theorem

In this section we shall prove the following theorem.

Theorem 2. *Let $P(\sigma)$ be a polynomial which does not vanish for real $\sigma \neq 0$. If $u(x)$ is a classical solution in R^n of $P(i(\partial/\partial x))u = 0$ satisfying*

(2.1) $$|u(x)| \leqslant C_\epsilon \exp [(c + \epsilon)|x|^\gamma] \qquad (0 < \gamma < 1)$$

for any $\epsilon > 0$, then $u(x)$ can be extended into an entire function $u(z)$ satisfying, for any $\epsilon > 0$,

(2.2) $$|u(z)| \leqslant C'_\epsilon \exp [(c_1 + \epsilon)|z|^\gamma] \qquad \left(c_1 = \frac{c}{\cos (\pi\gamma/2)}\right).$$

If $u(x) = 0(|x|^h)$ then from Theorem 1′, Sec. 1, Chap. 5 it follows that $u(z) = 0(|z|^{hn})$ and, hence, u is a polynomial of degree $\leqslant h$. This is another proof to Theorem 63, Sec. 15, Chap. 3.

In proving Theorem 2 we shall make use of the explicit form of generalized functions over spaces $S_{0,B}^{\alpha,A}$. We recall that if $\alpha > 1$ then $S_{0,B}^{\alpha,A}$ is a nontrivial space whose elements $\psi(\sigma)$ are functions vanishing outside $|\sigma_j| < B_j$ $(j = 1, \ldots, n)$ and the topology of $S_{0,B}^{\alpha,A}$ is given by the sequence of norms

(2.3) $$\|\psi\|_m = \sup_{\sigma\in R^n} \sup_q \frac{|D^q\psi(\sigma)|}{\left(A + \dfrac{1}{m}\right)^q q^{q\alpha}} \qquad (m = 1, 2, \ldots).$$

We shall prove that the sequence of norms

(2.4) $$\|\psi\|'_m = \left\{\sum_{0\leqslant|q|<\infty} \int_{G_B} \frac{|D^q\psi(\sigma)|^2}{\left[\left(A + \dfrac{1}{m}\right)^q q^{q\alpha}\right]^2} d\sigma\right\}^{1/2} \qquad (m = 1, 2, \ldots)$$

is equivalent to the sequence (2.3). Here,

$$G_B = \{\sigma; |\sigma_j| \leqslant B_j \text{ for } j = 1, \ldots, n\}.$$

Writing

$$\int_{G_B} \frac{|D^q\psi(\sigma)|^2}{\left[\left(A+\frac{1}{m}\right)^q q^{q\alpha}\right]^2} d\sigma = \left(\frac{A+\frac{1}{k}}{A+\frac{1}{m}}\right)^{2q} \int_{G_B} \frac{|D^q\psi(\sigma)|^2}{\left[\left(A+\frac{1}{k}\right)^q q^{q\alpha}\right]^2} d\sigma$$

and observing that the right side is bounded by $\theta^{2q}2^nB^n\|\psi\|_k^2$ where $\theta < 1$ if $k > m$, we obtain

$$(\|\psi\|_m')^2 \leqslant 2^nB^n\|\psi\|_k^2 \sum_q \theta^{2q} = B_1\|\psi\|_k^2 \qquad \text{if } k > m. \tag{2.5}$$

On the other hand,

$$|D^q\psi(\sigma)| \leqslant \int_{-B}^{\sigma} |D^{q+n}\psi(\sigma)|\, d\sigma \leqslant (2^nB^n)^{1/2}\left\{\int_{G_B} |D^{q+n}\psi(\sigma)|^2\, d\sigma\right\}^{1/2}$$

and, therefore,

$$\frac{|D^q\psi(\sigma)|}{\left(A+\frac{1}{k}\right)^q q^{q\alpha}} \leqslant (2^nB^n)^{1/2}\left(\frac{A+\frac{1}{m}}{A+\frac{1}{k}}\right)^q \frac{(q+n)^{(q+n)\alpha}}{q^{q\alpha}}\left(A+\frac{1}{m}\right)^n$$
$$\times\left\{\int_{G_B} \frac{|D^{q+n}\psi(\sigma)|^2}{\left[\left(A+\frac{1}{m}\right)^{q+n}(q+n)^{(q+n)\alpha}\right]^2} d\sigma\right\}^{1/2},$$

from which it follows that

$$\|\psi\|_k \leqslant B_2 \sup_q \{\theta^q(q+n)^{n\alpha}\}\|\psi\|_m' = B_3\|\psi\|_m' \qquad \text{if } m > k. \tag{2.6}$$

The inequalities (2.5), (2.6) show that the sequences (2.3), (2.4) are equivalent and, in particular, the topologies defined by the two sequences are equivalent.

Set $\Psi = S_{0,B}^{\alpha,A}$ and denote by Ψ_m the completion of Ψ with respect to the norm $\|\cdot\|_m'$. Let Ψ^m be the cartesian product $\Psi_1 \times \cdots \times \Psi_m$. Ψ_m and Ψ^m are Hilbert spaces. The elements ψ of Ψ can be identified with the elements $(\psi, \psi, \ldots, \psi)$ of Ψ^m. These elements form a subspace of Ψ^m which will be denoted by Ψ_0^m.

Let g be a continuous linear functional on Ψ. Then it is a continuous linear functional on Ψ provided with the norm $\|\cdot\|_m'$, for some $m \geqslant 1$, i.e., g is a continuous linear functional on the subspace Ψ_0^m of Ψ^m. Extending g into a continuous linear functional on Ψ^m, we then have the representation

$$(g, \psi) = (g, (\psi_1, \ldots, \psi_m)) = \sum_{k=1}^{m} (g_k, \bar{\psi}_k)_k$$

for some $g_k \in \Psi_k$, where

$$(h, \chi)_k = \sum_{0 \leqslant |q| < \infty} \int_{G_B} \frac{D^qh(\sigma)\cdot\overline{D^q\chi(\sigma)}}{\left[\left(A+\frac{1}{k}\right)^q q^{q\alpha}\right]^2} d\sigma \qquad \text{if } h \in \Psi, \chi \in \Psi. \tag{2.7}$$

Hence, if $\psi \in \Psi$,

$$(g, \psi) = \sum_{k=1}^{m} (g_k, \psi)_k. \tag{2.8}$$

In order to evaluate $(g_k, \psi)_k$ we introduce the Hilbert space H of all sequences $w = (w_q(\sigma); 0 \leqslant |q| < \infty)$ where the $w_j(\sigma)$ are measurable functions on G_B, and

$$\|w\| = \left\{ \sum_{0 \leqslant |q| < \infty} \int_{G_B} |w_q(\sigma)|^2 \, d\sigma \right\}^{1/2},$$

$$(w^{(1)}, w^{(2)}) = \sum_{0 \leqslant |q| < \infty} \int_{G_B} w_q^{(1)}(\sigma) \overline{w_q^{(2)}(\sigma)} \, d\sigma.$$

The space Ψ_k may be considered as a subspace of H via the correspondence

$$w \leftrightarrow \left\{ \frac{D^q \psi(\sigma)}{\left(A + \frac{1}{k}\right)^q q^{q\alpha}}; 0 \leqslant |q| < \infty \right\} \qquad \text{for } \psi \in \Psi.$$

The g_k can then be identified with a sequence $\{g_{kq}\}$ in H and (2.8) can be rewritten in the form

$$(g, \psi) = \sum_{k=1}^{m} \sum_{0 \leqslant |q| < \infty} \int_{G_B} \frac{g_{kq}(\sigma) \, D^q \psi(\sigma)}{\left(A + \frac{1}{k}\right)^q q^{q\alpha}} \, d\sigma,$$

where
$$\sum_{0 \leqslant |q| < \infty} \int_{G_B} |g_{kq}(\sigma)|^2 \, d\sigma < \infty.$$

Defining

$$g_q(\sigma) = \sum_{k=1}^{m} \left(\frac{A + \frac{1}{m}}{A + \frac{1}{k}} \right)^q g_{kq}(\sigma)$$

and noting that

$$\sum_{0 \leqslant |q| < \infty} \int_{G_B} |g_q(\sigma)|^2 \, d\sigma \leqslant m \sum_{0 \leqslant |q| < \infty} \sum_{k=1}^{m} \int_{G_B} |g_{kq}(\sigma)|^2 \, d\sigma < \infty$$

we obtain the following lemma.

Lemma 1. *Every continuous linear functional g over $S_{0,B}^{\alpha,A}$ $(\alpha > 1)$ can be represented in the form*

$$(g, \psi) = \sum_{0 \leqslant |q| < \infty} \int_{G_B} g_q(\sigma) \frac{D^q \psi(\sigma)}{\left(A + \frac{1}{m}\right)^q q^{q\alpha}} \, d\sigma, \tag{2.9}$$

where the $g_q(\sigma)$ belong to $L^2(G_B)$ and

$$\sum_{0 \leqslant |q| < \infty} \int_{G_B} |g_q(\sigma)|^2 \, d\sigma < \infty. \tag{2.10}$$

Conversely, every functional of the form (2.9) *is a continuous linear functional on* $S_{0,B}^{\alpha,A}$ *provided* (2.10) *holds.*

The proof of the second part of the lemma is obvious.

Proof of Theorem 2. Set $\alpha = 1/\gamma$, $a = \alpha/(eA^{1/\alpha})$ and choose A such that $a > c$. Since the functions $\varphi(x)$ of $S_{\alpha,A}$ satisfy

$$|x^k D^q\varphi(x)| \leqslant C_{q\delta} \exp\left[-(a-\delta)|x|^{1/\alpha}\right] \qquad \text{(for any } \delta > 0),$$

it follows that

$$(u, \varphi) = \int_{R^n} u(x)\varphi(x)\, dx$$

is a continuous linear functional on $S_{\alpha,A}$. Hence, $v = \mathfrak{F}u$ belongs to $(S^{\alpha,A})'$ and

$$P(\sigma)v(\sigma) = 0 \qquad \text{in } (S^{\alpha,A})'. \tag{2.11}$$

Let $\psi \in S^{\alpha,A}$ be such that $\psi = 0$ in some neighborhood V of 0 and ψ has a compact support N_0. Since $P(\sigma) \neq 0$ outside V, $1/P(\sigma)$ is an analytic function and, therefore,

$$\left|D^q\left(\frac{1}{P(\sigma)}\right)\right| \leqslant C_1C_2^q q^q \qquad \text{if } \sigma \in N_0,\ \sigma \notin V.$$

It is then easily verified that $\psi_0(\sigma) \equiv \psi(\sigma)/P(\sigma)$ belongs to $S^{\alpha,A}$ and we have

$$(v, \psi) = (v, P\psi_0) = (Pv, \psi_0) = 0. \tag{2.12}$$

Let $1 < \alpha_0 < \alpha$ and take $h(\sigma)$ to be a function in S^{α_0} which is equal to 1 in some neighborhood V of $\sigma = 0$ and which has a compact support. $h(\sigma)$ can be taken to be $\chi_0(\sigma/\mu)$ where χ_0 is defined in (7.16), Chap. 6 (with $\delta = \alpha_0$) and μ depends on V. If $\psi \in S^{\alpha,A}$, then the function $h(\sigma/m)\psi(\sigma)$ belongs to $S^{\alpha,A}$, and if ψ further vanishes in some neighborhood of 0, then, by (2.12),

$$\left(v, h\left(\frac{\sigma}{m}\right)\psi\right) = 0.$$

If we prove that, as $m \to \infty$,

$$h\left(\frac{\sigma}{m}\right)\psi(\sigma) \to \psi(\sigma) \qquad \text{in } S^{\alpha,A}, \tag{2.13}$$

then it follows that

$$(v, \psi) = 0 \qquad \text{if } \psi \in S^{\alpha,A} \text{ and } \psi = 0 \text{ in a neighborhood of } 0. \tag{2.14}$$

The proof of (2.13) is obtained by noting that $h(\sigma/m) \to \psi$ properly and that the sequence $\{h(\sigma/m);\ m = 1, 2, \ldots\}$ is bounded in each norm

$\|\cdot\|_{kp}$ of $S^{\alpha,A}$, and then applying step (f) of the proof of Theorem 1, Sec. 1, Chap. 9.

We choose B such that

$$a_0 \equiv \frac{\alpha}{e(A+B)^{1/\alpha}} > c. \tag{2.15}$$

Then $u \in (S_{\alpha,A+B})'$ and, consequently, $v \in (S^{\alpha,A+B})'$. The proof of (2.14) now yields the following:

$$(v, \psi) = 0 \quad \text{if } \psi \in S^{\alpha,A+B} \text{ and } \psi = 0 \text{ in some neighborhood of } 0. \tag{2.16}$$

We next choose m sufficiently large so that $h(m\sigma)$ vanishes outside G_B, and set $h_B(\sigma) = h(m\sigma)$. m will be fixed from now on. Since $h_B \in S_{0,B}^{\alpha,B}$, for any $\psi \in S^{\alpha,A}$ $h_B\psi$ belongs to $S_{0,B}^{\alpha,A+B}$ and $(1 - h_B)\psi$ belongs to $S^{\alpha,A+B}$ and vanishes in some neighboohood of 0. Using (2.16) we obtain

$$(v, h_B\psi) = (v, \psi) - (v, (1 - h_B)\psi) = (v, \psi). \tag{2.17}$$

By Lemma 1 we have (considering v as as an element of $(S_{0,B}^{\alpha,A+B})'$):

$$\begin{aligned}(v, \psi) = (v, h_B\psi) &= \sum_q \int_{G_B} g_q(\sigma) \frac{D^q(h_B\psi)}{\left(A + B + \dfrac{1}{m}\right)^q q^{q\alpha}}\, d\sigma \\ &= \sum_q \left(\frac{D^q v_{qB}}{\left(A + B + \dfrac{1}{m}\right)^q q^{q\alpha}}, h_B\psi\right) = \left(h_B \sum_q \frac{D^q v_{qB}}{\left(A + B + \dfrac{1}{m}\right)^q q^{q\alpha}}, \psi\right)\end{aligned} \tag{2.18}$$

where $v_{qB} = (-1)^{|q|} g_q$ and $\sum_q \int_{G_B} |v_{qB}(\sigma)|^2\, d\sigma < \infty$. Equation (2.18) can be rewritten in the form

$$v(\sigma) = h_B(\sigma) \sum_q \frac{D^q v_{qB}(\sigma)}{\left(A + B + \dfrac{1}{m}\right)^q q^{q\alpha}} \qquad \text{over } S^{\alpha,A}. \tag{2.19}$$

We may assume that m is such that

$$a_1 \equiv \frac{\alpha}{e\left(A + B + \dfrac{1}{m}\right)^{1/\alpha}} > c$$

since, by (2.15), $a_1 > c$ if m is sufficiently large.

We now take the inverse Fourier transforms of both sides of (2.19) and obtain, formally,

$$u(x) = k_B(x) * \sum_q (ix)^q \frac{u_{qB}(x)}{\left(A + B + \dfrac{1}{m}\right)^q q^{q\alpha}} \equiv k_B(x) * \Gamma(x), \tag{2.20}$$

where $\tilde{k}_B = h_B$, $\tilde{u}_{qB} = v_{qB}$, and

$$\sum_q \int_{R^n} |u_{qB}(x)|^2 \, dx = \frac{1}{(2\pi)^n} \sum_q \int_{G_B} |v_{qB}(\sigma)|^2 \, d\sigma < \infty.$$

In order to justify (2.20) we apply Theorem 9, Sec. 4, Chap. 4, to partial sums, noting that h_B is a multiplier in $S^{\alpha,A}$, and then pass to the limit.

We proceed to estimate $\Gamma(x)$. Since $u_{qB} = \mathfrak{F}^{-1}(v_{qB})$,

$$|u_{qB}(x)| \leqslant \frac{1}{(2\pi)^n} \int_{G_B} |v_{qB}(\sigma)| \, d\sigma \leqslant C_3 \left\{ \int_{G_B} |v_{qB}(\sigma)|^2 \, d\sigma \right\}^{1/2}. \tag{2.21}$$

By (1.19), Chap. 9, with

$$1 - r = 2\alpha, \qquad \eta = \frac{x_j^2}{e\left(A_j + B_j + \dfrac{1}{m}\right)^2},$$

we get

$$\sum_q \frac{|x^{2q}|}{\left(A + B + \dfrac{1}{m}\right)^{2q} q^{2q\alpha}} \leqslant C_4(\epsilon) \exp\left[2(a_1 + \epsilon)|x|^{1/\alpha}\right] \tag{2.22}$$

for any $\epsilon > 0$. Combining (2.21), (2.22) we obtain

$$\begin{aligned} |\Gamma(x)| &= \left| \sum_q (ix)^q \frac{u_{qB}(x)}{\left(A + B + \dfrac{1}{m}\right)^q q^{q\alpha}} \right| \\ &\leqslant \left\{ \sum_q \frac{|x^{2q}|}{\left(A + B + \dfrac{1}{m}\right)^{2q} q^{2q\alpha}} \right\}^{1/2} \{\sum_q |u_{qB}(x)|^2\}^{1/2} \\ &\leqslant C_5(\epsilon) \exp\left[(a_1 + \epsilon)|x|^{1/\alpha}\right]. \end{aligned} \tag{2.23}$$

Since $k_B(x)$ belongs to $S^{0,B}_{\alpha,B}$, it can be extended into an entire function $k_B(z)$ satisfying

$$|k_B(x + iy)| \leqslant C_6(\epsilon) \left[\exp\left[-(b - \epsilon)|x|^{1/\alpha}\right]\right] \left[\exp\left[(B + \epsilon)|y|\right]\right] \qquad \left(b = \frac{\alpha}{eB^{1/\alpha}}\right), \tag{2.24}$$

for any $\epsilon > 0$.

We now note that even though the convolution in (2.20) is taken in the abstract sense, it exists also as a classical convolution, in view of the inequalities (2.23) and (2.24) for $y = 0$. Furthermore, the abstract convolution is equal to the classical convolution, as one can verify by means of Lemma 3, Sec. 5, Chap. 7. If we define

$$u(x + iy) = \int_{R^n} k_B(x + iy - \xi)\Gamma(\xi) \, d\xi,$$

then this function coincides with $u(x)$ when $y = 0$.

Using (2.23), (2.24) and the inequality

$$(a_1 + \epsilon)|\xi|^{1/\alpha} - (b - \epsilon)|x - \xi|^{1/\alpha} \leqslant -\epsilon|\xi|^{1/\alpha} + (a_1 + 3\epsilon)|x|^{1/\alpha}$$

which holds if b is sufficiently large (i.e., if B is sufficiently small, which we may assume to be the case), we obtain

$$(2.25) \qquad |u(z)| \leqslant C_7(\epsilon)\,[\exp\,[(a_1 + 3\epsilon)|x|^{1/\alpha}]]\,e^{(B+\epsilon)|y|}.$$

Since $k_B(z)$ is an entire function of z, $u(z)$ is also an entire function in z, and the last inequality shows that $u(z)$ is of order $\leqslant 1$ and finite type $\leqslant B$. Furthermore, on the real axis,

$$(2.26) \qquad |u(x)| \leqslant C_7(\epsilon)\,\exp\,[(a_1 + 3\epsilon)|x|^{1/\alpha}].$$

We now remark that Theorem 1, Sec. 1, Chap. 5, and the result in Problem 2, Chap. 5, remain true if $\varphi_j = \pi/p_j$ provided $f(z)$ is of order $\leqslant p$ and minimal type, i.e., provided the b in (1.1), Chap. 5, can be taken arbitrarily small (C will depend on b). The proofs are very similar to the proofs in the case where $\varphi_j < \pi/p_j$. Since (2.26) holds and since $u(z)$ is of order $\leqslant 1$ and type $\leqslant B$, where B can be taken arbitrarily small, it follows by the modified version of Problem 2, Chap. 5, that (2.2) holds.

It will be shown at the end of Sec. 2, Chap. 11, that Theorem 2 is not true if $\gamma = 1$.

3. Fundamental Solutions of Equations with Constant Coefficients

We recall that a fundamental solution of a differential equation

$$(3.1) \qquad P\left(i\frac{\partial}{\partial x}\right)u = 0$$

is a distribution G satisfying in $K' \equiv (D')$ the equation

$$(3.2) \qquad P\left(i\frac{\partial}{\partial x}\right)G(x) = \delta(x).$$

Theorem 3. *For every differential equation with constant coefficients there exists a fundamental solution.*

Proof. (3.2) is equivalent to

$$(3.3) \qquad P(\sigma)E = 1 \qquad \text{in } Z'.$$

Consider first the case that

$$(3.4) \qquad P(s) = as_1^m + \sum_{k=0}^{m-1} P_k(s')s_1^k \qquad (a \neq 0),$$

where $s' = (s_2, \ldots, s_n)$. We shall try to find E in the form

$$(E(\sigma), \psi(\sigma)) = \int_T \frac{\psi(\sigma_1 + i\tau_1, \sigma')}{P(\sigma_1 + i\tau_1, \sigma')} \, d\sigma, \tag{3.5}$$

where T is a manifold to be defined below. The motivation for (3.5) is that if we take $T = R^n$, $\tau_1 = 0$, then, formally,

$$(P(\sigma)E(\sigma), \psi(\sigma)) = (E(\sigma), P(\sigma)\psi(\sigma)) = \int \frac{P(\sigma)\psi(\sigma)}{P(\sigma)} \, d\sigma \tag{3.6}$$
$$= \int \psi(\sigma) \, d\sigma$$

and (3.3) is satisfied.

For every fixed σ', the equation $P(s_1, \sigma') = 0$ has m roots $s_1^{(j)}$ ($j = 1, \ldots, n$) and

$$P(s_1, \sigma') = a(s_1 - s_1^{(1)}) \cdots (s_1 - s_1^{(m)}).$$

For at least one value $\tau_1 = k$ ($k = 0, 1, \ldots, m + 1$) all the imaginary parts of the $s_1^{(j)}$ differ from τ_1, in absolute value, by at least 1. Therefore

$$|P(s_1, \sigma')| \geqslant a \qquad \text{for some } \tau_1 = k \qquad (k = 0, 1, \ldots, m + 1), \tag{3.7}$$

for all values of σ_1. If σ' varies in a sufficiently small $(n - 1)$-dimensional rectangle Δ, then the corresponding roots $s_1^{(j)}$ still satisfy the inequality $|k - \text{Im}\,\{s_1^{(j)}\}| > 1$. Hence, (3.7) continues to hold for $\sigma' \in \Delta$, $-\infty < \sigma_1 < \infty$, $\tau_1 = k$. By the Heine-Borel theorem we can cover the σ' space by a sequence of rectangles Δ_j such that if

$$\sigma' \in \Delta_j, \ -\infty < \sigma_1 < \infty, \ \tau_1 = k_j \qquad (0 \leqslant k_j < m + 1),$$

then the inequality (3.7) holds. Defining

$$\Gamma_1 = \Delta_1, \quad \Gamma_2 = \Delta_2 - \Delta_1 \cap \Delta_2, \quad \Gamma_3 = \Delta_3 - \Delta_3 \cap (\Delta_1 \cup \Delta_2), \ \ldots,$$

we take T to be the manifold

$$\{\sigma' \in \Gamma_j, \ -\infty < \sigma_1 < \infty, \ \tau_1 = k_j; \ j = 1, 2, \ldots\}$$

and define E by (3.5).

We proceed to prove (3.3). Since every $\psi \in Z$ decreases on $s = \sigma$ faster than any power of $|\sigma|^{-1}$, provided τ remains in a bounded set, the integral in (3.5) is convergent and, in fact, one easily sees that E is a bounded linear functional on Z. Since $P(\sigma)$ is a multiplier in Z,

$$
\begin{aligned}
(P(\sigma)E(\sigma), \psi(\sigma)) &= (E(\sigma), P(\sigma)\psi(\sigma)) \\
&= \int_T P(\sigma_1 + i\tau_1, \sigma') \frac{\psi(\sigma_1 + i\tau_1, \sigma')}{P(\sigma_1 + i\tau_1, \sigma')}\, d\sigma \\
&= \int_T \psi(\sigma_1 + i\tau_1, \sigma')\, d\sigma \\
&= \sum_j \int_{\Gamma_j} \left[\int_{-\infty}^{\infty} \psi(\sigma_1 + ik_j, \sigma')\, d\sigma_1\right] d\sigma' \\
&= \sum_j \int_{\Gamma_j} \left[\int_{-\infty}^{\infty} \psi(\sigma_1, \sigma')\, d\sigma_1\right] d\sigma' = \int_{R^n} \psi(\sigma)\, d\sigma,
\end{aligned}
$$

which proves (3.3). The change of the domain of integration is permissible since $\psi(\sigma_1 + i\tau_1, \sigma') = 0(|\sigma_1|^{-\alpha})$ for any $\alpha > 0$ and fixed σ'.

Now let $P(s)$ be an arbitrary polynomial. We claim that there exists a linear transformation

$$
\sigma_j = \sum_{k=1}^{n} c_{jk}\eta_k \qquad (\det (c_{jk}) \neq 0) \tag{3.8}
$$

such that $\hat{P}(\eta) \equiv P(\sigma)$ is of the form (3.4), i.e., $\hat{P}(\eta) = \hat{a}\eta_1^m +$ lower powers in η_1, and $\hat{a} \neq 0$. Indeed, if $P_0(\sigma)$ is the principal part of $P(\sigma)$, then the coefficient $\hat{a}$ is equal to $P_0(c_1)$ where $c_1 = (c_{11}, \ldots, c_{n1})$. We therefore take c_1 to be any vector which does not lie on the manifold $P_0(\eta) = 0$ and then choose the other c_{jk} such that $\det (c_{jk}) \neq 0$.

By what we have already proven, there exists an $\hat{E}(\eta)$ in K' satisfying

$$
\hat{P}(\eta)\hat{E}(\eta) = 1 \qquad \text{in } Z'.
$$

Consider the functional $E(\sigma) = \hat{E}(C^{-1}\sigma)$, where C^{-1} is the inverse matrix of $C = (c_{jk})$. It should be mentioned that for any generalized function $f(x)$, $f(Mx)$ (M a matrix) is defined by

$$
(f(Mx), \varphi(Mx)) = \frac{1}{\det M} (f(x), \varphi(x)).
$$

We then have

$$
\begin{aligned}
(P(\sigma)E(\sigma), \psi(\sigma)) &= (\hat{P}(C^{-1}\sigma)\hat{E}(C^{-1}\sigma), \psi(\sigma)) \\
&= \frac{1}{\det (C^{-1})} (\hat{P}(\sigma)\hat{E}(\sigma), \psi(C\sigma)) \\
&= \frac{1}{\det (C^{-1})} (1, \psi(C\sigma)) = (1, \psi(\sigma))
\end{aligned}
$$

and the proof is completed.

Having proved that every differential equation with constant coefficients has a fundamental solution (in K'), the next problem is to prove, under appropriate conditions on $P(s)$, that G satisfies some differentiability and growth properties. The problem of differentiability will be treated in detail in the next chapter.

That G need not be a function is seen from the example

$$\frac{\partial^k u}{\partial x_1 \cdots \partial x_k} = 0, \qquad G(x) = Y_{x_1} \otimes \cdots \otimes Y_{x_k} \otimes \delta_{x_{k+1}} \otimes \cdots \otimes \delta_{x_n}$$

(see Problem 8, Chap. 3). Here $G(x)$ is not a function in any neighborhood of any portion of a hyperplane $x_m = 0$ $(m > k)$.

We state some results for elliptic equations (for proofs, see References 39 and 40). Let $P_0(i\,(\partial/\partial x))$ be the principal part of $P(i\,(\partial/\partial x))$. If $P_0(\xi) \neq 0$ for any real vector $\xi \neq 0$, then $P(i\,(\partial/\partial x))$ is called an *elliptic operator* and the equation (3.1) is said to be of *elliptic type*. Assume that the coefficients of P are real constants (m is then necessarily an even number). The fundamental solution $K(x)$ of (3.1) then has the form

$$K(x) = \begin{cases} A\left(\frac{x}{r}, r\right) r^{m-n} & \text{for } n \text{ odd}, \\ B\left(\frac{x}{r}, r\right) r^{m-n} + C\left(\frac{x}{r}, r\right) r^{m-n} \log r & \text{for } n \text{ even}, \end{cases} \tag{3.9}$$

where $A(\xi, r)$, $B(\xi, r)$, $C(\xi, r)$ are analytic functions in (ξ, r) in a neighborhood of $|\xi| = 1$, $r = 0$, and $C(x/r, r)r^{m-n}$ is a function $C(x)$ which is analytic in x at $x = 0$. If the elliptic operator is also homogeneous (i.e., it coincides with its principal part) then

$$K(x) = \begin{cases} A\left(\frac{x}{r}\right) r^{m-n} & \text{for } n \text{ odd or for } m < n, \\ B\left(\frac{x}{r}\right) r^{m-n} + C(x) \log r & \text{for } n \text{ even}, m \geqslant n, \end{cases} \tag{3.9'}$$

where $A(\xi)$, $B(\xi)$ are analytic functions on $|\xi| = 1$ and $C(x)$ is a polynomial in x of degree $m - n$.

Consider now the Cauchy problem

$$P\left(\frac{\partial}{\partial t}, \frac{\partial}{\partial x}\right) u(x, t) = 0, \tag{3.10}$$

$$\frac{\partial^\nu u(x, 0)}{\partial t^\nu} = u_\nu(x) \qquad (\nu = 0, 1, \ldots, m - 1), \tag{3.11}$$

where $P(\lambda, s)$ is a polynomial in λ of degree m. A *fundamental solution of the Cauchy problem* (3.10), (3.11) is a family of distributions $G_j(x, t)$ (j fixed, $0 \leqslant j \leqslant m - 1$) depending on a parameter t $(-\infty < t < \infty)$

and satisfying

$$(3.12)\qquad \begin{aligned} P\left(\frac{\partial}{\partial t}, \frac{\partial}{\partial x}\right) G_j(x, t) &= 0, \\ \frac{\partial^\nu G_j(x, 0)}{\partial t^\nu} &= \delta_{\nu j}\delta(x) \qquad (\nu = 0, 1, \ldots, m-1) \end{aligned}$$

in the sense of K'. One can formally express the solution of (3.10), (3.11) in terms of the G_j, i.e.,

$$u(x, t) = \sum_{j=0}^{m-1} G_j(x, t) * u_j(x).$$

Writing (3.10) in a matrix form

$$(3.13)\qquad \frac{\partial U(x, t)}{\partial t} = \hat{P}\left(i\frac{\partial}{\partial x}\right) U \qquad \left(U = \left(u, \frac{\partial u}{\partial t}, \ldots, \frac{\partial^{m-1} u}{\partial t^{m-1}}\right)\right),$$

the Green matrix of (3.13) (see Chap. 7, Sec. 4) is the diagonal matrix $(\delta_{kj}G_j)$.

There is a formal connection between the fundamental solutions $K(x, t)$ of (3.10) and the fundamental solutions $G_j(x, t)$ of the Cauchy problem (3.10), (3.11). Thus, if $m = 1$ and the coefficient of $\partial/\partial t$ in P is 1, then

$$K(x, t) = \begin{cases} 0 & \text{if } t < 0, \\ G_0(x, t) & \text{if } t \geqslant 0 \end{cases}$$

is a fundamental solution (3.10).

In Sec. 5 we shall construct in a closed form the fundamental solution for the Cauchy problem for hyperbolic equations. The results of Sec. 4 will be used.

4. Special Distributions and Radon's Problem

Consider the distribution x_+^λ defined by

$$(4.1)\qquad (x_+^\lambda, \varphi) = \int_0^\infty x^\lambda[\varphi(x) - \varphi(0) - x\varphi'(0) - \cdots - \frac{x^{m-1}}{(m-1)!}\varphi^{(m-1)}(0)]\,dx,$$

where $\varphi \in (D)$, x is one-dimensional and $-m \geqslant \operatorname{Re}\lambda > -m-1$, $\lambda \neq -m$. If $\operatorname{Re}\lambda > -1$ then x_+^λ coincides with the distribution of the function

$$x_+^\lambda = \begin{cases} x^\lambda & \text{if } x > 0, \\ 0 & \text{if } x \leqslant 0. \end{cases}$$

For fixed φ, the function (x_+^λ, φ) $(\text{Re}\,\lambda > -1)$ can be continued analytically into the whole complex plane as a meromorphic function whose poles occur at the points $\lambda = -1, -2, \ldots$. Indeed, if $\text{Re}\,\lambda > -1$ and m is any positive integer, then

$$(4.2)\qquad \int_0^\infty x^\lambda \varphi(x)\,dx = \int_0^1 x^\lambda \Big[\varphi(x) - \varphi(0) - x\varphi'(0) - \cdots - \frac{x^{m-1}}{(m-1)!}\varphi^{(m-1)}(0)\Big]\,dx + \int_1^\infty x^\lambda \varphi(x)\,dx + \sum_{k=1}^{m} \frac{\varphi^{(k-1)}(0)}{(k-1)!(\lambda+k)}$$

and the right side gives the desired analytic continuation in $\text{Re}\,\lambda > -m-1$.

If $-m \geqslant \text{Re}\,\lambda > -m-1$, $\lambda \neq -m$, then the right side of (4.2) is equal to the right side of (4.1). Hence, for any $\lambda \neq -m$ $(m = 1, 2, \ldots)$, (x_+^λ, φ) is obtained from the (x_+^μ, φ), $\text{Re}\,\mu > -1$, by analytic continuation.

One easily proves directly that

$$(4.3)\qquad \frac{dx_+^\lambda}{dx} = \lambda x_+^{\lambda-1} \qquad (\lambda \neq -1, -2, \ldots).$$

Another way to prove (4.3) is by noting that

$$\left(x_+^\lambda, \frac{d\varphi}{dx}\right) + \lambda(x_+^{\lambda-1}, \varphi)$$

is a meromorphic function of λ which vanishes for $\text{Re}\,\lambda > -1$; hence it vanishes for all $\lambda \neq -1, -2, \ldots$, and (4.3) follows.

We introduce the gamma function $\Gamma(\lambda)$ which is defined for $\text{Re}\,\lambda > 0$ by

$$\Gamma(\lambda) = \int_0^\infty x^{\lambda-1} e^{-x}\,dx.$$

For $\text{Re}\,\lambda \leqslant 0$, it is defined by analytic continuation. Applying the previous results we see that $\Gamma(\lambda) = (x_+^{\lambda-1}, e^{-x})$ and that

$$(4.4)\qquad \Gamma(\lambda) = \int_0^\infty x^{\lambda-1}\Big[e^{-x} - \sum_{k=0}^{m} (-1)^k \frac{x^k}{k!}\Big]\,dx \qquad \text{if } -m-1 < \text{Re}\,\lambda \leqslant -m,\ \lambda \neq -m,$$

$$(4.5)\qquad \Gamma(\lambda) = \int_0^1 x^{\lambda-1}\Big[e^{-x} - \sum_{k=0}^{m} (-1)^k \frac{x^k}{k!}\Big]\,dx + \int_1^\infty x^{\lambda-1}e^{-x}\,dx + \sum_{k=0}^{m} \frac{(-1)^k}{k!(k+\lambda)} \qquad \text{if } \text{Re}\,\lambda > -m-1.$$

From (4.5) it follows that $\Gamma(\lambda)$ is a meromorphic function whose singularities are simple poles located at the points $\lambda = -m$ $(m = 0, 1, 2, \ldots)$, the residues being $(-1)^m/m!$. Using the relation $\Gamma(\lambda+1) = \lambda\Gamma(\lambda)$, (4.3) and the well-known fact that $\Gamma(\lambda) \neq 0$ for any λ, we see that the distribution

$$f_+^\lambda = \frac{x_+^\lambda}{\Gamma(\lambda + 1)} \tag{4.6}$$

satisfies

$$\frac{d}{dx} f_+^\lambda = f_+^{\lambda-1} \qquad \text{if } \lambda \neq -1, -2, \ldots . \tag{4.7}$$

Since both (x_+^λ, φ) and $\Gamma(\lambda)$ have simple poles at $\lambda = -m$ $(m = 1, 2, \ldots)$, it follows that (f_+^λ, φ) remains bounded as $\lambda \to -m$ and hence its singularity at $\lambda = -m$ is removable. (f_+^λ, φ) is therefore an entire function of λ.

(x_-^λ, φ) for $\operatorname{Re} \lambda \leqslant -1$ is defined as the analytic continuation of (x_-^λ, φ) for $\operatorname{Re} \lambda > -1$, where x_-^λ is the function

$$x_-^\lambda = \begin{cases} 0 & \text{if } x \geqslant 0, \\ |x|^\lambda & \text{if } x < 0. \end{cases}$$

One easily verifies either directly or by analytic continuation that

$$(x_-^\lambda, \varphi(x)) = (x_+^\lambda, \varphi(-x)). \tag{4.8}$$

Furthermore, if $-m - 1 < \operatorname{Re} \lambda \leqslant -m, \lambda \neq -m$,

$$(x_-^\lambda, \varphi) = \int_0^\infty x^\lambda[\varphi(-x) - \varphi(0) + x\varphi'(0) - \cdots - (-1)^{m-1} \frac{x^{m-1}}{(m-1)!} \varphi^{(m-1)}(0)]\, dx. \tag{4.9}$$

The poles of (x_-^λ, φ) are located at the points $-1, -2, \ldots$, and they are simple poles.

We define

$$|x|^\lambda = x_+^\lambda + x_-^\lambda, \tag{4.10}$$

$$|x|^\lambda \operatorname{sgn} x = x_+^\lambda - x_-^\lambda. \tag{4.11}$$

The reader will easily verify that the residue of (x_+^λ, φ) at $\lambda = -m$ is $\varphi^{(m-1)}(0)/(m - 1)!$ and the residue of (x_-^λ, φ) at $\lambda = -m$ is $(-1)^{m-1}\varphi^{(m-1)}(0)/(m-1)!$. Hence $(|x|^\lambda, \varphi)$ has poles (which are simple) only at the points $\lambda = -2m + 1$ $(m = 1, 2, \ldots)$, with residue $2\varphi^{(2m)}(0)/(2m)!$. Analogously, $(|x|^\lambda \operatorname{sgn} x, \varphi)$ has poles (which are simple) only at the points $\lambda = -2m$ $(m = 1, 2, \ldots)$, with residue $2\varphi^{(2m-1)}(0)/(2m - 1)!$.

For $\lambda \to -2m$ we set $\lim |x|^\lambda = x^{-2m}$ and for $\lambda \to -2m - 1$ we set $\lim |x|^\lambda \operatorname{sgn} x = x^{-2m-1}$. Using (4.1), (4.9) and taking $\lambda \to -2m$,

$\lambda \to -2m - 1$ we obtain

$$(4.12) \qquad (x^{-2m}, \varphi) = \int_0^\infty x^{-2m} \Bigg\{ \varphi(x) + \varphi(-x) - 2\left[\varphi(0) + \frac{x^2}{2!}\varphi''(0) + \cdots + \frac{x^{2m-2}}{(2m-2)!}\varphi^{(2m-2)}(0)\right]\Bigg\}\, dx,$$

$$(4.13) \qquad (x^{-2m-1}, \varphi) = \int_0^\infty x^{-2m-1} \Bigg\{ \varphi(x) - \varphi(-x) - 2\left[x\varphi'(0) + \frac{x^3}{3!}\varphi^{(3)}(0) + \cdots + \frac{x^{2m-1}}{(2m-1)!}\varphi^{(2m-1)}(0)\right]\Bigg\}\, dx.$$

We denote a primitive of a distribution $T(x)$ by $\int T(x)\,dx$. Then,

$$\int x_+^\lambda \, dx = \frac{x_+^{\lambda+1}}{\lambda+1} + C_1(\lambda) \qquad \text{if } \lambda \neq -1, -2, \ldots,$$

$$\int x_-^\lambda \, dx = -\frac{x_-^{\lambda+1}}{\lambda+1} + C_2(\lambda) \qquad \text{if } \lambda \neq -1, -2, \ldots,$$

$$\int |x|^\lambda \, dx = \frac{|x|^{\lambda+1}\operatorname{sgn} x}{\lambda+1} + C_3(\lambda) \qquad \text{if } \lambda \neq -1, -3, \ldots,$$

$$\int |x|^\lambda \operatorname{sgn} x \, dx = \frac{|x|^{\lambda+1}}{\lambda+1} + C_4(\lambda) \qquad \text{if } \lambda \neq -1;\ \lambda \neq -2, -4, \ldots,$$

where the $C_j(\lambda)$ are constants. By iteration we also get

$$(4.14) \qquad \int \cdots \int |x|^\lambda \, dx^q = \frac{|x|^{\lambda+q}(\operatorname{sgn} x)^q}{(\lambda+1)\cdots(\lambda+q)} + Q_\lambda(x),$$

where $Q_\lambda(x)$ is a polynomial in x of degree $< q$. The integrand has poles at $\lambda = -1, -3, \ldots$. The first term on the right side of (4.14) has additional poles at $\lambda = -2k$, $k = 1, 2, \ldots, [q/2]$, which are simple poles with residue

$$-\frac{x^{q-2k}}{(2k-1)!(q-2k)!}.$$

Taking

$$(4.15) \qquad Q_\lambda(x) = \sum_{k=1}^{[q/2]} \frac{x^{q-2k}}{(2k-1)!(q-2k)!}\,\frac{1}{\lambda+2k}$$

we then eliminate the poles at $\lambda = -2k$, $k = 1, 2, \ldots, [q/2]$, on the right side of (4.14).

The distribution $x_+^\lambda \log^k x_+$ (k a positive integer) is defined for $-m-1 < \operatorname{Re}\lambda \leqslant -m$, $\lambda \neq -m$, by

$$(x_+^\lambda \log^k x_+, \varphi) = \int_0^\infty x^\lambda \log^k x \left[\varphi(x) - \varphi(0) - \cdots - \frac{x^{m-1}}{(m-1)!}\varphi^{(m-1)}(0)\right] dx.$$

Similarly one defines $x_-^\lambda \log^k x_-$, $|x|^\lambda \log^k |x|$. We leave for the reader to verify that

$$\frac{d^k}{d\lambda^k} x_+^\lambda = x_+^\lambda \log^k x_+ \qquad (\lambda \neq -1, -2, \ldots).$$

We now turn to the case $n > 1$ and begin by defining r^λ by

$$(r^\lambda, \varphi) = \int_{R^n} r^\lambda \varphi(x)\, dx \qquad (r = |x|) \tag{4.16}$$

for Re $\lambda > -n$. Denoting by Ω the unit sphere and by Ω_n its area, we formally have

$$\int_{R^n} r^\lambda \varphi(x)\, dx = \int_0^\infty r^\lambda \left\{ \int_\Omega \varphi(r\omega)\, d\omega \right\} r^{n-1}\, dr.$$

Developing φ by Taylor's formula one finds that

$$S_\varphi(r) \equiv \frac{1}{\Omega_n} \int_\Omega \varphi(r\omega)\, d\omega$$

is a C_c^∞ function if $\varphi \in C_c^\infty$ and that the odd derivatives of $S_\varphi(r)$ at $r = 0$ vanish. We now define

$$(r^\lambda, \varphi) = \Omega_n (r_+^{\lambda+n-1}, S_\varphi(r)) \tag{4.17}$$

for any $\lambda \neq -m - n + 1$, $m = 1, 2, \ldots$, where the right side is defined according to (4.1).

(r^λ, φ) is a meromorphic function whose poles are located at $\lambda = -n - m + 1$ ($m = 1, 2, \ldots$) and are simple, and the corresponding residues are

$$\Omega_n \frac{S_\varphi^{(m-1)}(0)}{(m-1)!}. \tag{4.18}$$

Since $S_\varphi^{(k)}(0) = 0$ if k is odd, the actual poles occur only at the points $\lambda = -n - 2m$ ($m = 0, 1, 2, \ldots$).

Noting that $\Gamma((\lambda + n)/2)$ has a simple pole with residue 2 at $\lambda = -n$, we get

$$2 \lim_{\lambda \to -n} \frac{(r^\lambda, \varphi)}{\Gamma\left(\dfrac{\lambda + n}{2}\right)} = \Omega_n S_\varphi(0) = \Omega_n \varphi(0).$$

Hence,

$$\delta(x) = \frac{2}{\Omega_n} \lim_{\lambda \to -n} \frac{r^\lambda}{\Gamma\left(\dfrac{\lambda + n}{2}\right)}. \tag{4.19}$$

This formula will be used in Sec. 5 in conjunction with the following formula:

$$(4.20)\qquad \frac{2r^\lambda}{\Gamma\left(\dfrac{\lambda+n}{2}\right)} = \frac{1}{\pi^{(n-1)/2}\Gamma\left(\dfrac{\lambda+1}{2}\right)}\int_\Omega |\omega\cdot x|^\lambda\, d\omega \qquad (\text{Re}\,\lambda > -1).$$

By analytic continuation, (4.20) is true for all $\lambda \neq -1, -3, \ldots$.

To prove (4.20) note first that the integral is of the form Cr^λ, where $C = C(n, \lambda)$ is a constant. To determine C, take $x_1 = \cdots = x_{n-1} = 0$ and use polar coordinates $(\theta_1, \ldots, \theta_{n-1}, r)$ so that

$$\omega_n = \cos\theta_{n-1}, \qquad d\omega = \sin^{n-2}\theta_{n-1}\, d\omega_{n-1},$$

where $d\omega_{n-1}$ is the element of area in the unit sphere in R^{n-1}. We get

$$C = \int_\Omega |\omega_n|^\lambda\, d\omega = 2\Omega_{n-1}\int_0^{\pi/2} \sin^{n-2}\theta\cos^\lambda\theta\, d\theta = \frac{2\pi^{(n-1)/2}\Gamma\left(\dfrac{\lambda+1}{2}\right)}{\Gamma\left(\dfrac{\lambda+n}{2}\right)},$$

from which (4.20) follows.

We conclude this section by solving *Radon's problem*, i.e., the problem of representing a function φ at any point x in terms of averages of φ and its derivatives on hyperplanes. We shall derive the following formulas:

$$(4.21)\qquad \varphi(0) = \frac{(-1)^{(n-1)/2}}{2(2\pi)^{n-1}}\int_\Omega d\omega \int_{\omega\cdot x=0} \frac{\partial^{n-1}\varphi(x)}{\partial\nu^{n-1}}\, d\sigma \qquad \text{for odd } n,$$

where $d\sigma$ is the element of area on the hyperplane $\omega\cdot x = 0$ and $\partial/\partial\nu$ is the derivative in the direction orthogonal to $\omega\cdot x = 0$, and

$$(4.22)\qquad \varphi(0) = \frac{(-1)^{n/2}(n-1)!}{(2\pi)^n}\int_\Omega ((\omega\cdot x)^{-n}, \psi(\omega\cdot x; \omega))\, d\omega \quad \text{for even } n,$$

where $(\xi^{-n}, \psi(\xi))$ is defined according to (4.12),

$$(4.23)\qquad \psi(\xi;\omega) = \int_{\omega\cdot x=\xi} \varphi(x)\, d\sigma_\xi,$$

and $d\sigma_\xi$ is the element of area on $\omega\cdot x = \xi$.

To prove (4.21), note that from the previous calculations of the residues of $|x|^\lambda$ (x one-dimensional) and of $\Gamma(\lambda)$ one gets

$$(4.24)\qquad \lim_{\lambda\to -n} \frac{|x|^\lambda}{\Gamma\left(\dfrac{\lambda+1}{2}\right)} = (-1)^{(n-1)/2}\frac{\left(\dfrac{n-1}{2}\right)!}{(n-1)!}\,\delta^{(n-1)}(x)$$

(x is one-dimensional)

for any odd number n. Combining this with (4.19) and with (4.20), as

$\lambda \to -n$, we obtain

$$\delta(x) = \frac{(-1)^{(n-1)/2}\left(\dfrac{n-1}{2}\right)!}{\pi^{(n-1)/2}(n-1)!\Omega_n} \int_\Omega \delta^{(n-1)}(\omega \cdot x)\, d\omega.$$

Since $\Omega_n = 2\pi^{n/2}/\Gamma(n/2)$, we get

$$\delta(x) = \frac{(-1)^{(n-1)/2}}{2(2\pi)^{n-1}} \int_\Omega \delta^{(n-1)}(\omega \cdot x)\, d\omega. \tag{4.25}$$

Equation (4.21) now follows by applying both sides of (4.25) to $\varphi(x)$.

To prove (4.22) we shall use the obvious relation

$$\lim_{\lambda \to -n} \frac{|x|^\lambda}{\Gamma\left(\dfrac{\lambda+1}{2}\right)} = \frac{1}{\Gamma\left(-\dfrac{n}{2}+\dfrac{1}{2}\right)} x^{-n} \qquad (x \text{ is one-dimensional}).$$

Combining it with (4.19) and with (4.20), as $\lambda \to -n$, and proceeding as before, we arrive at

$$\delta(x) = \frac{(-1)^{n/2}(n-1)!}{(2\pi)^n} \int_\Omega (\omega \cdot x)^{-n}\, d\omega, \tag{4.26}$$

and (4.22) follows by applying both sides of (4.26) to $\varphi(x)$.

5. Fundamental Solutions for Hyperbolic Equations

A differential equation of order m $P(\partial/\partial t, \partial/\partial x)u(x, t) = 0$ is said to be of *hyperbolic type* if for any real vector $\xi \neq 0$ the equation $P_0(\lambda, \xi) = 0$ has m real and distinct roots, where P_0 is the principal part of P. This definition is more restrictive than the one for which the results of Chap. 7, Sec. 7 were proven. We shall assume that P is homogeneous, i.e., $P = P_0$, and derive explicit formulas for the fundamental solutions of the Cauchy problem (see (3.12)). It will be enough to construct G_{m-1}, since the other G_j can be constructed in the same way. Setting, for brevity, $G = G_{m-1}$, our problem is to find explicit formulas for the solution of the system

$$P\left(\frac{\partial}{\partial t}, \frac{\partial}{\partial x}\right) G(x, t) = 0, \tag{5.1}$$

$$\frac{\partial^k G(x, 0)}{\partial t^k} = 0 \quad \text{if } 0 \leqslant k \leqslant m-2, \qquad \frac{\partial^{m-1} G(x, 0)}{\partial t^{m-1}} = \delta(x). \tag{5.2}$$

Formula (4.19) suggests that we may first solve (5.1) under the initial conditions

$$\begin{aligned} \frac{\partial^k G(x, 0)}{\partial t^k} &= 0 \quad \text{if } 0 \leqslant k \leqslant m-2, \\ \frac{\partial^{m-1} G(x, 0)}{\partial t^{m-1}} &= \frac{2r^\lambda}{\Omega_n \Gamma\left(\dfrac{\lambda+n}{2}\right)}, \end{aligned} \tag{5.3}$$

and then take $\lambda \to -n$. The solution of (5.1), (5.3) will be denoted by $G(x, t; \lambda)$.

From (4.20) it follows that if $G_\omega(x, t; \lambda)$ satisfies (5.1) and

$$\begin{aligned} \left.\frac{\partial^k G_\omega}{\partial t^k}\right|_{t=0} &= 0 \qquad \text{if } 0 \leqslant k \leqslant m-2, \\ \left.\frac{\partial^{m-1} G_\omega}{\partial t^{m-1}}\right|_{t=0} &= \frac{|x \cdot \omega|^\lambda}{\Omega_n \pi^{(n-1)/2} \Gamma\left(\dfrac{\lambda+1}{2}\right)}, \end{aligned} \tag{5.4}$$

then

$$G(x, t; \lambda) = \int_\Omega G_\omega(x, t; \lambda)\, d\omega. \tag{5.5}$$

Observe now that if $\omega = (\omega_1, \ldots, \omega_n)$ is a real unit vector and if v_j $(j = 1, \ldots, m)$ are the m real and distinct solutions of the algebraic equation $P(v, \omega) = 0$, then, for any function f, $f(x \cdot \omega + v_j t)$ is a solution of (5.1). We try to find G_ω in the form

$$G_\omega(x, t; \lambda) = \sum_{j=1}^{m} c_j f(x \cdot \omega + v_j t). \tag{5.6}$$

The initial conditions (5.4) reduce to the equations

$$\sum_{j=1}^{m} c_j v_j^k = 0 \qquad (0 \leqslant k \leqslant m-2), \qquad \sum_{j=1}^{m} c_j v_j^{m-1} = 1, \tag{5.7}$$

$$f^{(m-1)}(\xi) = \frac{|\xi|^\lambda}{\Omega_n \pi^{(n-1)/2} \Gamma\left(\dfrac{\lambda+1}{2}\right)}, \tag{5.8}$$

where ξ is one-dimensional. A solution of (5.8) is given by (compare (4.14), (4.15))

$$f(\xi) = \frac{1}{\Omega_n \pi^{(n-1)/2} \Gamma\left(\dfrac{\lambda+1}{2}\right)} \left\{ \frac{|\xi|^{\lambda+m-1} (\operatorname{sgn} \xi)^{m-1}}{(\lambda+1) \cdots (\lambda+m-1)} + Q_\lambda(\xi) \right\}, \tag{5.9}$$

where

$$Q_\lambda(\xi) = \sum_{k=1}^{[(m-1)/2]} \frac{\xi^{m-2k-1}}{(2k-1)!(m-2k-1)!(\lambda+2k)}. \tag{5.10}$$

Hence,

$$\begin{aligned} G_\omega(x, t; \lambda) = {} & \frac{1}{\Omega_n \pi^{(n-1)/2} \Gamma\left(\dfrac{\lambda+1}{2}\right)} \\ & \times \sum_{j=1}^{m} c_j \left\{ \frac{|x \cdot \omega + v_j t|^{\lambda+m-1} [\operatorname{sgn}(x \cdot \omega + v_j t)]^{m-1}}{(\lambda+1)(\lambda+2) \cdots (\lambda+m-1)} + Q_\lambda(x \cdot \omega + v_j t) \right\}. \end{aligned} \tag{5.11}$$

In order to obtain the fundamental solution $G(t, x)$, one substitutes G_ω from (5.11) into (5.5) and takes $\lambda \to -n$. It can be verified without difficulty that the limit indeed exists, in the sense of $(D') \equiv K'$.

It is more advantageous in applications to replace the integration over the unit sphere Ω by integration over a manifold which is related to the differential equation. The following calculations are aimed at replacing Ω by the manifold

$$H(\xi) \equiv P(1, \xi) = 0 \qquad (\xi \text{ real}). \tag{5.12}$$

For any ω, there are n distinct real solutions $v_j = v_j(\omega)$ of $P(v, \omega) = 0$. Index them such that

$$v_1(\omega) > v_2(\omega) > \cdots > v_m(\omega). \tag{5.13}$$

Since $P(-v_j(\omega), -\omega) = 0$ and, similarly, $P(v_j(-\omega), -\omega) = 0$ for all j, the set $\{-v_j(\omega)\}$ coincides with the set $\{v_j(-\omega)\}$. Since

$$-v_1(\omega) < -v_2(\omega) < \cdots < -v_m(\omega),$$

whereas (5.13) holds also for ω replaced by $-\omega$, it follows that

$$v_k(-\omega) = -v_{m-k+1}(\omega) \qquad \text{for } k = 1, \ldots, m. \tag{5.14}$$

Consider first the case that m is even. As ω varies on Ω, the $v_k = v_k(\omega)$ generate $m/2$ sheets S_k $(S_{m-k+1} = S_k)$ and the points $\xi = (\xi_1, \ldots, \xi_n)$, where $\xi_j = \omega_j/v_k$ (k fixed), generate a sheet T_k of the surface $H(\xi) = 0$ $(T_{m-k+1} = T_k)$. The surface $H(\xi) = 0$ consists of $m/2$ sheets $T_1, \ldots, T_{m/2}$.

Denoting by $d\sigma$ the element of surface area of $H(\xi) = 0$ and denoting by φ the angle between the normal to Ω at ω and the normal to $H(\xi) = 0$ at ξ, we have

$$d\omega = \frac{|\cos \varphi|\, d\sigma}{|\xi|^{n-1}} = \frac{|\xi \cdot \operatorname{grad} H|\, d\sigma}{|\xi|^n |\operatorname{grad} H|}. \tag{5.15}$$

For simplicity it will now be assumed that

$$P(1, 0) = 1. \tag{5.16}$$

Then,

$$\begin{aligned} c_j &= \frac{1}{(v_j - v_1) \cdots (v_j - v_{j-1})(v_j - v_{j+1}) \cdots (v_j - v_m)} \\ &= \frac{1}{\left[\dfrac{\partial P}{\partial v}\right]}\Bigg|_{v=v_j} \\ &= -\frac{v_j}{\sum \omega_i \dfrac{\partial P}{\partial \omega_i}\Big|_{v=v_j}}, \end{aligned}$$

where in the last equation use has been made of the relation

$$v_j \frac{\partial P}{\partial v}\bigg|_{v=v_j} + \sum_i \omega_i \frac{\partial P}{\partial \omega_i}\bigg|_{v=v_j} = mP\bigg|_{v=v_j} = 0,$$

which follows from the fact that $P(v, \omega)$ is homogeneous of degree m. Replacing ω_i by $v_j\xi_i$ ($\omega \in \Omega$, $\xi \in T_j$) and $P(v, \omega)$ by $v^m P(1, \omega/v)$ we get

$$c_j = -\frac{v_j^{1-m}}{\xi\cdot\operatorname{grad} H} = -\frac{(\operatorname{sgn} v_j)^{m-1}|\xi|^{m-1}}{\xi\cdot\operatorname{grad} H}. \tag{5.17}$$

We also have

$$\begin{aligned} &|x\cdot\omega + v_j t|^{\lambda+m-1}[\operatorname{sgn}(x\cdot\omega + v_j t)]^{m-1} \\ &\quad = |v_j|^{\lambda+m-1}(\operatorname{sgn} v_j)^{m-1}|x\cdot\xi + t|^{\lambda+m-1}[\operatorname{sgn}(x\cdot\xi + t)]^{m-1}. \end{aligned} \tag{5.18}$$

Substituting in the jth term of the sum of (5.11) for c_j and $x\cdot\omega + v_j t$ the expressions derived in (5.17) and (5.18), respectively, and replacing $d\omega$ by the right side of (5.15) with $\xi \in T_j$, we obtain, since each sheet T_k occurs twice and m is even,

$$\begin{aligned} G(x, t; \lambda) = &-\frac{2}{\Omega_n \pi^{(n-1)/2}\Gamma\left(\frac{\lambda+1}{2}\right)} \\ &\times \int_{H=0} \left\{\frac{|\xi|^{-\lambda-n}|x\cdot\xi + t|^{\lambda+m-1}\operatorname{sgn}(x\cdot\xi + t)}{(\lambda+1)(\lambda+2)\cdots(\lambda+m-1)} + Q\right\} d\Sigma, \end{aligned} \tag{5.19}$$

where

$$d\Sigma = \frac{d\sigma}{|\operatorname{grad} H|\operatorname{sgn}(\xi\cdot\operatorname{grad} H)} \tag{5.20}$$

and

$$Q = (\operatorname{sgn} v_j)|\xi|^{m-n-1}Q_\lambda(x\cdot\omega + v_j t)$$

on T_j.

Taking $\lambda \to -n$ and recalling the definition of Q_λ in (5.10) we obtain, for n odd, $m \geqslant n + 1$,

$$\begin{aligned} G(x, t) = &\frac{(-1)^{(n+1)/2}}{2(2\pi)^{n-1}(m-n-1)!} \\ &\times \int_{H=0} (x\cdot\xi + t)^{m-n-1}[\operatorname{sgn}(x\cdot\xi + t)]\, d\Sigma. \end{aligned} \tag{5.21}$$

For n even, $m \geqslant n + 1$, the first term in the integrand of (5.19) has a pole at $\lambda = -n$, but this singularity is cancelled by the singularity of Q at $\lambda = -n$. Taking $\lambda \to -n$, we obtain, after throwing away a polynomial of degree $< m$,

$$\begin{aligned} G(x, t) = &\frac{2(-1)^{n/2}}{(2\pi)^n(m-n-1)!} \\ &\times \int_{H=0} (x\cdot\xi + t)^{m-n-1}\log\left|\frac{x\cdot\xi + t}{x\cdot\xi}\right| d\Sigma. \end{aligned} \tag{5.22}$$

Formulas (5.21), (5.22) are due to Herglotz and Petrowski.

If $m < n + 1$ then we obtain, upon taking $\lambda \to -n$ in (5.19),

$$G(x, t) = \frac{(-1)^{(n+1)/2}}{(2\pi)^{n-1}} \int_{H=0} \delta^{(n-m)}(x \cdot \xi + t)\, d\Sigma \tag{5.23}$$

for odd n, and

$$G(x, t) = \frac{(-1)^{(n+2)/2}(n - m)!}{(2\pi)^n} \int_{H=0} (x \cdot \xi + t)^{m-n-1}\, d\Sigma \tag{5.24}$$

for even n.

The case where m is odd can be treated in the same way; the resulting formulas are slightly different.

PROBLEMS

1. A *fundamental matrix* G of the differential system

$$\sum_{k=1}^{N} P_{jk}\left(i\frac{\partial}{\partial x}\right) u_k = 0 \qquad (1 \leqslant j \leqslant N)$$

is a matrix $G = (G_{jk})$ satisfying

$$\sum_{k=1}^{N} P_{jk}\left(i\frac{\partial}{\partial x}\right) G_{kh}(x) = \begin{cases} 0 & \text{if } j \neq h, \\ \delta(x) & \text{if } j = h, \end{cases} \qquad \text{in } K'.$$

Taking Fourier transforms and using Theorem 3, prove that if $\det (P_{jk}(\sigma)) \not\equiv 0$ then there exists a fundamental matrix.

2. Give a formal procedure by which the Cauchy problem (3.10), (3.11) can be solved by solving only problems where $u_0 \equiv \cdots \equiv u_{m-2} \equiv 0$. [*Hint:* If $u_\nu(x) \equiv 0$ for $\nu \neq m - 2$, $u_{m-2}(x) = g(x)$, one proceeds as follows: let $u_0(x, t)$ be a solution of (3.10), (3.11), where $u_\nu \equiv 0$ if $\nu \neq m - 1$ and $u_{m-1} = g$. Set $u_1(x, t) = \partial u_0(x, t)/\partial t$ and let $u_2(x, t)$ be a solution of (3.10), (3.11), where $u_\nu \equiv 0$ if $\nu \neq m - 1$ and $u_{m-1} = [\partial^{m-1}u_1/\partial t^{m-1}]_{t=0}$. Then $u_3(x, t) = u_1(x, t) - u_2(x, t)$ satisfies (3.10), (3.11) with $u_\nu(x) \equiv 0$ if $\nu \neq m - 2$, $u_{m-2}(x) = g(x)$.]

3. Let $g(x) = \sum_{\nu=1}^{N} \dfrac{A_\nu}{(x - a)^{\lambda_\nu}} + \dfrac{B}{x - a} + h(x)$ where $\operatorname{Re}\{\lambda_\nu\} \geqslant 1$, $\lambda_\nu \neq 1$, and $h(x)$ is integrable in $a \leqslant x \leqslant b$. The *finite part* of the integral $\int_a^b g(x)\, dx$ is defined by

$$\text{Fp.} \int_a^b g(x)\, dx = \lim_{\epsilon \to 0} \left[\int_{a+\epsilon}^b g(x)\, dx - I(\epsilon) - B \log \frac{1}{\epsilon}\right]$$
$$= -\sum_{\nu=1}^{N} \frac{A_\nu}{\lambda_\nu - 1}\left(\frac{1}{b - a}\right)^{\lambda_\nu - 1} + B \log (b - a) + \int_a^b h(x)\, dx,$$

where

$$I(\epsilon) = \sum_{\nu=1}^{N} \frac{A_\nu}{\lambda_\nu - 1}\left(\frac{1}{\epsilon}\right)^{\lambda_\nu - 1}.$$

Set $F(\lambda) = \int_a^b g(x)(x-a)^\lambda\, dx$ for $\operatorname{Re}\lambda > \max \operatorname{Re}\{\lambda_\nu\}$. Prove that, if $B = 0$, $F(\lambda)$ can be continued as a meromorphic function whose poles occur at $\lambda = \lambda_\nu - 1$, and

$$F(0) = \text{Fp.} \int_a^b g(x)\, dx.$$

It then follows that

$$(x_+^\mu, \varphi) = \text{Fp.} \int_0^\infty x^\mu \varphi(x)\, dx \qquad (\mu \neq -1, -2, \ldots).$$

4. Prove: $\Delta(r^{\lambda+2}) = (\lambda+2)(\lambda+n)r^\lambda$,

$$r^\lambda = \frac{\Delta^k(r^{\lambda+2k})}{(\lambda+2)(\lambda+4)\cdots(\lambda+2k)(\lambda+n)(\lambda+n+2)\cdots(\lambda+n+2k-2)}.$$

5. Using $(\Delta^k r^{\lambda+2k}, \varphi) = (r^{\lambda+2k}, \Delta^k\varphi)$ and Problem 4, show that the residue of (r^λ, φ) at $\lambda = -n-2k$ is

$$\frac{\Omega_n \Delta^k\varphi(0)}{2^k k!\, n(n+2)\cdots(n+2k-2)}.$$

6. Prove that if $\varphi(x)$ is an analytic function in a neighborhood of $x = 0$ then

$$S_\varphi(r) \equiv \frac{1}{\Omega_n}\int_\Omega \varphi(r\omega)\, d\omega = \Gamma\left(\frac{n}{2}\right) \sum_{k=0}^\infty \left(\frac{r}{2}\right)^{2k} \frac{\Delta^k\varphi(0)}{k!\,\Gamma\left(k+\dfrac{n}{2}\right)}$$

(Pizetti's formula).

[*Hint:* Compare the residues of (r^λ, φ) as given by Problem 5 and by (4.18).]

7. Using Problem 5 and $(r^\lambda, e^{-r^2}) = \frac{1}{2}\Omega_n\Gamma((\lambda+n)/2)$, show that

$$\lim_{\lambda\to -n-2k} \frac{2}{\Omega_n} \frac{r^\lambda}{\Gamma\left(\dfrac{\lambda+n}{2}\right)} = (-1)^k \frac{\Delta^k\delta(x)}{2^k k!\, n\cdots(n+2k-2)}.$$

(This is a generalization of (4.19).)

8. Prove the formula

$$\mathfrak{F}(|x|^\lambda) = -2\sin\frac{\lambda\pi}{2}\,\Gamma(\lambda+1)|\sigma|^{-\lambda-1}$$

for $n = 1$, $\lambda \neq 0, \pm 1, \pm 2, \ldots$.

CHAPTER 11

DIFFERENTIABILITY OF SOLUTIONS OF PARTIAL DIFFERENTIAL EQUATIONS

In this chapter we give necessary and sufficient conditions on a polynomial $P(s)$ in order that all the local solutions of $P(D)u = 0$ be C^∞ functions, analytic functions, etc. The fundamental solution constructed in Chap. 10, Sec. 3, will play an important role. We shall also solve in R^n the equation $P(D)u = f$.

1. Hypoelliptic Equations and their Fundamental Solutions

Consider the differential equation

$$P(D)u(x) = 0. \tag{1.1}$$

By a solution u in an open set $\Omega \subset R^n$ we mean (in this chapter) a distribution u defined on Ω (i.e., $u \in (D'_\Omega)$) and satisfying (1.1) in the sense that

$$(u, P(-D)\varphi) = 0 \qquad \text{for any } \varphi \in (D_\Omega).$$

By a *local* solution we mean a solution in some open set Ω.

Definition. Equation (1.1) is said to be *hypoelliptic* (and $P(s)$ is then called a *hypoelliptic polynomial*) if every local solution of (1.1) is a C^∞ function.

Let u be a solution of (1.1) in Ω and let Ω_0 be a bounded domain whose closure $\overline{\Omega_0}$ lies in Ω. Let β be a C_c^∞ function which equals 1 on Ω_0 and 0 outside Ω. βu is a solution of (1.1) in Ω_0, and it is also a distribution whose support is a compact set lying in Ω. Next let α be a C_c^∞ function, which is equal to 1 in some neighborhood of 0 and whose support is contained in a neighborhood U of 0. Denoting by $G(x)$ a fundamental solution of (1.1) (it was constructed in Chap. 10, Sec. 3), we shall prove the following useful lemma.

Lemma 1. *If W is a domain such that $W + U \subset \Omega_0$ then, on W,*

$$u(x) = P(D)[(1 - \alpha(x))G(x)] * \beta(x)u(x). \tag{1.2}$$

Proof. If $\varphi \in (D_W)$ then

$$(P(D)(\alpha G) * \beta u, \varphi) = (\alpha G * P(D)(\beta u), \varphi) = (P(D)(\beta u), \psi),$$

where $\psi = \alpha G * \varphi$. By Theorem 41, Sec. 11, Chap. 3, ψ is a C^∞ function and, by Theorem 38, Sec. 10, Chap. 3, its support lies in $W + U \subset \Omega_0$. Since on Ω_0 $\beta u = u$ and $P(D)u = 0$, $P(D)(\alpha G) * \beta u = 0$ on (D_W). Hence, on D_W,

$$(1.3) \qquad P(D)[(1 - \alpha)G] * \beta u = P(D)G * \beta u - P(D)(\alpha G) * \beta u = \delta * \beta u = u.$$

Since the support of $P(D)[(1 - \alpha)G]$ lies in U, on W the left side of (1.3) depends on the values of βu in $W - U$ only (by Theorem 39, Sec, 10, Chap. 3). Hence we have the following corollary.

Corollary. *If $W - U \subset \Omega_0$, then, on W,*

$$(1.2') \qquad u(x) = P(D)[(1 - \alpha(x))G(x)] * u(x).$$

Theorem 1. *If $G(x)$ is a C^∞ function for $x \neq 0$, then P is hypoelliptic.* The converse is trivial since $P(D)G(x) = 0$ for $x \neq 0$.

Proof. Let u be a solution in Ω. It suffices to prove that u is a C^∞ function in any bounded domain Ω_0, where $\overline{\Omega_0} \subset \Omega$. Using the representation (1.2) (or (1.2′)) and applying Theorem 41, Sec. 11, Chap. 3, we find that u is a C^∞ function in W. Since U can be taken arbitrarily small, u is a C^∞ function in Ω_0.

Theorem 2. *If P is hypoelliptic, then, for any $\Omega \subset R^n$, if $u \in (D'_\Omega)$ and $P(D)u$ is a C^∞ function in Ω then u is a C^∞ function in Ω.*

Proof. Observe that if $W \pm U \subset \Omega_0$ then (1.3) implies, for $x \in W$,

$$u(x) = P(D)[(1 - \alpha(x))G(x)] * u(x) + \alpha(x)G(x) * P(D)u(x),$$

and the convolutions depend on the values of $u(x)$ and of $P(D)u(x)$ in Ω_0 only. The assertion now immediately follows.

The following lemma will be useful in deriving necessary conditions for polynomials to be hypoelliptic.

Lemma 2. *Let P be hypoelliptic, let u be a solution in Ω of (1.1), and let U, W, Ω_0 be bounded domains such that $\overline{\Omega_0} \subset \Omega$, $0 \in U$, $W \pm U \subset \Omega_0$. Then, for any q there exists a constant C independent of u such that*

$$(1.4) \qquad \max_{x \in W} |D^q u(x)| \leqslant C \max_{x \in \Omega_0} |u(x)|.$$

The converse is also true and its proof follows from Theorem 4 below, and from the proof of the first part of Theorem 4.

Proof. The function $A(x) = P(D)[(1 - \alpha(x))G(x)]$ is a C_c^∞ function whose support lies in U. From (1.2′) it follows that, if $x \in W$,

$$|D^q u(x)| \leqslant C \max_{x \in \Omega_0} |u(x)|,$$

where $C = \max_{x \in W} \int |D^q A(x - \xi)|\, d\xi$.

Theorem 3. *Let Ω be a bounded domain and denote its boundary by Γ. If u is a solution of* (1.1) *in some neighborhood $\hat{\Omega}$ of $\bar{\Omega} = \Omega \cup \Gamma$ and if u is a C^∞ function in some neighborhood $\hat{\Gamma}$ of Γ ($\hat{\Gamma} \subset \hat{\Omega}$) then u is a C^∞ function in Ω.*

Proof. Set $\hat{u} = u\gamma$ where γ is a C_c^∞ function which equals 1 in Ω and 0 in the exterior of $\hat{\Gamma}$. Then $P(D)\hat{u} = \hat{f}$ in R^n where $f \in C_c^\infty$. Set

$$D_R = \{x; |x| \leqslant R\}, \qquad S_R = \{x; R \leqslant |x| \leqslant R + 1\}.$$

We use the formula

$$u(x) = P(D)[(1 - \alpha(x))G(x)] * \hat{u}(x) + \alpha(x)G(x) * \hat{f}(x) \qquad (x \in \Omega)$$

where $\alpha = 1$ in D_R, $\alpha = 0$ outside D_{R+1}. The second term on the right side is a C_c^∞ function. The support of the first term is contained in $S_R + \hat{\Omega}$ and it therefore does not intersect Ω if R is sufficiently large. It follows that $u(x)$ is a C^∞ function in Ω.

If (1.1) is hypoelliptic, then (1.2) can be written in the form

$$u(x) = \int (1 - \alpha(\xi))G(\xi)P(D_x)w(x - \xi)\, d\xi \qquad (w = \beta u).$$

Taking $\alpha \to \chi_U$, where χ_U is the characteristic function of U, we obtain

$$u(x) = \int_{\hat{U}} G(\xi)P(D_x)w(x - \xi)\, d\xi,$$

where $\hat{U}$ is the complement of U in R^n. Replacing $P(D_x)w(x - \xi)$ by $P(-D_\xi)w(x - \xi)$ and integrating by parts, one gets

$$u(x) = \sum_k \int_\Gamma c_k(\xi)Q_k(-D_\xi)w(x - \xi)\, d\xi,$$

where P_k, Q_k are polynomials of degree $< m$, m being the order of the differential equation (1.1), $c_k(\xi) = P_k(D_\xi)G(\xi)$, and Γ is the boundary of U. Making, in the integrand, the substitution $x - \xi = \eta$ which maps Γ into $x - \Gamma$, and noting that U and Γ can be taken to be dependent on x, so that $x - \Gamma$ can be replaced by $x^0 - \Gamma$ (provided $x^0 + \bar{U} \subset \Omega_0$ and $|x - x^0|$ is sufficiently small), we obtain

$$(1.5) \qquad u(x) = \sum_k \int_{x^0 - \Gamma} c_k(x - \eta)Q_k(D_\eta)u(\eta)\, d\eta \qquad (\text{if } x^0 \pm \bar{U} \subset \Omega_0),$$

provided $|x - x^0|$ is sufficiently small. Applying any derivative D^q to both sides of (1.5) one gets

$$D^q u(x) = \sum_k \int_{x^0 - \Gamma} D_x^q c_k(x - \eta) \cdot Q_k(D_\eta) u(\eta) \, d\eta. \tag{1.6}$$

This formula enables one to estimate derivatives of u in terms of derivatives of the fundamental solutions.

2. Conditions for Hypoellipticity

In this section we derive necessary and sufficient conditions for hypoellipticity in terms of the complex zeroes of $P(s) = 0$. More direct conditions, depending on the behavior of $P(\sigma)$ (σ real), will be given in the next section. It will be somewhat more convenient to write the differential equations in the form

$$P\left(i \frac{\partial}{\partial x}\right) u(x) = 0 \tag{2.1}$$

rather than in the form (1.1). We denote by $N(P)$ the set of all complex solutions s of the equation $P(s) = 0$.

Theorem 4. *If (2.1) is hypoelliptic then for any constant $A > 0$ there exists a constant $B > 0$ such that all the points $s = \sigma + i\tau$ of $N(P)$ satisfy the inequality*

$$|\tau| \geqslant A \log |\sigma| - B. \tag{2.2}$$

Conversely, if for every positive constant A there exists a positive constant B such that (2.2) is satisfied for all $s = \sigma + i\tau$ in $N(P)$, then (2.1) is hypoelliptic.

Proof. If $s \in N(P)$ then

$$P\left(i \frac{\partial}{\partial x}\right) e^{-is \cdot x} = P(s) e^{-ix \cdot s} = 0.$$

Assuming that (2.1) is hypoelliptic and using (1.4) with $|q| = 1$, $u = e^{-is \cdot x}$, and W, Ω_0 being two concentric balls of radii r and R, respectively, one obtains

$$|s| \leqslant C_1 e^{(R-r)|\tau|} \text{ (in this chapter, } \exp[a|\tau|^b] \equiv \exp[a(\sum_{j=1}^n \tau_j^2)^{b/2}]).$$

Since $|\sigma| \leqslant |s|$, it follows that $\log |\sigma| \leqslant \log C_1 + (R - r)|\tau|$. We thus arrive at (2.2) with $A = 1/(R - r)$. Since $R - r$ can be taken arbitrarily small, (2.2) holds for any given A, with some B depending on A.

In order to prove the converse we need the following:

Lemma 3. *Let $f(t)$ ($t > 0$) be a continuously differentiable function such that $f(t) \to \infty$ if $t \to \infty$ and let $f'(t)$ be bounded for $t \geqslant t_0 > 0$. If $N(P)$ lies* *in the region* $|\tau| \geqslant f(|\sigma|)$, *then there exists a positive constant* C' *such that*

$$|P(s)| \geqslant C' \qquad \text{if } s \in N(P),\ |\tau| \leqslant \tfrac{1}{2}f(|\sigma|). \tag{2.3}$$

Proof. We first assume that

$$P(s) = a_0 s_1^m + \sum_{k=1}^{m} a_k(\hat{s})s_1^{m-k} \qquad (\hat{s} = (s_2, \ldots, s_n),\ a_0 \neq 0). \tag{2.4}$$

For $\hat{s}$ fixed, there exist n roots $s_1 = \lambda_j(\hat{s})$ of $P(s) = 0$. Set $\lambda_j = \xi_j + i\eta_j$, $|\hat{\sigma}|^2 = \sigma_2^2 + \cdots + \sigma_n^2$, $|\hat{\tau}|^2 = \tau_2^2 + \cdots + \tau_n^2$. By assumption,

$$(\eta_j^2 + |\hat{\tau}|^2)^{1/2} \geqslant f(\sqrt{\xi_j^2 + |\hat{\sigma}|^2}) \qquad (j = 1, \ldots, m). \tag{2.5}$$

Let $|\tau| \leqslant \frac{1}{2}f(|\sigma|)$ and write

$$|P(s)| = |a_0| \prod_{j=1}^{m} |s_1 - \lambda_j| = |a_0| \left[\prod_{j=1}^{m} ((\sigma_1 - \xi_j)^2 + (\tau_1 - \eta_j)^2)\right]^{1/2}.$$

If we show that each of the m factors $|s_1 - \lambda_j|$ is $\geqslant$ const. > 0, then the proof is completed. Suppose that this is not true. Then there exist sequences $\{\lambda_h^\nu\}$, $\{s^\nu\}$ (h fixed, $\lambda_h^\nu = \lambda_h(\hat{s}^\nu)$) such that $|\tau^\nu| \leqslant \frac{1}{2}f(|\sigma^\nu|)$, and

$$(\sigma_1^\nu - \xi_h^\nu)^2 + (\tau_1^\nu - \eta_h^\nu)^2 \to 0 \qquad \text{as } \nu \to \infty \tag{2.6}$$

$$(\lambda_h^\nu = \xi_h^\nu + i\eta_h^\nu,\ s_1^\nu = \tau_1^\nu + i\sigma_1^\nu).$$

Obviously $|\sigma^\nu|^2 = (\sigma_1^\nu)^2 + |\hat{\sigma}^\nu|^2 \to \infty$ as $\nu \to \infty$.

Using (2.6) we get

$$|[(\tau_1^\nu)^2 + |\hat{\tau}^\nu|^2]^{1/2} - [(\eta_h^\nu)^2 + |\hat{\tau}^\nu|^2]^{1/2}| \leqslant |\tau_1^\nu - \eta_h^\nu| \to 0 \qquad \text{as } \nu \to \infty, \tag{2.7}$$

$$|[(\sigma_1^\nu)^2 + |\hat{\sigma}^\nu|^2]^{1/2} - [(\xi_h^\nu)^2 + |\hat{\sigma}^\nu|^2]^{1/2}| \leqslant |\sigma_1^\nu - \xi_h^\nu| \to 0 \qquad \text{as } \nu \to \infty. \tag{2.8}$$

By (2.5), (2.7),

$$f(\sqrt{(\xi_h^\nu)^2 + |\hat{\sigma}^\nu|^2}) \leqslant [(\tau_1^\nu)^2 + |\hat{\tau}^\nu|^2]^{1/2} + \epsilon \leqslant \tfrac{1}{2}f(|\sigma^\nu|) + \epsilon \tag{2.9}$$

for any $\epsilon > 0$, if ν is sufficiently large. From (2.8) and the assumption that $f'(t)$ is bounded, say by H, it follows that

$$f(\sqrt{(\xi_h^\nu)^2 + |\hat{\sigma}^\nu|^2}) \geqslant f(|\sigma^\nu|) - H\epsilon \tag{2.10}$$

if ν is sufficiently large, depending on ϵ. Use has been made of the fact that $|\sigma^\nu| \to \infty$ (and hence, by (2.8), also $(\xi_h^\nu)^2 + |\hat{\sigma}^\nu|^2 \to \infty$) since $f'(t)$ is bounded only if $t \geqslant t_0$. Combining (2.9), (2.10) we obtain

$$f(|\sigma^\nu|) \leqslant \tfrac{1}{2}f(|\sigma^\nu|) + (H + 1)\epsilon,$$

which is impossible for large ν since $f(|\sigma^\nu|) \to \infty$ as $\nu \to \infty$.

If (2.4) is not assumed, then it can be achieved by a transformation $s \to Cs$, as in the proof of Theorem 3, Sec. 3, Chap. 10. Since C may be taken to be an orthogonal matrix, the transformation preserves distances.

Hence both the assumptions and assertion of the lemma remain unaffected by the transformation. The proof of the lemma is therefore completed.

We return to the proof of the second part of Theorem 4. From the last paragraph it follows that we may assume that $P(s)$ has the form (2.4). From Theorem 1 it follows that it suffices to prove that $G(x)$ is a C^∞ function for $x \neq 0$. This will be done by modifying the construction of G which was given in Chap. 10, Sec. 3. Let V be any bounded domain in R^n. If we prove that, for any p,

$$(G(-x), \varphi(x)) = \int \hat{G}_p(x)\varphi(x)\,dx \qquad \text{for any } \varphi \in (D_V),$$

where $\hat{G}_p(x)$ is a C^p function, then the proof is completed. Without loss of generality we may assume that V is contained in the domain $x_j > a$ ($j = 1, \ldots, n$), where $a > 0$, since otherwise, assuming, as we may, that V is a sufficiently small domain containing a fixed point $\neq 0$, we first perform an appropriate orthogonal transformation (which does not affect the form (2.4)).

By Sec. 3, Chap. 10,

$$(2.11) \qquad (E, \psi) = \int_T \frac{\psi(s)\,ds}{P(s)} \qquad (\psi = \mathfrak{F}\varphi,\ E = \mathfrak{F}G).$$

For the sake of clarity we consider the case $n = 2$. The proof for $n > 2$ follows by obvious modifications. We divide the σ plane into nine regions Ω_j by $\sigma_1 = \pm\mu$, $\sigma_2 = \pm\mu$ and denote by Ω_1 the bounded square. μ is chosen sufficiently large such that outside Ω_1

$$(2.12) \qquad A \log|\sigma| - B > 2 \max_{s \in T} |\tau|,$$

where A, B are the constants for which (2.2) is assumed to hold. Set

$$(2.13) \qquad (E_j, \psi) = \int_{T_j} \frac{\psi(s)\,ds}{P(s)} \qquad (j = 1, \ldots, 9),$$

where T_j is the projection of Ω_j on T. We shall prove that, for any given p,

$$(2.14) \qquad (E_j, \psi) = \int_V G_j(x)\varphi(x)\,dx,$$

where G_j is a C^p function in V. Since $G(-x)$ coincides on V with $(2\pi)^{-n} \Sigma\, G_j(x)$, the proof of the theorem is then completed.

Since

$$(E_1, \psi) = \int_{T_1} \frac{\psi(s)\,ds}{P(s)} = \int_{T_1} \frac{1}{P(s)} \left\{ \int_V \varphi(x)^{ix\cdot s}\,dx \right\} ds$$
$$= \int_V \left\{ \int_{T_1} \frac{e^{ix\cdot s}}{P(s)}\,ds \right\} \varphi(x)\,dx,$$

$G_1(x) = \int_{T_1} [e^{ix\cdot s}/P(s)]\,ds$ is a C^∞ function.

Suppose that T_2 is defined by $s = \sigma_1 + i\tau_1$, $\mu < \sigma_1 < \infty$, $s_2 = \sigma_2$, $-\mu < \sigma_2 < \mu$. In order to proceed as in the previous case, one has to justify the change of order of integration. Before being able to justify this, we modify the domain of integration by replacing, for s_2 fixed, τ_1 by $\frac{1}{2}(A \log \sigma_1 - B)$ $(\mu < \sigma_1 < \infty)$. The difference between the two integrals is (by Cauchy's theorem)

$$(2.15) \qquad -\int_{-\mu}^{\mu}\int_{Q(\mu)}^{Q_1(\mu)} \frac{\psi(s_1, s_2)}{P(s_1, s_2)}\, ds_1\, ds_2 + \lim_{\nu\to\infty} \int_{-\mu}^{\mu}\int_{Q(\nu)}^{Q_1(\nu)} \frac{\psi(s_1, s_2)}{P(s_1, s_2)}\, ds_1\, ds_2 \equiv I_1 + I_2,$$

where the integration with respect to s_1 is along vertical segments connecting Q to Q_1.

I_1 is an expression of the form (E_1, ψ) and therefore

$$I_1 = \int_V G_{12}(x)\varphi(x)\, dx,$$

where G_{12} is a C^∞ function. $I_2 = 0$ since (using Lemma 3)

$$\int_{Q(\nu)}^{Q_1(\nu)} \left| \frac{\psi(s_1, s_2)}{P(s_1, s_2)}\, ds_1 \right| \leqslant C'' \max |\psi(s_1, s_2)| \overline{Q(\nu)Q_1(\nu)}$$
$$\leqslant C_k \left.\frac{\exp[b|\tau_1|] \log|\sigma_1|}{|\sigma_1|^k}\right|_{\sigma_1=\nu} \leqslant C_k \frac{\nu^{bA/2} \log \nu}{\nu^k}$$

for any $k > 0$; C_k depends on s_2, but if $s_2 = \sigma_2$ and $-\mu < \sigma_2 < \mu$, then C_k can be taken to be independent of s_2.

It thus remains to consider

$$\int_{T'_2} \frac{\psi(s)\, ds}{P(s)},$$

where T'_2 is the manifold

$$s_1 = \sigma_1 + i\tau_1, \quad \mu < \sigma_1 < \infty, \quad 2\tau_1 = A \log \sigma_1 - B; \quad s_2 = \sigma_2, \quad -\mu < \sigma_2 < \mu.$$

One can formally write, as in the case $j = 1$,

$$(2.16) \qquad \int_{T'_2} \frac{\psi(s)\, ds}{P(s)} = \int_{T'_2} \frac{1}{P(s)} \left\{ \int_V e^{ix\cdot s}\varphi(x)\, dx \right\} ds = \int_V \varphi(x) \left\{ \int_{T'_2} \frac{e^{ix\cdot s}\, ds}{P(s)} \right\} dx.$$

If we prove the uniform convergence of the inner integral on the right side of (2.16), then the change of order of integration is justified. The proof of convergence follows immediately from the inequalities

$$|e^{ix\cdot s}| = e^{-x\cdot\tau} \leqslant H_1 e^{-x_1\tau_1} \leqslant H_1 \exp[-a\tau_1] \leqslant H_2 \exp[-\tfrac{1}{2}aA \log \sigma_1] = H_2 \sigma_1^{-aA/2},$$

$$|ds| = |ds_1|\,|ds_2| \leqslant \left(1 + \frac{A}{2\sigma_1}\right) d\sigma_1\, d\sigma_2, \qquad |P(s)| \geqslant C'.$$

Use has been made of Lemma 3 and of the fact that $x_1 > a$ if $x \in V$.

Having proved that

$$\int_{T'_2} \frac{\psi(s)\, ds}{P(s)} = \int G_{22}(x)\varphi(x)\, dx, \qquad \text{where } G_{22}(x) = \int_{T'_2} \frac{e^{ix\cdot s}\, ds}{P(s)},$$

we next find that

$$D^q G_{22}(x) = \int_{T'_2} \frac{(is)^q e^{ix\cdot s}}{P(s)}\, ds \qquad \left(|q| < \frac{aA}{2} - 1\right)$$

and the integral is uniformly convergent, since the integrand is bounded by

$$H_3 \sigma_1^{-aA/2} \sigma_1^{|q|}.$$

Hence G_{22}, and consequently $G_2(x)$ in (2.14), is of differentiability class C^p for any $p < (aA/2) - 1$.

Consider next the case $j = 3$ and assume that Ω_3 is defined by $\mu < \sigma_1 < \infty$, $\mu < \sigma_2 < \infty$. We first modify T_3 into T'_3 defined by

$$s_1 = \sigma_1 + i\tau_1, \quad \mu < \sigma_1 < \infty, \quad 2\tau_1 = A \log \sigma_1 - B; \quad s_2 = \sigma_2, \quad \mu < \sigma_2 < \infty.$$

The difference between the two integrals is (compare (2.15))

$$-\int_\mu^\infty \int_{Q(\mu)}^{Q_1(\mu)} \frac{\psi(s_1, s_2)}{P(s_1, s_2)}\, ds_1\, ds_2 + \lim_{\nu\to\infty} \int_\mu^\infty \int_{Q(\nu)}^{Q_1(\nu)} \frac{\psi(s_1, s_2)}{P(s_1, s_2)}\, ds_1\, ds_2 \equiv J_1 + J_2.$$

Using Lemma 3 it follows that

$$|J_2| \leqslant C_{kh} \lim_{\nu\to\infty} \frac{\nu^{bA/2} \log \nu}{\nu^k} \int_\mu^\infty \frac{d\sigma_2}{(1 + |\sigma_2|)^h}$$

for any $k > 0$, $h > 0$. Taking $k > bA/2$, $h > 1$ we conclude that $J_2 = 0$. As for J_1, it can be handled in the same way as (E_2, ψ). Hence,

$$J_1 = \int G_{31}(x)\varphi(x)\, dx,$$

where G_{31} is of differentiability class C^p for any $p < (aA/2) - 1$.

It remains to consider the case $j = 3$ with T_3 replaced by T'_3. We next replace T'_3 by T''_3 defined by

$$\mu < \sigma_1 < \infty, \qquad 2\tau_1 = A \log \sigma_1 - B,$$
$$\mu < \sigma_2 < \infty, \qquad 2\tau_2 = A \log \sigma_2 - B.$$

Again, the difference between the two integrals is of the form $\int G_{32}(x)\varphi(x)\,dx$ where $G_{32}(x)$ is of differentiability class C^p for any $p < (aA/2) - 1$. With T''_3 as the domain of integration, we may change the order of integration as in (2.16), making use of Lemma 3 and of the inequalities

$$|e^{ix\cdot s}| \leqslant H_4\sigma_1^{-aA/2}\sigma_2^{-aA/2}, \qquad |ds| \leqslant \left(1 + \frac{A}{2\sigma_1}\right)\left(1 + \frac{A}{2\sigma_2}\right) d\sigma_1\, d\sigma_2.$$

It also follows from these inequalities that

$$\int_{T''_3} \frac{e^{ix\cdot s}}{P(s)}\, ds$$

is of differentiability class C^p for any $p < (aA/2) - 1$.

Since any of the other integrals over the Ω_j $(j = 4, \ldots, 9)$ is either of the type of $j = 2$ or the type of $j = 3$, we conclude that all the $G_j(x)$ in (2.14) are of differentiability class C^p, on V, for any $p < (aA/2) - 1$. Since A can be taken arbitrarily large, $G(x)$ is a C^∞ function and the proof of the theorem is completed.

Definition. A function $u(x)$ defined in a domain $V \subset R^n$ and satisfying the inequalities

$$|D^q u(x)| \leqslant B_0 B_1^q q^{q\beta} \qquad (0 \leqslant |q| < \infty)$$

for some constants B_0, B_1 and for a given $\beta \geqslant 1$, is said to belong to *Gevrey's class* G_β in V.

If the fundamental solution $G(x)$ belongs to G_β in any domain V which does not contain the origin, then by using (1.6) one can easily estimate the derivatives of any solution u and thus prove that u is locally of class G_β, i.e., u is of class G_β in every neighborhood W whose closure is contained in a domain where u is a solution.

Definition. If every local solution is locally of class G_β, then P is said to be β-hypoelliptic.

Thus, if $G(x)$ is locally of class G_β for $x \neq 0$ then P is β-hypoelliptic. The converse is trivial.

Theorem 5. *A necessary and sufficient condition for* P *to be* β*-hypoelliptic is that all the points* $s = \sigma + i\tau$ *of the manifold* $N(P) = \{s; P(s) = 0\}$

satisfy

$$|\tau| \geqslant A|\sigma|^{1/\beta} - B \tag{2.17}$$

for some positive constants A, B.

Proof. Assume that P is β-hypoelliptic and let $V \subset V' \subset V''$ be three concentric balls about the origin, having radii r, r', R. Using (1.6) with $x = x^0 \in V$, Γ being the boundary of V', we get

$$|D^q u(x)| \leqslant \sum_{k \leqslant m} \int_{x-\Gamma} |D^q c_k(x-\eta)| \, [\max_{\eta \in V'} |Q_k(D_\eta)u(\eta)|] \, d\eta.$$

Recalling that the c_k are derivatives of G, we have

$$\int_{x-\Gamma} |D^q c_k(x-\eta)| \, d\eta \leqslant A_0 A_1^q q^{q\beta}.$$

The Q_k are differential operators of order $< m$. Using (1.4) one gets

$$\max_{\eta \in V'} |Q_k(D_\eta)u(\eta)| \leqslant A_2 \max_{\eta \in V''} |u(\eta)|.$$

Combining these estimates we arrive at

$$\max_{|x| \leqslant r} |D^q u(x)| \leqslant A_3 A_1^q q^{q\beta} \max_{|x| \leqslant R} |u(x)|.$$

Taking $u = e^{-ix \cdot s}$, where $P(s) = 0$, we obtain

$$|s^q| e^{r|\tau|} \leqslant A_3 A_1^q q^{q\beta} e^{R|\tau|}.$$

Hence,

$$e^{(r-R)|\tau|} \leqslant A_4 \frac{A_1^q q^{q\beta}}{|s^q|}.$$

Taking the minimum with respect to q we get

$$e^{(r-R)|\tau|} \leqslant A_4 \exp[-b|s|^{1/\beta}],$$

from which it follows that

$$(R - r)|\tau| \geqslant b|s|^{1/\beta} - \log A_4 \geqslant b|\sigma|^{1/\beta} - \log A_4,$$

and (2.17) is proved.

To prove the converse it suffices, in view of the remark preceding Theorem 5, to prove that $G(x)$ is locally of class G_β if $x \neq 0$. We proceed as in the proof of Theorem 4 except that in the final expressions $\int \frac{e^{ix \cdot s}}{P(s)} ds$ we modify the domain of integration by replacing $\log |\sigma_j|$ by $|\sigma_j|^{1/\beta}$. Thus, for the integral $\int_{T'_2} \frac{e^{ix \cdot s}}{P(s)} ds$ on the right side of (2.16) we replace T'_2 by $\tilde{T}'_2$ defined by

$$\mu < \sigma_1 < \infty, \quad \tau_1 = \tfrac{1}{2}(A\sigma_1^{1/\beta} - B); \quad s_2 = \sigma_2, \quad -\mu < \sigma_2 < \mu.$$

The difference

$$\int_{T'_2} \frac{e^{ix\cdot s}}{P(s)}\, ds - \int_{\tilde{T}'_2} \frac{e^{ix\cdot s}}{P(s)}\, ds$$

is, by Cauchy's theorem, an integral of the type obtained for $j = 1$. The estimate for $e^{ix\cdot s}$ $(s \in \tilde{T}'_2)$ is now

$$|e^{ix\cdot s}| = \exp\,[-\,x_1\tau_1] \leqslant A_5 \exp\left[\frac{-aA}{2}\,\sigma_1^{1/\beta}\right] = A_5 \exp\,[-a'\sigma_1^{1/\beta}]$$
$$\left(a' = \frac{aA}{2}\right).$$

It follows that

$$\left| D_1^{q_1} \int_{\tilde{T}'_2} \frac{e^{ix\cdot s}}{P(s)}\, ds \right| \leqslant A_6 \int_{\mu}^{\infty} \sigma_1^{q_1} \exp\,[-a'\sigma_1^{1/\beta}]\, d\sigma_1$$
$$\leqslant A_6 \int_0^{\infty} \sigma_1^{q_1} \exp\,[-a'\sigma_1^{1/\beta}]\, d\sigma_1.$$

Substituting $\xi_1 = a'\sigma_1^{1/\beta}$, we get the bound $A_7 A_8^{q_1}\Gamma(q_1\beta)$. One also gets

$$\left| D_1^{q_1} D_2^{q_2} \int_{\tilde{T}'_2} \frac{e^{ix\cdot s}}{P(s)}\, ds \right| \leqslant A_9 A_8^{q_1} A_{10}^{q_2}\Gamma(q_1\beta) \leqslant A_9 A_{11}^{q} q_1^{q_1\beta} q_2^{q_2\beta}.$$

The integrals of the other Ω_j can be handled similarly. This completes the proof of the theorem.

We next prove the following:

Theorem 6. *If* $|\sigma| \to \infty$, $\sigma + i\tau \in N(P)$ *imply* $|\tau| \to \infty$, *then* P *is hypoelliptic. If* P *is hypoelliptic then* P *is also* β*-hypoelliptic for some* $\beta \geqslant 1$.

In view of Theorems 4, 5, Theorem 6 is a consequence of the following lemma.

Lemma 4. *If* $|\sigma| \to \infty$, $\sigma + i\tau \in N(P)$ *imply* $|\tau| \to \infty$, *then there exist positive constants* A, σ_0, γ *such that if* $\sigma + i\tau \in N(P)$ *and* $|\sigma| \geqslant \sigma_0$ *then* $|\tau| \geqslant A|\sigma|^{\gamma}$.

Proof. By an argument similar to that given in the proof of Lemma 9, Sec. 14, Chap. 7, one shows that the function

$$M_n(\sigma_n) \equiv \inf_{(\sigma_1,\ldots,\sigma_{n-1})} d(\sigma, N(P))$$

(where $d(\sigma, N(P))$ is the distance from σ to $N(P)$) is an algebraic function of σ_n for large σ_n. By the Puiseux expansion,

$$M_n(\sigma_n) = \rho_n\sigma_n^{\gamma_n}(1 + o(1)) \qquad \text{as } \sigma_n \to \infty. \tag{2.18}$$

We next claim that $M_n(\sigma_n) \to \infty$ as $\sigma_n \to \infty$. Indeed, for any given $H > 0$ there exists $C(H) > 0$ such that if $|\tau| < H$ and $\sigma_n > C(H)$ (and consequently $|\sigma| > C(H)$) then $P(\sigma + i\tau) \neq 0$ for any $\sigma_1, \ldots, \sigma_{n-1}$. This implies that $M(\sigma_n) > H$ if $\sigma_n > C(H) + H$, i.e., $M(\sigma_n) \to \infty$ as $\sigma_n \to \infty$. Hence, $\rho_n > 0$, $\gamma_n > 0$ in (2.18).

In the same way we define $M_j(\sigma_j)$ (for $j = 1, 2, \ldots, n-1$) and prove that

$$M_j(\sigma_j) = \rho_j \sigma_j^{\gamma_j}(1 + o(1)) \qquad \text{as } \sigma_j \to \infty,$$

where $\rho_j > 0$, $\gamma_j > 0$. The same can be proved for $\sigma_j \to -\infty$ ($\rho_j \sigma_j^{\gamma_j}$ is replaced by $\nu_j|\sigma_j|^{\delta_j}$, $\nu_j > 0$, $\delta_j > 0$). If follows that

$$d(\sigma, N(P)) \geqslant \rho'|\sigma_j|^{\gamma} \qquad \text{for } j = 1, \ldots, n,$$

where $\rho' > 0$, $\gamma > 0$, provided $|\sigma_j| \geqslant \sigma_{j0}$. Hence, for some $A > 0$, $\sigma_0 > 0$,

$$d(\sigma, N(P)) \geqslant A|\sigma|^{\gamma} \qquad \text{if } |\sigma| \geqslant \sigma_0. \tag{2.19}$$

In order to complete the proof of the lemma, observe that if

$$P(\sigma + i\tau) = 0$$

then $d(\sigma, N(P)) \leqslant |\tau|$, and use (2.19).

As an application of the first part of Theorem 6 we shall prove the following result.

Theorem 7. *Assume that there exists a domain $\Omega \subset R^n$ such that all the solutions of* (2.1) *in Ω which belong to some class $C^h(\Omega)$ belong also to $C^{h+1}(\Omega)$. Then P is hypoelliptic.*

Proof. Let X be the space of all the $C^h(\Omega)$ functions which are solutions of (2.1) in Ω and which have a finite norm

$$\|u\| = \sup_{x \in \Omega} \{\exp[-|x|^2]\} \sum_{|\alpha| \leqslant h} |D^\alpha u(x)|.$$

X is a Banach space. Let Ω_0 be a bounded domain whose closure is contained in Ω and denote by X_0 the space of $C^{h+1}(\Omega_0)$ functions having a finite norm

$$\|u\|' = \sup_{x \in \Omega_0} \sum_{|\alpha| \leqslant h+1} |D^\alpha u(x)|.$$

X_0 is also a Banach space. By assumption, if $u \in X$ then $u \in X_0$. Consider now the mapping $Tu = u$ of X into X_0. T is closed, i.e., if $u_k \to \hat{u}$ in X, $Tu_k \to \bar{u}$ in X_0, then $T\hat{u}$ exists and $T\hat{u} = \bar{u}$. We now use the *closed graph theorem* which states that *any closed linear mapping of a Banach space X into a Banach space X_0 is a continuous mapping.* It follows that

$$\|u\|' \leqslant B\|u\| \qquad (B \text{ a constant}).$$

Taking $u(x) = e^{-ix\cdot s}$, where $P(s) = 0$, we obtain

$$|s| \sup_{x\in\Omega_0} e^{x\cdot\tau} \leqslant B' \sup_{x\in\Omega} \exp\,[-|x|^2 + x\cdot\tau].$$

The left side is $\geqslant B_1|s|$. Using the inequalities

$$-\,|x|^2 + x\cdot\tau \leqslant -\,|x|^2 + |x|\,|\tau| \leqslant \frac{|\tau|^2}{4}$$

we obtain $|\sigma| \leqslant |s| \leqslant B_2 \exp\,[|\tau|^2/4]$, from which it follows that $|\tau| \to \infty$ if $|\sigma| \to \infty$. Hence, by Theorem 6, P is hypoelliptic.

Note that if we replace $\exp\,[-|x|^2]$ in the definition of $\|u\|$ by $\exp\,[-|x|^\delta]$ for any $\delta > 1$ then we can still arrive at the conclusion that $|\tau| \to \infty$ as $|\sigma| \to \infty$. On the other hand, if $\exp\,[-|x|^2]$ is replaced by $e^{-\mu|x|}$ ($\mu > 0$), then the above proof yields

$$|\sigma| \leqslant B_3 e^{-\mu|x| + x\cdot\tau} \leqslant B_3$$

if $\mu > |\tau|$. Hence $P(\sigma + i\tau) \neq 0$ if $|\sigma| > B_3$. If μ can be taken arbitrarily large then it follows that if $P(\sigma + i\tau) = 0$, $|\sigma| \to \infty$, then $|\tau| \to \infty$. We have thus proved the following:

Corollary. *Assume that for any* $\mu > 0$, $B' > 0$ *all the solutions of* (2.1) *in* Ω *of class* C^h *which satisfy*

$$|u(x)| \leqslant B' e^{\mu|x|} \tag{2.20}$$

are of class C^{h+1}. *Then P is hypoelliptic.*

Using this corollary one can show that Theorem 2, Sec. 2, Chap. 10, is not true for $\gamma = 1$ for any P which is not hypoelliptic. Indeed, if the theorem were true, then all the solutions of any class C^h (h larger than the order of P) satisfying (2.20) in R^n were analytic and, by the corollary, P would be hypoelliptic. An example of a polynomial P which satisfies the assumptions of Theorem 2, Sec. 2, Chap. 10, but which is not hypoelliptic is given by

$$P(\sigma) = (\sigma_1^2 + \sigma_2^2)(\sigma_2^2 + 1).$$

3. Conditions for Hypoellipticity (Continued)

In this section we give necessary and sufficient conditions for hypoellipticity in terms of the behaviour of $P(\sigma)$ as $|\sigma| \to \infty$. The main results are stated in the following theorem.

Theorem 8. *P is hypoelliptic if and only if one of the following (equivalent) conditions is satisfied:*

(a) *for any real* θ

$$\frac{P(\sigma + \theta)}{P(\sigma)} \to 1 \qquad as\ |\sigma| \to \infty; \tag{3.1}$$

(b) *for any* $q > 0$

$$\frac{D^q P(\sigma)}{P(\sigma)} \to 0 \qquad as\ |\sigma| \to \infty. \tag{3.2}$$

The proof depends upon three lemmas which we proceed to establish.

Lemma 5. *If for any* $c > 0$ *there exists an* $A > 0$ *such that there are no roots of* $P(s)$ *in the domain* $|\text{Im } s| < c$, $|\text{Re } s| > A$, *then for any real* θ

$$\frac{P(\sigma + \theta)}{P(\sigma)} \to 1 \qquad as\ |\sigma| \to \infty. \tag{3.3}$$

Proof. We may assume that $\theta = (1, 0, \ldots, 0)$ since otherwise we first perform an appropriate linear transformation. If $|\sigma| > A + c$ and $P(\lambda) = 0$ then either $|\text{Im } \lambda| < c$, in which case $|\text{Re } \lambda| > A$, or $|\text{Im } \lambda| > c$. In either case $|\sigma - \lambda| > c$. Writing $P(\sigma)$ in the form

$$P(\sigma) = P_0(\hat{\sigma}) \prod_{j=1}^{k} (\sigma_1 - \lambda_j(\hat{\sigma})) \qquad (\hat{\sigma} = (\sigma_2, \ldots, \sigma_n),\ k \equiv k(\hat{\sigma}) \leqslant m),$$

it follows that

$$\frac{P(\sigma + \theta)}{P(\sigma)} = \prod_{j=1}^{k} \frac{\sigma_1 + 1 - \lambda_j}{\sigma_1 - \lambda_j} = \prod_{j=1}^{k} \left(1 + \frac{1}{\sigma_1 - \lambda_j}\right) = 1 + R,$$

where (since $P(\lambda_j, \hat{\sigma}) = 0$)

$$|R| \leqslant \frac{m}{c} + \binom{m}{2}\frac{1}{c^2} + \cdots + \frac{1}{c^m} \leqslant \frac{m}{c}\left(1 + \frac{1}{c}\right)^{m-1} < \epsilon$$

for any $\epsilon > 0$, if c is sufficiently large. This proves (3.3).

Lemma 6. *If* (3.3) *holds for any real* θ, *then, for any* $q > 0$,

$$\frac{D^q P(\sigma)}{P(\sigma)} \to 0 \qquad as\ |\sigma| \to \infty. \tag{3.4}$$

Proof. By Taylor's formula

$$\frac{P(\sigma + \theta)}{P(\sigma)} = 1 + \sum_{0<|q|\leqslant m} \frac{D^q P(\sigma)}{P(\sigma)} \frac{\theta^q}{q!}. \tag{3.5}$$

Taking $|\sigma| \to \infty$ and using (3.3) it follows that

$$\sum_{0<|q|\leqslant m} \frac{D^q P(\sigma)}{P(\sigma)} \frac{\theta^q}{q!} \to 0 \qquad \text{as } |\sigma| \to \infty.$$

Since this is true for any real θ, (3.4) is easily obtained.

Noting that for some q, $|q| = m$, $D^q P(\sigma) = \text{const.} \neq 0$, we obtain the following:

Corollary. *If* (3.3) *holds for all real* θ *then* $P(\sigma) \to \infty$ *as* $|\sigma| \to \infty$.

Lemma 7. *If for any* $q > 0$ (3.4) *is satisfied then for any* $c > 0$ *there exists an* $A > 0$ *such that there are no roots of* $P(s)$ *in the domain* $|\text{Im } s| < c$, $|\text{Re } s| > A$.

Proof. If the assertion is not true then there exists a sequence $\{\lambda_k\}$ of roots of $P(s)$ such that

$$|\text{Im } \lambda_k| < c, \ |\text{Re } \lambda_k| \equiv r_k \to \infty \qquad \text{as } k \to \infty.$$

The complex ball with center λ_k and radius c intersects R^n in a ball of positive radius and, since $P \not\equiv 0$, for at least one point $\sigma^{(k)}$ in that intersection $P(\sigma^{(k)}) \neq 0$. Setting $\mu_k - \sigma^{(k)} = \theta_k$ we have

$$(3.6) \qquad \frac{P(\sigma^{(k)} + \theta_k)}{P(\sigma_k)} = \frac{P(\lambda_k)}{P(\sigma_k)} = 0 \quad \text{and} \quad |\theta_k| < 1.$$

But since (3.4) is satisfied for any $q > 0$, (3.5) implies that if $|\theta| < 1$, θ complex, then

$$\frac{P(\sigma + \theta)}{P(\sigma)} \to 1 \qquad \text{as } |\sigma| \to \infty,$$

uniformly with respect to θ. This contradicts (3.6).

Combining Lemmas 5–7 and Theorems 4, 6, the proof of Theorem 8 follows.

Corollary (*to Theorem* 8). *If*

$$(3.7) \qquad \frac{D_j P(\sigma)}{P(\sigma)} \to 0 \qquad \textit{as } |\sigma| \to \infty \qquad (j = 1, \ldots, n),$$

then P *is hypoelliptic.*

Proof. Since for any real θ, $0 < \theta < 1$,

$$\log \frac{P(\sigma + \theta)}{P(\sigma)} = \log P(\sigma + \theta) - \log P(\sigma) = \sum_{j=1}^{n} \frac{\theta_j D_j P(\sigma + t\theta)}{P(\sigma + t\theta)} \to 0$$

as $|\sigma| \to \infty$, it follows that $P(\sigma + \theta)/P(\sigma) \to 1$. Now apply Theorem 8.

We conclude this section with some remarks concerning systems of differential equations:

$$(3.8) \qquad \sum_{k=1}^{N} P_{jk}\left(i \frac{\partial}{\partial x}\right) u_k(x) = 0 \qquad (j = 1, \ldots, N).$$

The concept of hypoellipticity is defined as for one equation. The construction of fundamental solutions is similar to that of one equation (see

Problem 1, Chap. 10). The representation formulas of Sec. 1 and the results of Secs. 2, 3 also remain true with minor modifications except that the role of $P(s)$ is now given to $\hat{P}(s) \equiv \det (P_{jk}(s))$. The necessary and sufficient conditions for hypoellipticity are the same conditions as given in Secs. 2, 3 with $P(s)$ replaced by $\hat{P}(s)$.

4. Examples of Hypoelliptic Equations

Theorem 9. *A polynomial $P(s)$ is elliptic if and only if it is 1-hypoelliptic, i.e., if and only if the points s of $N(P)$ satisfy*

$$|\tau| > A|\sigma| - B \tag{4.1}$$

for some positive constants A, B.

Proof. Let $P_0(s)$ be the principal part of $P(s)$, and set

$$P(s) = P_0(s) + P_1(s).$$

Assume first that P is elliptic, i.e., $P_0(\sigma) \neq 0$ if $\sigma \neq 0$. If (4.1) is not true then there exists a sequence $\{s^{(\nu)}\}$ such that

$$s^{(\nu)} = \sigma^{(\nu)} + i\tau^{(\nu)}, \quad P(s^{(\nu)}) = 0, \quad |\tau^{(\nu)}| < \frac{1}{\nu} |\sigma^{(\nu)}|, \quad |\sigma^{(\nu)}| \to \infty.$$

By Taylor's formula,

$$P(\sigma^{(\nu)} + i\tau^{(\nu)}) = P_0(\sigma^{(\nu)}) + \sum_{0<|q|\leqslant m} D^q P_0(\sigma^{(\nu)}) \frac{(i\tau^{(\nu)})^q}{q!} + \sum_{0\leqslant|q|\leqslant m-1} D^q P_1(\sigma^{(\nu)}) \frac{(i\tau^{(\nu)})^q}{q!}. \tag{4.2}$$

Since $P_0(\sigma)$ is homogeneous of degree m, $|P_0(\sigma)| \geqslant C|\sigma|^m$ $(C > 0)$. We also have, for large $|\sigma|$,

$$|D^q P_0(\sigma)| \leqslant C_q|\sigma|^{m-|q|},$$

$$|D^q P_1(\sigma)| \leqslant C'_q|\sigma|^{m-|q|-1}.$$

Using these inequalities and the inequality $|\tau^{(\nu)}| < |\sigma^{(\nu)}|/\nu$ to evaluate the right side of (4.2), we obtain

$$|P(\sigma^{(\nu)} + i\tau^{(\nu)})| > \tfrac{1}{2} C|\sigma^{(\nu)}|^m > 0$$

if ν is sufficiently large, which is impossible since $P(\sigma^{(\nu)} + i\tau^{(\nu)}) = 0$.

Suppose conversely that (4.1) is satisfied. If P is not elliptic then $P_0(\sigma^0) = 0$ for some $\sigma^0 \neq 0$. Since $P \not\equiv 0$, $P_0(\sigma^1) \neq 0$ for some σ^1. We may assume that $\sigma^0 = (1, 0, \ldots, 0)$, $\sigma^1 = (0, 1, 0, \ldots, 0)$ since otherwise we first perform a linear transformation which maps σ^0, σ^1 into $(1, 0, \ldots, 0)$ and $(0, 1, 0, \ldots, 0)$ respectively.

We shall prove that there exists a curve $s_2 = s_2(s_1)$ extending to infinity such that $(s_1, s_2(s_1))$ are roots of the polynomial

$$Q(s_1, s_2) \equiv P(s_1, s_2, 0, \ldots, 0)$$

and

$$|\tau| \leqslant \hat{A}|\sigma|^\gamma \qquad \text{for some } \gamma < 1, \hat{A} > 0. \tag{4.3}$$

Let $Q = Q_0 + Q_1$, where Q_0 is the principal part of Q. Since

$$Q_0(1, 0) = 0,$$

also $Q_0(s_1, 0) \equiv 0$. $Q_1(s_1, 0) \not\equiv 0$, for otherwise $P(s_1, 0, \ldots, 0) \equiv 0$ and we obtain a contradiction to (4.1). Hence there exist in Q_1 terms independent of s_2. Denote by i_0 the highest of the exponents of s_1 which occur in these terms. Clearly $i_0 < m$.

The solutions of $Q(s_1, s_2) = 0$, for large $|s|$, are given in terms of the Puiseux series and can therefore be represented by a finite number of curves

$$s_2 = Ms_1^\gamma E \qquad (\gamma > 0), \tag{4.4}$$

where M, γ are constants and $E \to 1$ as $|s| \to \infty$. $M \not\equiv 0$ since $Q(s_1, 0) \not\equiv 0$. We now have to recall (see for instance, Reference 33, pp. 105–109) the method by which one finds the γ's. Setting

$$Q_0(s_1, s_2) = \sum_{k=0}^{m-1} a_k s_1^k s_2^{m-k}, \qquad Q_1(s_1, s_2) = \sum_{j=0}^{m-1} \sum_{k=0}^{j} b_{kj} s_1^k s_2^{j-k},$$

we have

$$Q(s_1, Ms_1^\gamma E) = \sum_{k=0}^{m-1} a_k M^{m-k} s_1^{k+\gamma(m-k)} E^{m-k} + \sum_{j=0}^{m-1} \sum_{k=0}^{j} b_{kj} M^{j-k} s_1^{k+\gamma(j-k)} E^{j-k} = 0. \tag{4.5}$$

Denote by $N(\gamma)$ the exponents of s_1, and by $N_0(\gamma)$ the maximum of the $N(\gamma)$. Clearly $\gamma \leqslant 1$ in (4.4), for if $\gamma > 1$ then from (4.5) it follows (taking $s_1 \to \infty$) that $a_0 = Q(0, 1) = 0$. As γ increases from 0 to 1 there will occur a finite number of values $\gamma = \gamma_h$ at which two or more of the $N(\gamma)$ will be equal to $N_0(\gamma)$. In any interval (γ_h, γ_{h+1}) there is only one term in (4.5) for which $N(\gamma) = N_0(\gamma)$. The numbers γ_h are the values of γ for which (4.4) is valid.

Note that there is exactly one term for which $N(\gamma) \equiv m\gamma$ (as $a_0 \neq 0$) and exactly one term for which $N(\gamma) \equiv i_0$. The corresponding γ-lines intersect at $\gamma = i_0/m < 1$. Since $N_0(0) = i_0$ $N_0(1) = m$, it follows that the first of the points γ_h is $\leqslant i_0/m < 1$.

Having proved that there exists a curve (4.4) with $\gamma < 1$ for which $Q(s_1, s_2) = 0$, and taking $s_1 = \sigma_1$, we obtain

$$|\tau| = |\tau_2| \leqslant |s_2| \leqslant \hat{A}|\sigma_1|^\gamma \leqslant \hat{A}|\sigma|^\gamma$$

for $|\sigma|$ sufficiently large, which contradicts (4.1).

In treating parabolic equations it is more convenient to consider P as a polynomial in $n + 1$ variables. Let then

$$P(\lambda, s) = a_0\lambda^p + \sum_{k=1}^{p} a_k(s)\lambda^{p-k}, \tag{4.6}$$

where P is a polynomial in (λ, s) of degree m, and let p_0 be the reduced order of P. Assume that a_0 is a constant, $a_0 \neq 0$, and denote the roots of $P(\lambda, s) = 0$, for fixed s, by $\lambda_j = \lambda_j(s) = \xi_j(s) + \eta_j(s)$ $(j = 1, \ldots, p)$.

Theorem 10. *Let h, μ be positive numbers, $\mu \leqslant 1$, and assume that if s lies in the region $|\tau| \leqslant C_0|\sigma|^\mu$ $(C_0 > 0)$ and $|\sigma| \geqslant C_1$ then*

$$\frac{|\eta_j(s)|}{|\sigma|^h} \geqslant C_2 > 0 \qquad (j = 1, \ldots, p). \tag{4.7}$$

Then P is γ-hypoelliptic, where $\gamma = \max\,(p_0/\mu, 1/\mu, p_0/h^2)$.

Proof. Let $\bar{s} = (\lambda, s)$ satisfy $P(\lambda, s) = 0$ and set $\operatorname{Re} \bar{s} = (\xi, \sigma)$, $\operatorname{Im} \bar{s} = (\eta, \tau)$. In view of Theorem 5, Sec. 2, if suffices to show that

$$|\operatorname{Im} \bar{s}| > C|\operatorname{Re} \bar{s}|^{1/\gamma} \qquad \text{if } |\operatorname{Re} \bar{s}| \geqslant C' \tag{4.8}$$

for some positive constants C, C'. Consider first the case where $|\tau| > C_0|\sigma|^\mu$. Then,

$$\begin{aligned} |\operatorname{Im} \bar{s}|^2 &= |\tau|^2 + \eta^2 > C_0^2|\sigma|^{2\mu} = C_0^2(|\operatorname{Re} \bar{s}|^2 - \xi^2)^\mu \\ &= C_0^2|\operatorname{Re} \bar{s}|^{2\mu}\left(1 - \left(\frac{\xi}{|\operatorname{Re} \bar{s}|}\right)^2\right)^\mu. \end{aligned} \tag{4.9}$$

Next, by Theorem 4, Sec. 2, Chap. 7,

$$\xi^2 + \eta^2 \leqslant C_3(|\sigma|^2 + |\tau|^2)^{p_0} \leqslant C_4(|\tau|^{2/\mu} + |\tau|^2)^{p_0} \leqslant C_5|\tau|^{2p_0/\mu},$$

and therefore, for some $C_6 > 0$,

$$\begin{aligned} |\operatorname{Im} \bar{s}|^2 &= |\tau|^2 + \eta^2 \geqslant |\tau|^2 \geqslant C_6(\xi^2 + \eta^2)^{\mu/p_0} \geqslant C_6(\xi^2)^{\mu/p_0} \\ &= C_6(|\operatorname{Re} \bar{s}|^2 - |\sigma|^2)^{\mu/p_0} \\ &= C_6|\operatorname{Re} \bar{s}|^{2\mu/p_0}\left(1 - \left(\frac{|\sigma|}{|\operatorname{Re} \bar{s}|}\right)^2\right)^{\mu/p_0}. \end{aligned} \tag{4.10}$$

Since $|\sigma|^2 + \xi^2 = |\operatorname{Re} \bar{s}|^2$, either $|\sigma| \leqslant |\operatorname{Re} \bar{s}|/\sqrt{2}$ or $|\xi| \leqslant |\operatorname{Re} \bar{s}|/\sqrt{2}$.

Applying, correspondingly, either (4.10) or (4.9) we conclude that

$$(4.11)\qquad |\text{Im } \bar{s}| \geqslant C_7|\text{Re } \bar{s}|^\delta \qquad \text{where } \delta = \min\left(\mu, \frac{\mu}{p_0}\right),\ C_7 > 0.$$

Consider next the case where $|\tau| \leqslant C_0|\sigma|^\mu$. If $|\sigma| \geqslant C_1$ then by Theorem 4, Sec. 2, Chap. 7, and by (4.7),

$$(4.12)\qquad \begin{aligned} |\text{Im } \bar{s}|^2 &= |\tau|^2 + \eta^2 \geqslant |\tau|^2 + (C_2)^2|\sigma|^{2h} \geqslant C_8(|\sigma|^2 + |\tau|^2)^h \\ &\geqslant C_9(\xi^2 + \eta^2)^{h/p_0} \geqslant C_9[\xi^2 + (C_2)^2|\sigma|^{2h}]^{h/p_0} \\ &\geqslant C_{10}(\xi^2 + |\sigma|^2)^{h^2/p_0} \\ &= C_{10}|\text{Re } \bar{s}|^{2h^2/p_0} \qquad (C_{10} > 0), \end{aligned}$$

provided $h \leqslant 1$. If $h \geqslant 1$,

$$(4.13)\qquad \begin{aligned} |\text{Im } \bar{s}|^2 &= |\tau|^2 + \eta^2 \geqslant |\tau|^2 + (C_2)^2|\sigma|^{2h} \geqslant |\tau|^2 + (C_2)^2|\sigma|^2 \\ &\geqslant C_{11}(\xi^2 + \eta^2)^{1/p_0} \geqslant C_{11}[\xi^2 + (C_2)^2|\sigma|^{2h}]^{1/p_0} \\ &\geqslant C_{12}(\xi^2 + |\sigma|^2)^{1/p_0} = C_{12}|\text{Re } \bar{s}|^{2/p_0} \qquad (C_{12} > 0). \end{aligned}$$

From (4.12), (4.13) it follows that (4.11) holds with $\delta = \min(h^2/p_0, 1/p_0)$. Hence, (4.8) holds with $\gamma = \max(p_0/\mu, 1/\mu, p_0/h^2)$.

Consider the differential equation

$$(4.14)\qquad P\left(i\frac{\partial}{\partial t}, i\frac{\partial}{\partial x}\right)u(x, t) = 0,$$

where $P(\lambda, s)$ is given by (4.6), and assume that it is parabolic in the Petrowski sense; i.e., for all real σ, $|\sigma| = 1$, the eigenvalues $\lambda_j(\sigma)$ of the principal part of $P(\lambda, s)$ satisfy

$$\text{Im }\{\lambda_j(\sigma)\} \leqslant -\delta,$$

where δ is a positive constant. If we rewrite (4.14) as a system, then (by Theorem 1, Sec. 1, Chap. 7) the present definition coincides with the definition given in Chap. 7, Sec. 6. It therefore follows (by Theorems 11, 12, Sec. 6, Chap. 7, and by Problem 1, Chap. 7) that $h = p = p_0 > 1$, $\mu = 1$. From Theorem 10 we conclude:

Theorem 11. *If* (4.14) *is parabolic in Petrowski's sense, then it is p-hypoelliptic.*

We next prove a converse of Theorem 10.

Theorem 12. *If* $P(\lambda, s)$ *is* γ*-hypoelliptic then there exists a region* $|\tau| \leqslant C_0|\sigma|^{1/\gamma}$ $(C_0 > 0)$, $|\sigma| \geqslant C_1$, *such that*

$$(4.15)\qquad \frac{|\eta_j(s)|}{|\sigma|^{1/\gamma}} \geqslant C_2 > 0 \qquad (j = 1, \ldots, p).$$

Proof. Since P is γ-hypoelliptic, if $\bar{s} = (\lambda, s)$, $P(\lambda, s) = 0$, $|\text{Re}\, \bar{s}| \geq C_1$, then $|\text{Im}\, \bar{s}| \geq C|\text{Re}\, \bar{s}|^{1/\gamma}$ $(C > 0)$. Let $|\tau| \leq C_0|\sigma|^{1/\gamma}$, $|\sigma| \geq C_1$. Then $|\text{Re}\, \bar{s}|^2 = (\text{Re}\, \lambda)^2 + |\sigma|^2 \geq C_1$ and hence

$$(\text{Im}\, \lambda)^2 + |\tau|^2 \geq C^2[(\text{Re}\, \lambda)^2 + |\sigma|^2]^{1/\gamma}.$$

It follows that

$$\begin{aligned}(\text{Im}\, \lambda)^2 &\geq C^2[(\text{Re}\, \lambda)^2 + |\sigma|^2]^{1/\gamma} - C_0^2|\sigma|^{2/\gamma} \\ &\geq C^2|\sigma|^{2/\gamma} - C_0^2|\sigma|^{2/\gamma} \geq \tfrac{1}{2}\, C^2|\sigma|^{2/\gamma}\end{aligned}$$

provided $C_0 < C/\sqrt{2}$. This completes the proof of (4.15).

From Theorem 12 it follows that if the equation (4.14) is γ-hypoelliptic and correctly posed then it is a parabolic equation (not necessarily in Petrowski's sense).

5. Nonhomogeneous Equations

Let $P(D)$ be a differential operator of order m and denote by $P(D)\Phi$ the space $\{P(D)\varphi;\, \varphi \in \Phi\}$. Denote by K'_r the space of distributions of finite order $\leq r$, by H the space of entire functions (i.e., the functions whose Taylor's series about 0 is convergent for all $x \in R^n$), and by H_ρ the space of all entire functions of order $\leq \rho$. The following results are valid:

(5.1) $$P(D)K' = K',$$

(5.2) $$P(D)K'_r \supset K'_t \qquad (r = t + \text{const.}),$$

(5.3) $$P(D)C^r \supset C^t \qquad (r = t + \text{const.}),$$

(5.4) $$P(D)C^\infty = C^\infty,$$

(5.5) $$P(D)H = H,$$

(5.6) $$P(D)H_\rho = H_\rho \qquad (\rho > 1),$$

(5.7) $$P(D)S' = S'.$$

Note that in proving a statement of the form $P(D)\Phi = \Phi_0$ one has to prove (a) $P(D) \subset \Phi_0$, and (b) for any $f \in \Phi_0$ there exists a $u \in \Phi$ such that $P(D)u = f$. The proof of (a), for (5.1)–(5.7), is trivial. As for (b), introducing a fundamental solution G one may try to find u in the form $u = G * f$ since, formally, $P(D)(G * f) = \delta * f = f$. A proof of each of the statements (5.1)–(5.6) can indeed be given along this line, but in the construction of G (given in Chap. 10, Sec. 3) one takes a different contour of integration. The manner in which this choice of the domain of integration is made depends upon the particular statement which is to be proved. (The fundamental solutions corresponding to different contours are, in general, different from each other.) Furthermore, in proving the statements (5.1)–(5.4) one first breaks f into a sum Σf_ν (using, for instance, a partition of unity) and for each term $G * f_\nu$ one takes an appropriate domain of integration which depends also on ν. This method can also be used to derive other formulas—for instance,

$$P(D)(S_{\alpha,A})' = (S_{\alpha,A})' \qquad (0 \leqslant \alpha \leqslant 1), \tag{5.8}$$

$$P(D)(W_{M,a})' = (W_{M,a})', \tag{5.9}$$

provided for any $\epsilon > 0$, $C > 0$, $M_j(\epsilon t) > C\mu_j(t)$ for $j = 1, \ldots, n$ and for all t sufficiently large.

Equation (5.7) is equivalent to the problem of division of tempered distributions by polynomials (see Problems 2, 5) and the known proofs are entirely different from the known proofs of (5.1)–(5.6).

We proceed to give detailed proofs of (5.1)–(5.4).

Proof of (5.4). We may assume that $P(s)$ has the form (3.4), Chap. 10. From the proof of Theorem 3, Sec. 3, Chap. 10, it is clear that $|P(s)| \geqslant C' > 0$ on the manifold $\hat{T}$ defined by

$$\{\sigma \in R^n, \quad \tau_1 = \gamma_1 + k(\sigma'), \quad \tau_j = \gamma_j \qquad (j = 2, \ldots, n)\}, \tag{5.10}$$

where $\sigma' = (\sigma_2, \ldots, \sigma_n)$, γ_j are constants and $k(\sigma')$ is an appropriate step function satisfying $\gamma_0 < k(\sigma') < \gamma_0'$; C' depends only on $\gamma_0' - \gamma_0$ and P. Furthermore, if $F(s)$ is any entire function satisfying (for τ varying in bounded sets B)

$$\prod_{j=1}^{n} (1 + |s_j|)^2 |F(s)| \leqslant C'' \qquad (C'' = C''(B)),$$

then

$$\int_{\hat{T}} F(s)\, d\sigma = \int_{R^n} F(\sigma)\, d\sigma. \tag{5.11}$$

Let $g(x)$ be a C^∞ function which is positive for $-\frac{1}{2} \leqslant x_j < \frac{1}{2}$ $(j = 1, \ldots, n)$ and which vanishes for x outside $-1 \leqslant x_j \leqslant 1$ $(j = 1, \ldots, n)$. Set

$$\alpha_\nu(x) = \frac{g(x - \nu)}{\sum_\mu g(x - \mu)},$$

where $\nu = (\nu_1, \ldots, \nu_n)$, $\mu = (\mu_1, \ldots, \mu_n)$ and the ν_j, μ_j run over all the integers. Then $\Sigma\, \alpha_\nu(x) \equiv 1$. Given any $f \in C^\infty$, define $f_\nu(x) = f(x)\alpha_\nu(x)$. $f_\nu(x)$ has its support in the cube $|x_j - \nu_j| \leqslant 1$ $(j = 1, \ldots, n)$ and $f = \Sigma f_\nu$. Let $C_{\alpha\nu}$ be constants for which

$$|D^\alpha f_\nu(x)| \leqslant C_{\alpha\nu} \qquad (0 \leqslant |\alpha| < \infty). \tag{5.12}$$

Defining

$$B_h = 1 + \max_{0 \leqslant \alpha_k \leqslant h, 0 \leqslant \nu_k \leqslant h, 1 \leqslant k \leqslant n} C_{\alpha\nu}, \qquad \delta_j = \frac{1}{B_j(2 + j^2)},$$

and setting $\delta_\nu = \delta_{\nu_1} \cdots \delta_{\nu_n}$, it follows that $\delta_j < 1$ and

$$\Sigma\, C_{\alpha\nu}\delta_\nu < \infty. \tag{5.13}$$

Since $\delta_j < 1$, $t_j = -\log \delta_j$ are positive numbers. We now construct, for each ν, a manifold $\hat{T}_\nu$ of the form (5.10) with

$$\gamma_0 = 0,\ \gamma_0' = 1 \quad \text{if } \nu_1 \geqslant 0; \qquad \gamma_0 = -1,\ \gamma_0' = 0 \qquad \text{if } \nu_1 < 0;$$

$$\gamma_j = t_{\nu_j}(\operatorname{sgn} \nu_j) \qquad (j = 1, \ldots, n),$$

so that $|P(s)| \geqslant C' > 0$ on $\hat{T}_\nu$, where C' is independent of ν. Using (5.12) one finds that the Fourier transform $\tilde{f}_\nu(s) = \int f_\nu(x) e^{ix\cdot s}\, dx$ of $f_\nu(x)e^{-x\cdot\tau}$ satisfies, for $s \in \hat{T}_\nu$,

$$|s^\alpha \tilde{f}_\nu(s)| \leqslant C_1 C_{\alpha\nu} \exp \{- \sum_{j=1}^n t_{\nu_j}(|\nu_j| - 1)\}.$$

Hence,

$$\begin{aligned} |s^\alpha|\, |\tilde{f}_\nu(s)e^{-ix\cdot s}| &\leqslant C_2 C_{\alpha\nu} \exp \{ \sum_{j=1}^n |x_j|(t_{\nu_j} + 1) - \sum_{j=1}^n t_{\nu_j}(|\nu_j| - 1)\} \\ &\leqslant C_3(x) C_{\alpha\nu}\delta_\nu \end{aligned} \tag{5.14}$$

provided $2 + |x_j| < |\nu_j|$. By increasing $C_3(x)$, if necessary, we conclude that (5.14) holds for all x belonging to any bounded set and $C_3(x)$ is independent of ν. One can therefore differentiate the integral

$$\int_{\hat{T}_\nu} \frac{\tilde{f}_\nu(s)}{P(s)} e^{-ix\cdot s}\, d\sigma$$

any number of times by differentiating the integrand. Using (5.13) we further conclude that the function

$$u(x) = \frac{1}{(2\pi)^n} \sum_\nu \int_{\hat{T}_\nu} \frac{\tilde{f}_\nu(s)}{P(s)} e^{-ix\cdot s}\, d\sigma \tag{5.15}$$

is a C^∞ function and

$$P\left(i\frac{\partial}{\partial x}\right) u(x) = \frac{1}{(2\pi)^n} \sum_\nu \int_{\hat{T}_\nu} \tilde{f}_\nu(s) e^{-ix\cdot s}\, d\sigma.$$

Applying the remark concerning (5.11), the right side reduces to

$$\frac{1}{(2\pi)^n} \sum_\nu \int_{R^n} \tilde{f}_\nu(\sigma)e^{-ix\cdot\sigma}\, d\sigma = \sum_\nu f_\nu(x) = f(x).$$

Hence $P(iD)u = f$. This completes the proof.

Incidentally one can easily show that, conversely, if u is a C^∞ solution of $P(iD)u = f$ then it is given by (5.15).

The proof of (5.3) is obtained by slightly modifying the previous proof.

Proof of (5.2). It will be slightly more convenient to prove (5.2) (and also (5.1)) for $P(-iD)$ instead of $P(D)$. Write $f \in K_t'$ in the form

$$f = \sum_{|q| \leqslant M} (-1)^q D^q \hat{f}_q \qquad (M = t + \text{const.}),$$

where the $\hat{f}_q$ are continuous functions, and set $f_{q\nu} = \hat{f}_q \alpha_\nu$. Let C_ν be positive constants such that

$$|f_{q\nu}| \leqslant C_\nu \qquad \text{for } |q| \leqslant M,$$

and let δ_j be constants such that

$$0 < \delta_j < 1, \qquad \sum C_\nu \delta_\nu < \infty \qquad (\delta_\nu = \delta_{\nu_1} \cdots \delta_{\nu_n}). \tag{5.16}$$

$\hat{T}_\nu$ is defined as in the proof of (5.4), except that sgn ν_j is replaced by $-$sgn ν_j. Setting

$$F_\nu(s) = \sum_{|q| \leqslant M} (-i)^q s^q \int_{R^n} f_{q\nu}(x) e^{-ix \cdot s}\, dx$$

we define u by

$$(u, \varphi) = \frac{1}{(2\pi)^n} \sum_\nu \int_{\hat{T}_\nu} \frac{F_\nu(s)\psi(s)}{P(s)}\, d\sigma, \tag{5.17}$$

where $\varphi \in K$, $\psi = \mathfrak{F}\varphi$.

Since $\mathfrak{F}(D^q\varphi) = i^{-q}\sigma^q\psi$ and since, for any $k \geqslant 0$, $s \in \hat{T}_\nu$,

$$\prod_{j=1}^{n} (1 + |s_j|)^k |F_\nu(s)\psi(s)| \leqslant \hat{C}_k C_\nu \delta_\nu, \tag{5.18}$$

it follows, using (5.16), that u is a distribution. Furthermore, from the proof of (5.18) one also sees that u is of order $\leqslant r = t + \text{const.}$ Finally,

$$\begin{aligned}(u, P(iD)\varphi) &= \frac{1}{(2\pi)^n} \sum_\nu \int_{\hat{T}_\nu} F_\nu(s)\psi(s)\, ds \\ &= \frac{1}{(2\pi)^n} \sum_\nu \int_{R^n} F_\nu(\sigma)\psi(\sigma)\, d\sigma \\ &= \Big(\sum_q \sum_\nu (-1)^q D^q f_{q\nu}, \varphi\Big) = (f, \varphi),\end{aligned}$$

i.e., $P(-iD)u = f$.

We finally describe the proof of (5.1), leaving some details to the reader.

Proof of (5.1). Let $f \in K'$ and set $f_\nu = \alpha_\nu f$. Denoting by Ω_ν the cube $-\frac{3}{2} \leqslant x_j - \nu_j \leqslant \frac{3}{2}$ $(j = 1, \ldots, n)$, we can write

$$f_\nu = \sum_{|q| \leqslant M(\nu)} (-1)^{|q|} D^q f_{\nu q} \qquad \text{in } K',$$

where the support of $f_{\nu q}$ lies in Ω_ν. Clearly

$$(f, \varphi) = \sum_\nu \sum_{|q| \leqslant M(\nu)} \int_{R^n} f_{\nu q}(x) \, D^q \varphi(x) \, dx \qquad \text{if } \varphi \in K.$$

Let

$$|f_{\nu q}| \leqslant C_\nu \qquad \text{for } |q| \leqslant M(\nu),$$

and define u by

$$(5.19) \qquad (u, \varphi) = \frac{1}{(2\pi)^n} \sum_\nu \int_{\hat{T}_\nu} \frac{F_\nu(s)\psi(s)}{P(s)} \, d\sigma,$$

where $$F_\nu(s) = \sum_{|q| \leqslant M(\nu)} (-i)^q s^q \int_{R^n} f_{\nu q}(x) e^{-ix \cdot s} \, dx.$$

We claim that u is a distribution satisfying $P(-iD)u = f$.

The $\hat{T}_\nu$ will now be defined differently from before. The definition of the $\hat{T}_\nu$ for a bounded set of ν's is insignificant. For the sake of definiteness we define $\hat{T}_\nu$ to be the manifold T of Sec. 3, Chap. 10, if $|\nu| < H$ (H will be determined later). Next we cover R^n by a finite number of (overlapping) circular cones K_j having the origin for vertex. K_1 is a cone whose axis is the ray $x_1 = x_2 = \cdots = x_n > 0$ and whose solid angle is sufficiently small such that each of its rays has positive components. The other K_j are taken to be isomorphic to K_1. If H is sufficiently large, then we can divide the Ω_ν ($|\nu| \geqslant H$) into disjoint sets W_j such that $\Omega_\nu \subset K_j$ if $\Omega_\nu \in W_j$. For $\Omega_\nu \in W_1$ we define $\hat{T}_\nu$ by

$$\tau_1 = \gamma_1 + k(\sigma'), \qquad \tau_j = \gamma_j \qquad (2 \leqslant j \leqslant n),$$

where

$$\gamma_0 = -1, \quad \gamma_0' = 0, \quad \gamma_j < 0, \quad |\gamma_j| = M(\nu) \log (1 + |s_j|) - \log \delta_{\nu_j} \qquad (1 \leqslant j \leqslant n),$$

and the δ_j are chosen in such a way that

$$(5.20) \qquad 0 < \delta_j < 1, \qquad \sum_{\Omega_\nu \in W_1} C_\nu \delta_\nu e^{|\nu| M(\nu)} < \infty.$$

If $k(\sigma')$ is an appropriately chosen step function, $-1 < k(\sigma') < 0$, then $|P(s)| \geqslant C' > 0$, where C' is independent of ν.

If $\varphi \in K$, $\psi = \mathfrak{F}\varphi$, then one finds that

$$\prod_{j=1}^n (1 + |s_j|)^2 |F_\nu(s)\psi(s)| \leqslant C(\psi) C_\nu \delta_\nu e^{|\nu| M(\nu)} \qquad (s \in \hat{T}_\nu),$$

where $C(\psi) \to 0$ if $\varphi \to 0$ in the topology of K. For the ν with $\Omega_\nu \in W_j$ ($j \geqslant 2$) we first perform an orthogonal transformation which maps K_j onto K_1 and then proceed as before. Using (5.20) and the analogous inequalities for the W_j with $j \geqslant 2$, it follows that u is a distribution. Finally, the proof that $P(-iD)u = f$ follows similarly to the proof for (5.2), noting that on $\hat{T}_\nu$, $\tau_j = 0[\log (1 + |s_j|)]$ and therefore

$$|s|^q|F_\nu(s)\psi(s)| \leqslant A'_{\nu q}e^{A|\tau|} \leqslant A''_{\nu q}|s|^{A'}$$

for any $q \geqslant 0$, from which it follows that

$$\int_{\hat{T}_\nu} F_\nu(s)\psi(s)\, ds = \int_{R^n} F_\nu(\sigma)\psi(\sigma)\, d\sigma.$$

One can verify without difficulty that, conversely, every solution u of $P(-iD)u = f$ which belongs to some K'_r can be written in the form (5.17), and every solution u in K' of $P(-iD)u = f$ can be written in the form (5.19).

PROBLEMS

1. Prove that if P is γ-hypoelliptic and $P(D)u$ is locally of class G_γ, then u is locally of class G_γ. [*Hint:* Compare the proof of Theorem 2.]
2. The *division problem* is the following: Given a distribution S and a C^∞ function H, find a distribution T such that $HT = S$. Prove that if $n = 1$, $H = x$ there exists a solution of the equation $HT = S$ and the difference between any two solutions is $C\delta$, C a constant. [*Hint:* Let (Δ) be the space of functions $\chi(x) = x\psi(x)$, where $\psi \in (D)$, and let φ_0 be a fixed function of (D) satisfying $\varphi_0(0) = 1$. Decompose any $\varphi \in (D)$ into $\lambda\varphi_0 + \chi$, where $\lambda = \varphi(0)$, and proceed similarly to the proof of Theorem 12, Sec. 4, Chap. 3. In proving uniqueness, use Theorem 30, Sec. 7, Chap. 3.]
3. Solve the division problem $x^pT = S$ ($n = 1$) by repeated applications of Problem 2.
4. Let $H(x)$ be a C^∞ function having isolated zeroes a_j and assume that $H(x) = (x - a_j)^{h_j}(1 + o(1))$ as $x \to a_j$, h_j being positive integers. Given S, solve $HT = S$. [*Hint:* Write $S = \Sigma S_j$, where the support of S_j contains a_j but no other a_i, $i \neq j$. Solve $HT_j = S_j$ and take $T = \Sigma T_j$.]
5. Let $P(\sigma)$ ($\not\equiv 0$) be a polynomial in n variables. Prove that each of the following statements implies the one following it: (a) The mapping $\varphi(\sigma) \to P(\sigma)\varphi(\sigma)$ from S into S has a continuous inverse; (b) $P(D)S' = S'$ or, equivalently, $P(\sigma)S' = S'$; (c) there exists a $G \in S'$ such that $P(D)G = \delta$, or, equivalently, there exists an $E \in S'$ such that $P(\sigma)E(\sigma) = 1$.
6. Prove that an inequality of the form

$$|\tau| \geqslant A|\sigma|^\delta \qquad (A > 0),$$

cannot hold for all $s \in N(P)$, $|s| \geqslant B$ (for any positive constant B) if $\delta > 1$.

BIBLIOGRAPHICAL REMARKS

Chapter 1. For the general theory of linear topological spaces, see Bourbaki [8]; see also a survey article of Dieudonné [9]. Countably normed spaces were introduced by Gelfand and Shilov [23]. Detailed results appear in [25]. Most of the results concerning complete countably normed spaces and perfect spaces are valid also for more general spaces, namely, for strict inductive limits of Fréchet spaces and for Montel spaces, respectively. They were derived by Dieudonné and Schwartz [10] and others (see [9]). Union spaces were introduced in [25].

Chapters 2, 3, 4. The spaces $K\{M_p\}$, $Z\{M_p\}$ were introduced by Shilov [25]. The concept of generalized functions was first introduced by Sobolev [61] in his study of the Cauchy problem for hyperbolic equations. The theory of distributions was developed in great detail by L. Schwartz [54]. Some of his methods extend to other types of generalized functions, such as $(K\{M_p\})'$. The structure theorems of Chap. 2, Sec. 5, were given by Gelfand and Shilov [25]; for the spaces S and $K(N) \equiv (D_N)$ they were previously derived by Schwartz [54]. The results of Chap. 3 constitute the main theory of distributions of Schwartz, as given in [54]. Theorem 3 was previously known. Theorem 53, for $A = \Delta$, is due to F. Riesz. The introduction of the spaces $K_r\{M_p\}$ and the structure of their generalized functions are due to A. Friedman. The results of Chap. 4 are extensions, given by Gelfand and Shilov [25], of the analogous results for distributions.

Chapter 5. The W spaces were introduced by Gurevich [30], [31]. Other similar spaces were previously introduced by Gelfand and Shilov [22]. Theorems 2–6 were given by Gelfand and Shilov [25]. A. Friedman and M. Schiffer noted that Theorems 3, 3′ are false if $n > 1$, by constructing the counterexample following Theorem 3′. The construction given in the proof of Theorem 7 is based on a general construction given by Levine [46] in his proof of Theorem 17. The remaining results of the chapter are taken from [25], [26] with one exception; namely, Theorem 20 is given in [25] only for $p = 1$.

Chapter 6. Theorem 5 is due to Schwartz [54, vol. 2]. Theorems 2, 3, 6 are due to Gelfand and Shilov [22], [25], [26]. Theorem 7′ is due to Eskin [14]. Theorem 7 and the results of Secs. 5 and 7 were derived by Friedman [18]. Theorem 9 is due to Kostuchenko [26]. Theorem 10 was obtained by Friedman in [15]. The result of Problem 3 is due to Ehrenpreis [11].

Chapter 7. Cauchy's problem for systems with constant coefficients was treated by Petrowski [52] who studied correctly posed systems and systems parabolic in the sense of Petrowski. His results for correctly posed systems were improved by Lantze [44]. In [22] Gelfand and Shilov developed the theory of Fourier transforms of generalized functions and proved uniqueness of the Cauchy problem for any system. Kostuchenko and Shilov [42] proved existence for correctly posed systems, for $\mu > 0$. Shilov [56] introduced the concept of parabolic equations used in Sec. 6 and solved the Cauchy problem. Kostuchenko solved incorrectly posed systems (see [56]). Some mildly incorrectly posed systems were previously treated by Lantze [44] by the method of [52].

In [26] all the cases are treated in detail by the method of [22] (which was used also in [42], [56]) except the case of incorrectly posed systems with $1 \leqslant h < p_0$. For any system, the initial values are assumed to belong to an appropriate class and the generalized solution $u(x, t)$ is then shown to be a function (in the usual sense) in x, for each fixed t. Recently Friedman [18] proved that (under the same initial conditions as in [26]) $u(x, t)$ is in fact a function, in the usual sense, in (x, t) and, consequently, the Cauchy problem has a solution in the classical sense. His method, which is given in the present treatment, differs from that of [26] in the analysis of the abstract convolution $G * u_0$ (Secs. 5–10). The question of extending Theorem 16′ to $1 \leqslant h < p_0$ is unsolved.

Theorems 7 and 9 for $\beta = \infty$ were first obtained by Gelfand and Shilov [22] for $p_0' - \epsilon$. The improvement to p_0' was given by Babenko [3] and Gurevich [30], [31]. The extension to $1 < \beta < \infty$ is due to Friedman. Theorem 9 is known to hold also for $\beta = 1$ for systems parabolic in Petrowski's sense. For other systems this is unknown. Remark 1 following Theorem 9′ is due to Gurevich. Zolotarev [65] obtained similar sufficient conditions for uniqueness, namely, $|u(x, t)| \leqslant 0[\exp\{\Phi(|x|)\}]$, where $\Phi(\lambda)$ is the conjugate of $F(\lambda)$ and

$$\int [F(\lambda^{1/p_0})/\lambda^2]\, d\lambda = \infty.$$

The present treatment of uniqueness is that of [26] (also [22]). Theorem 1 is well known. Gelfond [27, p. 42] derived formulas (2.5)–(2.9). Borok [7] introduced the concept of reduced order and derived (2.17) for $\Lambda(s)$. The result of Sec. 4 is given in [26]. The treatment of the existence theory in Secs. 5–10 is due to Friedman [18] with the following exceptions: Theorems 11, 12 were essentially obtained in [26]. The estimates (6.10), (6.11) were derived by Zhitomirski [64] by the method of Problem 2. The results of Sec. 12 are due to Gurevich (see [26]). The results of Sec. 13 were obtained by Friedman [17]. Some related problems were treated by Kaminin [41].

Theorem 24 was given by Seidenberg in [55]. A more detailed proof is given in [28]. Lemmas 9, 10 are due to Gorin [28]. The proofs follow a method previously used by Hörmander [34]. Theorem 25 was obtained by Borok [5] and Theorem 26 is due to Friedman [16]. For other inverse theorems, see [6], [26].

Hyperbolic equations with constant coefficients were treated by Gårding [20] who obtained, by more involved analysis, stronger results on the support of Green's function. He also derived an inverse theorem. The Cauchy problem for hyperbolic equations with variable coefficients was solved by Petrowski [51], Leray [45], and Gårding [21].

A uniqueness theorem for the heat equation was first given by Tychonov [63] and Täcklind [62]. Tychonov showed that $p_0' = 2$ cannot be replaced by $p_0' + \epsilon$ (the example in Remark 2, Sec. 3, was given by him in the case $p = 2$, $q = 1$), and Täcklind gave a more precise bound on the uniqueness class. Petrowski's uniqueness and existence theorems for parabolic systems with coefficients depending only on t [52] were extended to (Petrowski's parabolic) systems with variable coefficients by Ladyzhenskaya [43], Eidelman [13], and others. Zolotarev [66] derived a more precise bound on the uniqueness classes (in the case of variable coefficients), namely, $u(x, t) = O[\exp\{|x|h(|x|)\}]$, where h is monotone increasing and $\int dr/[h(r)]^{p-1} = \infty$. For constant coefficients this condition is also necessary. Zolotarev's results contain those of Täcklind. The Cauchy problem for parabolic equations, in the sense of Sec. 6, with variable coefficients, is unsolved.

Chapter 8. The results of Secs. 1, 3, 4, 5 were obtained by Friedman [16]. Some results in the case when condition (A) is satisfied were derived by Iskenderov [38] by the method of Petrowski [52].

Chapter 9. S spaces were introduced by Shilov [57] and are treated in detail in [25]. They were used, in [26], in solving the Cauchy problem. Babenko [3] gave necessary and sufficient conditions for nontriviality of the more general S spaces.

Chapter 10. Theorems 1, 2 are due to Shilov [26], [59]. The existence of fundamental solutions for differential equations with constant coefficients was first proved by Ehrenpreis [12, part I] and Malgrange [49]. The present construction was essentially given by Hörmander [34]. Radon's problem was solved by John (see [40]). He also gave a new derivation of Herglotz-Petrowski formulas. The present treatment of Radon's problem is due to Hachaturov [32]. The construction of the fundamental solution for hyperbolic equations is taken from [24] and is related to the method of John. Finite parts of integrals were introduced by Hadamard and used by him and by M. Riesz [53]. The connection with distributions was developed by Schwartz [54]. Many useful formulas were derived in [24].

Chapter 11. The concept of hypoellipticity is due to Schwartz [54]. Theorem 3 was found by Agranowich (see [2]). The results of Secs. 2, 3 are due to Hörmander [34]. (The corollary to Theorem 7 was given by Grushin [29].) The present treatment (including Secs. 1, 4) was essentially given by Shilov [58] and it contains some simplifications over the original work of Hörmander. An extension of some of the results to systems (M equations in N unknowns) was given by Hörmander [35]. Hypoelliptic convolution equations were treated by Ehrenpreis [12, part IV] and Hörmander [37].

Ehrenpreis [12] proved (5.1), (5.2), (5.4), (5.5), (5.6). Malgrange [49] proved (5.2), (5.3), (5.4). Shirota [60] used the method outlined in Sec. 5 to prove (5.1), (5.4). That method was also used by Agranowich [1] to derive (5.8), (5.9). In a survey article [2] Agranowich simplified the proofs of Shirota and gave also proofs of (5.2), (5.5), (5.6). The division problem for analytic functions was solved by Lojasiewicz [47]; for polynomials, a simpler proof was independently given by Hörmander [36].

BIBLIOGRAPHY

[1] M. S. Agranowich, "Existence of solutions of partial differential equations with constant coefficients in some classes of functions," *Vestnik Moscow Univ.*, no. 3 (1959), pp. 3–13.

[2] ————, "On partial differential equations with constant coefficients," *Uspehi Math. Nauk SSSR*, **16**:2 (1961), pp. 27–93.

[3] K. I. Babenko, "On a new problem of quasi-analyticity and on Fourier transforms of entire functions," *Trudy Moscow Math. Obshch.*, **5** (1956), 523–542.

[4] S. Bochner and K. Chandrasekharan, "Fourier transforms," *Annals of Mathematics Studies*, no. 19 (Princeton, 1949).

[5] V. M. Borok, "The solution of Cauchy's problem for certain types of linear partial differential equations," *Math Sbornik*, **36** (1955), 281–298.

[6] ————, "On one characteristic property of parabolic systems," *Doklady Akad. Nauk SSSR* (N.S.), **110** (1956), 903–905.

[7] ————, "Reducing systems of linear partial differential equations with constant coefficients to canonical form," *Doklady Akad. Nauk SSSR* (N.S.), **115** (1957), 13–16.

[8] N. Bourbaki, *Éléments de mathématique*, vol. 5, *Espaces vectorielles topologiques* (Paris: Hermann), Chaps. 1–2 (1953), 3–5 (1955).

[9] J. A. Dieudonné, "Recent developments in the theory of locally convex spaces," *Bull. Amer. Math. Soc.*, **59** (1953), 495–512.

[10] J. A. Dieudonné and L. Schwartz, "La dualité dans les espaces ($\mathfrak{F}$) *et* ($\mathfrak{L}\mathfrak{F}$)," *Ann. Inst. Fourier Grenoble*, **1** (1949–50), 61–101.

[11] L. Ehrenpreis, "Mean periodic functions," I, *Amer. J. Math.*, **77** (1955), 293–328.

[12] ————, "Solutions of some problems of division," *Amer. J. Math.*, I, **76** (1954), 883–903; II, **77** (1955), 286–292; III, **78** (1956), 685–715; IV, **82** (1960), 522–588.

[13] S. D. Eidelman, "On fundamental solutions for parabolic systems," *Math. Sbornik*, **38** (1956), 51–92.

[14] G. I. Eskin, "Generalization of the Paley-Wiener-Schwartz theorem," *Uspehi Math. Nauk SSSR*, **16**:1 (1961), 185–188.

[15] A. Friedman, "On the regularity of the solutions of nonlinear elliptic and parabolic systems of partial differential equations," *J. Math. and Mech.*, **7** (1958), 43–59.

[16] ————, "The Cauchy problem in several time variables," *J. Math. and Mech.*, **11** (1962), 859–889.

[17] ————, "A difference-differential scheme for the general Cauchy problem," *J. Math. and Mech.*, **11** (1962), 891–905.

[18] ————, "Existence of smooth solutions of the Cauchy problem for differential systems of any type," *J. Math. and Mech.*, **12** (1963), to appear.

[19] F. R. Gantmaher, *The theory of matrices* (New York: Chelsea, 1959), vol. 1.

[20] L. Garding, "Linear hyperbolic partial differential equations with constant coefficients," *Acta Math.*, **85** (1951), 1–62.

[21] ————, *Cauchy's problem for hyperbolic equations,* Lectures, University of Chicago, 1957.

[22] I. M. Gelfand and G. E. Shilov, "Fourier transforms of rapidly increasing functions and the questions of uniqueness of solutions of Cauchy's problem," *Uspehi Math. Nauk SSSR*, **8**:6 (1953), 3–54.

[23] ————, "Quelques applications de la théorie des fonctions généralisées," *J. de Math. Pure et Appl.*, **35** (1956), 383–412.

[24] ————, *Generalized functions,* vol. 1, *Generalized functions and operations on them* (2nd ed.; Moscow, 1959).

[25] ————, *Generalized functions,* vol. 2, *Spaces of fundamental functions and generalized functions* (Moscow, 1958).

[26] ————, *Generalized functions,* vol. 3, *Some questions in the theory of differential equations* (Moscow, 1958).

[27] A. O. Gelfond, *Calculus of finite differences* (2nd ed.; Moscow, 1959) (German edition, Berlin, 1958).

[28] E. A. Gorin, "Asymptotic properties of polynomials and algebraic functions of several variables," *Uspehi Math. Nauk SSSR*, **16**:1 (1961), 91–118.

[29] V. V. Grushin, "On a property of hypoelliptic equations," *Doklady Akad. Nauk SSSR* (N.S.), **137** (1961), 768–771.

[30] B. L. Gurevich, "New types of spaces of fundamental and generalized functions and Cauchy's problem for systems of finite difference equations," *Doklady Akad. Nauk SSSR* (N.S.), **99** (1954), 893–895.

[31] ————, "New types of fundamental and generalized spaces and Cauchy's problem for systems of difference equations involving differential operations," *Doklady Akad. Nauk SSSR* (N.S.), **108** (1956), 1001–1003.

[32] A. A. Hachaturov, "Defining the value of measures in domains of an n-dimensional space by their values on all half spaces," *Uspehi Math. Nauk SSSR*, **9**:3 (1954), 205–212.

[33] E. Hille, *Analytic function theory* (Ginn and Company, 1962), vol. 2.

[34] L. Hörmander, "On the theory of general partial differential operators," *Acta Math.*, **94** (1955), 161–248.

[35] ————, "Differentiability properties of solutions of systems of differential equations," *Arkiv för Matematik*, **3** (1958), 527–535.

[36] ———, "On the division problem of distributions by polynomials," *Arkiv för Matematik*, **3** (1958), 555–568.

[37] ———, "Hypoelliptic convolution equations," *Math. Scand.*, **9** (1961), 178–184.

[38] S. A. Iskenderov, "On the generalized Cauchy problem for a system," *Akad. Nauk Azerbaidzan, SSR., Trudy Inst. Fiz. Mat.*, no. 4–5 (1952), 106-127.

[39] F. John, "General properties of solutions of linear elliptic differential equations," *Proc. Sympos. Spectral Theory and Differential Problems* (Stillwater, Oklahoma, 1951), pp. 113–175.

[40] ———, *Plane waves and spherical means* (New York: Interscience Publishers, 1955).

[41] L. I. Kaminin, "Solutions of the Cauchy problem for infinite systems of ordinary differential equations," *Vestnik Moscow Univ.*, **4**:6 (1956), 3–10.

[42] A. G. Kostuchenko and G. E. Shilov, "On the solution of Cauchy's problem for regular systems of linear partial differential equations," *Uspehi Math. Nauk SSSR*, **9**:3 (1954), 141–148.

[43] O. A. Ladyzhenskaya, "On uniqueness of solutions of the Cauchy problem for linear parabolic equations," *Math. Sbornik*, **27** (1950), 175–184.

[44] V. E. Lantze, "On the Cauchy problem in the domain of functions of real variables," *Ukraine Math. J.*, **1** (1949), 42–63.

[45] J. Leray, *Hyperbolic differential equations* (Princeton, N.J.: Institute for Advanced Study, 1952).

[46] B. Ya. Levine, *Distribution of zeroes of entire functions* (Moscow, 1956).

[47] S. Lojasiewicz, "Sur le probleme de la division," *Studia Math.*, **18** (1959), 87–136.

[48] W. Magnus and F. Oberhettinger, *Formulas and theorems for the special functions of mathematical physics* (New York: Chelsea, 1949).

[49] B. Malgrange, "Existence et approximations des solutions de équations aux dérivées partielles et des équations de convolution," *Ann. Inst. Fourier Grenoble*, **6** (1955–56), 271–354.

[50] S. Mandelbrojt, *Séries de Fourier et classes quasi-analytiques de fonctions* (Paris: Gauthier-Villars, 1935).

[51] I. G. Petrowski, "Über das Cauchyche Problem für Systeme von partiellen Differentialgleichungen," *Rec. Math. (Math. Sbornik)*, **2** (1937), 814–868.

[52] ———, "On the Cauchy problem for systems of linear partial differential equations in the domain of nonanalytic functions," *Bull. Moscow Univ.*, Sec. A, I, **7** (1938).

[53] M. Riesz, "L'integrale de Riemann-Liouville et le problème de Cauchy," *Acta Math.*, **81** (1949), 1–223.

[54] L. Schwartz, *Théorie des distributions* (2nd ed.; Paris: Hermann), vol. 1 (1957), vol. 2 (1959).

[55] A. Seidenberg, "A new decision problem for elementary algebra," *Ann. of Math.*, **60** (1954), 365–374.

[56] G. E. Shilov, "On the condition of correctness of the Cauchy problem for systems of partial differential equations with constant coefficients," *Uspehi Math. Nauk SSSR*, **10**:4 (1955), 89–100.

[57] ————, "On a problem of quasi-analyticity," *Doklady Akad. Nauk SSSR* (N.S.), **102** (1955), 893–896.

[58] ————, "Local properties of solutions of partial differential equations with constant coefficients," *Uspehi Math. Nauk SSSR*, **14**:5 (1959), 3–44.

[59] ————, "An analogue of a theorem of Laurent Schwartz," *Izvestya Vis'shih Uchebnik Zavedenie Matematica*, no. 4 (1961), 137–147.

[60] T. Shirota, "On solutions of partial differential equations with parameter," *Proc. Japan Acad.*, **32** (1956), 401–405.

[61] S. L. Sobolev, "A new method for solving the Cauchy problem," *Math. Sbornik*, **1** (1936), 39–72.

[62] Täcklind, "Sur les classes quasi-analytiques des solutions des équations aux dérivées partielles du type parabolique," *Nova Acta Soc. Sci. Upsaliensis*, **10** (1936), 1–57.

[63] A. N. Tychonov, "Uniqueness theorems for the heat equation," *Math. Sbornik*, **42** (1935), 199–216.

[64] Y. I. Zhitomirski, "The Cauchy problem for one type of systems of partial differential equations, parabolic in the sense of G. E. Shilov, with variable coefficients," *Izvest. Akad. Nauk SSSR*, **23** (1959), 925–932.

[65] G. N. Zolotarev, "On upper estimates for the uniqueness classes of solutions of the Cauchy problem for systems of partial differential equations," *Nauchni Doklady Vis'shih Shkoli Fys. Mat.*, no. 2 (1958), 37–40.

[66] ————, "On uniqueness of the Cauchy problem for parabolic systems in the sense of I. G. Petrowski," *Izvestia Vis'shih Uchebnik Zavedenie Matematica*, no. 2 (1958), 118–135.

INDEX FOR SPACES

INDEX

D

E

F

G

H

I

L

M

N

O

P

A CATALOG OF SELECTED

DOVER BOOKS

IN SCIENCE AND MATHEMATICS

Astronomy

BURNHAM'S CELESTIAL HANDBOOK, Robert Burnham, Jr. Thorough guide to the stars beyond our solar system. Exhaustive treatment. Alphabetical by constellation: Andromeda to Cetus in Vol. 1; Chamaeleon to Orion in Vol. 2; and Pavo to Vulpecula in Vol. 3. Hundreds of illustrations. Index in Vol. 3. 2,000pp. 6⅛ x 9¼.
Vol. I: 23567-X
Vol. II: 23568-8
Vol. III: 23673-0

EXPLORING THE MOON THROUGH BINOCULARS AND SMALL TELESCOPES, Ernest H. Cherrington, Jr. Informative, profusely illustrated guide to locating and identifying craters, rills, seas, mountains, other lunar features. Newly revised and updated with special section of new photos. Over 100 photos and diagrams. 240pp. 8¼ x 11. 24491-1

THE EXTRATERRESTRIAL LIFE DEBATE, 1750–1900, Michael J. Crowe. First detailed, scholarly study in English of the many ideas that developed from 1750 to 1900 regarding the existence of intelligent extraterrestrial life. Examines ideas of Kant, Herschel, Voltaire, Percival Lowell, many other scientists and thinkers. 16 illustrations. 704pp. 5⅜ x 8½. 40675-X

THEORIES OF THE WORLD FROM ANTIQUITY TO THE COPERNICAN REVOLUTION, Michael J. Crowe. Newly revised edition of an accessible, enlightening book recreates the change from an earth-centered to a sun-centered conception of the solar system. 242pp. 5⅜ x 8½. 41444-2

A HISTORY OF ASTRONOMY, A. Pannekoek. Well-balanced, carefully reasoned study covers such topics as Ptolemaic theory, work of Copernicus, Kepler, Newton, Eddington's work on stars, much more. Illustrated. References. 521pp. 5⅜ x 8½.
65994-1

A COMPLETE MANUAL OF AMATEUR ASTRONOMY: Tools and Techniques for Astronomical Observations, P. Clay Sherrod with Thomas L. Koed. Concise, highly readable book discusses: selecting, setting up and maintaining a telescope; amateur studies of the sun; lunar topography and occultations; observations of Mars, Jupiter, Saturn, the minor planets and the stars; an introduction to photoelectric photometry; more. 1981 ed. 124 figures. 26 halftones. 37 tables. 335pp. 6½ x 9¼.
42820-6

AMATEUR ASTRONOMER'S HANDBOOK, J. B. Sidgwick. Timeless, comprehensive coverage of telescopes, mirrors, lenses, mountings, telescope drives, micrometers, spectroscopes, more. 189 illustrations. 576pp. 5⅝ x 8¼. (Available in U.S. only.)
24034-7

STARS AND RELATIVITY, Ya. B. Zel'dovich and I. D. Novikov. Vol. 1 of *Relativistic Astrophysics* by famed Russian scientists. General relativity, properties of matter under astrophysical conditions, stars, and stellar systems. Deep physical insights, clear presentation. 1971 edition. References. 544pp. 5⅝ x 8¼. 69424-0

Chemistry

THE SCEPTICAL CHYMIST: The Classic 1661 Text, Robert Boyle. Boyle defines the term "element," asserting that all natural phenomena can be explained by the motion and organization of primary particles. 1911 ed. viii+232pp. 5⅜ x 8½. 42825-7

RADIOACTIVE SUBSTANCES, Marie Curie. Here is the celebrated scientist's doctoral thesis, the prelude to her receipt of the 1903 Nobel Prize. Curie discusses establishing atomic character of radioactivity found in compounds of uranium and thorium; extraction from pitchblende of polonium and radium; isolation of pure radium chloride; determination of atomic weight of radium; plus electric, photographic, luminous, heat, color effects of radioactivity. ii+94pp. 5⅜ x 8½. 42550-9

CHEMICAL MAGIC, Leonard A. Ford. Second Edition, Revised by E. Winston Grundmeier. Over 100 unusual stunts demonstrating cold fire, dust explosions, much more. Text explains scientific principles and stresses safety precautions. 128pp. 5⅜ x 8½. 67628-5

THE DEVELOPMENT OF MODERN CHEMISTRY, Aaron J. Ihde. Authoritative history of chemistry from ancient Greek theory to 20th-century innovation. Covers major chemists and their discoveries. 209 illustrations. 14 tables. Bibliographies. Indices. Appendices. 851pp. 5⅜ x 8½. 64235-6

CATALYSIS IN CHEMISTRY AND ENZYMOLOGY, William P. Jencks. Exceptionally clear coverage of mechanisms for catalysis, forces in aqueous solution, carbonyl- and acyl-group reactions, practical kinetics, more. 864pp. 5⅜ x 8½. 65460-5

ELEMENTS OF CHEMISTRY, Antoine Lavoisier. Monumental classic by founder of modern chemistry in remarkable reprint of rare 1790 Kerr translation. A must for every student of chemistry or the history of science. 539pp. 5⅜ x 8½. 64624-6

THE HISTORICAL BACKGROUND OF CHEMISTRY, Henry M. Leicester. Evolution of ideas, not individual biography. Concentrates on formulation of a coherent set of chemical laws. 260pp. 5⅜ x 8½. 61053-5

A SHORT HISTORY OF CHEMISTRY, J. R. Partington. Classic exposition explores origins of chemistry, alchemy, early medical chemistry, nature of atmosphere, theory of valency, laws and structure of atomic theory, much more. 428pp. 5⅜ x 8½. (Available in U.S. only.) 65977-1

GENERAL CHEMISTRY, Linus Pauling. Revised 3rd edition of classic first-year text by Nobel laureate. Atomic and molecular structure, quantum mechanics, statistical mechanics, thermodynamics correlated with descriptive chemistry. Problems. 992pp. 5⅜ x 8½. 65622-5

FROM ALCHEMY TO CHEMISTRY, John Read. Broad, humanistic treatment focuses on great figures of chemistry and ideas that revolutionized the science. 50 illustrations. 240pp. 5⅜ x 8½. 28690-8

Engineering

DE RE METALLICA, Georgius Agricola. The famous Hoover translation of greatest treatise on technological chemistry, engineering, geology, mining of early modern times (1556). All 289 original woodcuts. 638pp. 6¾ x 11. 60006-8

FUNDAMENTALS OF ASTRODYNAMICS, Roger Bate et al. Modern approach developed by U.S. Air Force Academy. Designed as a first course. Problems, exercises. Numerous illustrations. 455pp. 5⅜ x 8½. 60061-0

DYNAMICS OF FLUIDS IN POROUS MEDIA, Jacob Bear. For advanced students of ground water hydrology, soil mechanics and physics, drainage and irrigation engineering, and more. 335 illustrations. Exercises, with answers. 784pp. 6⅛ x 9¼. 65675-6

THEORY OF VISCOELASTICITY (Second Edition), Richard M. Christensen. Complete, consistent description of the linear theory of the viscoelastic behavior of materials. Problem-solving techniques discussed. 1982 edition. 29 figures. xiv+364pp. 6⅛ x 9¼. 42880-X

MECHANICS, J. P. Den Hartog. A classic introductory text or refresher. Hundreds of applications and design problems illuminate fundamentals of trusses, loaded beams and cables, etc. 334 answered problems. 462pp. 5⅜ x 8½. 60754-2

MECHANICAL VIBRATIONS, J. P. Den Hartog. Classic textbook offers lucid explanations and illustrative models, applying theories of vibrations to a variety of practical industrial engineering problems. Numerous figures. 233 problems, solutions. Appendix. Index. Preface. 436pp. 5⅜ x 8½. 64785-4

STRENGTH OF MATERIALS, J. P. Den Hartog. Full, clear treatment of basic material (tension, torsion, bending, etc.) plus advanced material on engineering methods, applications. 350 answered problems. 323pp. 5⅜ x 8½. 60755-0

A HISTORY OF MECHANICS, René Dugas. Monumental study of mechanical principles from antiquity to quantum mechanics. Contributions of ancient Greeks, Galileo, Leonardo, Kepler, Lagrange, many others. 671pp. 5⅜ x 8½. 65632-2

STABILITY THEORY AND ITS APPLICATIONS TO STRUCTURAL MECHANICS, Clive L. Dym. Self-contained text focuses on Koiter postbuckling analyses, with mathematical notions of stability of motion. Basing minimum energy principles for static stability upon dynamic concepts of stability of motion, it develops asymptotic buckling and postbuckling analyses from potential energy considerations, with applications to columns, plates, and arches. 1974 ed. 208pp. 5⅜ x 8½. 42541-X

METAL FATIGUE, N. E. Frost, K. J. Marsh, and L. P. Pook. Definitive, clearly written, and well-illustrated volume addresses all aspects of the subject, from the historical development of understanding metal fatigue to vital concepts of the cyclic stress that causes a crack to grow. Includes 7 appendixes. 544pp. 5⅜ x 8½. 40927-9

Mathematics

FUNCTIONAL ANALYSIS (Second Corrected Edition), George Bachman and Lawrence Narici. Excellent treatment of subject geared toward students with background in linear algebra, advanced calculus, physics, and engineering. Text covers introduction to inner-product spaces, normed, metric spaces, and topological spaces; complete orthonormal sets, the Hahn-Banach Theorem and its consequences, and many other related subjects. 1966 ed. 544pp. 6⅛ x 9¼. 40251-7

ASYMPTOTIC EXPANSIONS OF INTEGRALS, Norman Bleistein & Richard A. Handelsman. Best introduction to important field with applications in a variety of scientific disciplines. New preface. Problems. Diagrams. Tables. Bibliography. Index. 448pp. 5⅜ x 8½. 65082-0

VECTOR AND TENSOR ANALYSIS WITH APPLICATIONS, A. I. Borisenko and I. E. Tarapov. Concise introduction. Worked-out problems, solutions, exercises. 257pp. 5⅝ x 8¼. 63833-2

THE ABSOLUTE DIFFERENTIAL CALCULUS (CALCULUS OF TENSORS), Tullio Levi-Civita. Great 20th-century mathematician's classic work on material necessary for mathematical grasp of theory of relativity. 452pp. 5⅝ x 8¼. 63401-9

AN INTRODUCTION TO ORDINARY DIFFERENTIAL EQUATIONS, Earl A. Coddington. A thorough and systematic first course in elementary differential equations for undergraduates in mathematics and science, with many exercises and problems (with answers). Index. 304pp. 5⅜ x 8½. 65942-9

FOURIER SERIES AND ORTHOGONAL FUNCTIONS, Harry F. Davis. An incisive text combining theory and practical example to introduce Fourier series, orthogonal functions and applications of the Fourier method to boundary-value problems. 570 exercises. Answers and notes. 416pp. 5⅜ x 8½. 65973-9

COMPUTABILITY AND UNSOLVABILITY, Martin Davis. Classic graduate-level introduction to theory of computability, usually referred to as theory of recurrent functions. New preface and appendix. 288pp. 5⅜ x 8½. 61471-9

ASYMPTOTIC METHODS IN ANALYSIS, N. G. de Bruijn. An inexpensive, comprehensive guide to asymptotic methods–the pioneering work that teaches by explaining worked examples in detail. Index. 224pp. 5⅜ x 8½ 64221-6

APPLIED COMPLEX VARIABLES, John W. Dettman. Step-by-step coverage of fundamentals of analytic function theory–plus lucid exposition of five important applications: Potential Theory; Ordinary Differential Equations; Fourier Transforms; Laplace Transforms; Asymptotic Expansions. 66 figures. Exercises at chapter ends. 512pp. 5⅜ x 8½. 64670-X

INTRODUCTION TO LINEAR ALGEBRA AND DIFFERENTIAL EQUATIONS, John W. Dettman. Excellent text covers complex numbers, determinants, orthonormal bases, Laplace transforms, much more. Exercises with solutions. Undergraduate level. 416pp. 5⅜ x 8½. 65191-6

Math–Decision Theory, Statistics, Probability

ELEMENTARY DECISION THEORY, Herman Chernoff and Lincoln E. Moses. Clear introduction to statistics and statistical theory covers data processing, probability and random variables, testing hypotheses, much more. Exercises. 364pp. 5⅜ x 8½. 65218-1

STATISTICS MANUAL, Edwin L. Crow et al. Comprehensive, practical collection of classical and modern methods prepared by U.S. Naval Ordnance Test Station. Stress on use. Basics of statistics assumed. 288pp. 5⅜ x 8½. 60599-X

SOME THEORY OF SAMPLING, William Edwards Deming. Analysis of the problems, theory, and design of sampling techniques for social scientists, industrial managers, and others who find statistics important at work. 61 tables. 90 figures. xvii +602pp. 5⅜ x 8½. 64684-X

LINEAR PROGRAMMING AND ECONOMIC ANALYSIS, Robert Dorfman, Paul A. Samuelson and Robert M. Solow. First comprehensive treatment of linear programming in standard economic analysis. Game theory, modern welfare economics, Leontief input-output, more. 525pp. 5⅜ x 8½. 65491-5

PROBABILITY: An Introduction, Samuel Goldberg. Excellent basic text covers set theory, probability theory for finite sample spaces, binomial theorem, much more. 360 problems. Bibliographies. 322pp. 5⅜ x 8½. 65252-1

GAMES AND DECISIONS: Introduction and Critical Survey, R. Duncan Luce and Howard Raiffa. Superb nontechnical introduction to game theory, primarily applied to social sciences. Utility theory, zero-sum games, n-person games, decision-making, much more. Bibliography. 509pp. 5⅜ x 8½. 65943-7

INTRODUCTION TO THE THEORY OF GAMES, J. C. C. McKinsey. This comprehensive overview of the mathematical theory of games illustrates applications to situations involving conflicts of interest, including economic, social, political, and military contexts. Appropriate for advanced undergraduate and graduate courses; advanced calculus a prerequisite. 1952 ed. x+372pp. 5⅜ x 8½. 42811-7

FIFTY CHALLENGING PROBLEMS IN PROBABILITY WITH SOLUTIONS, Frederick Mosteller. Remarkable puzzlers, graded in difficulty, illustrate elementary and advanced aspects of probability. Detailed solutions. 88pp. 5⅜ x 8½. 65355-2

PROBABILITY THEORY: A Concise Course, Y. A. Rozanov. Highly readable, self-contained introduction covers combination of events, dependent events, Bernoulli trials, etc. 148pp. 5⅜ x 8¼. 63544-9

STATISTICAL METHOD FROM THE VIEWPOINT OF QUALITY CONTROL, Walter A. Shewhart. Important text explains regulation of variables, uses of statistical control to achieve quality control in industry, agriculture, other areas. 192pp. 5⅜ x 8½. 65232-7

Math–Geometry and Topology

ELEMENTARY CONCEPTS OF TOPOLOGY, Paul Alexandroff. Elegant, intuitive approach to topology from set-theoretic topology to Betti groups; how concepts of topology are useful in math and physics. 25 figures. 57pp. 5⅜ x 8½. 60747-X

COMBINATORIAL TOPOLOGY, P. S. Alexandrov. Clearly written, well-organized, three-part text begins by dealing with certain classic problems without using the formal techniques of homology theory and advances to the central concept, the Betti groups. Numerous detailed examples. 654pp. 5⅜ x 8½. 40179-0

EXPERIMENTS IN TOPOLOGY, Stephen Barr. Classic, lively explanation of one of the byways of mathematics. Klein bottles, Moebius strips, projective planes, map coloring, problem of the Koenigsberg bridges, much more, described with clarity and wit. 43 figures. 210pp. 5⅜ x 8½. 25933-1

CONFORMAL MAPPING ON RIEMANN SURFACES, Harvey Cohn. Lucid, insightful book presents ideal coverage of subject. 334 exercises make book perfect for self-study. 55 figures. 352pp. 5⅝ x 8¼. 64025-6

THE GEOMETRY OF RENÉ DESCARTES, René Descartes. The great work founded analytical geometry. Original French text, Descartes's own diagrams, together with definitive Smith-Latham translation. 244pp. 5⅜ x 8½. 60068-8

PRACTICAL CONIC SECTIONS: The Geometric Properties of Ellipses, Parabolas and Hyperbolas, J. W. Downs. This text shows how to create ellipses, parabolas, and hyperbolas. It also presents historical background on their ancient origins and describes the reflective properties and roles of curves in design applications. 1993 ed. 98 figures. xii+100pp. 6½ x 9¼. 42876-1

THE THIRTEEN BOOKS OF EUCLID'S ELEMENTS, translated with introduction and commentary by Thomas L. Heath. Definitive edition. Textual and linguistic notes, mathematical analysis. 2,500 years of critical commentary. Unabridged. 1,414pp. 5⅜ x 8½.. Three-vol. set. Vol. I: 60088-2 Vol. II: 60089-0 Vol. III: 60090-4

GEOMETRY OF COMPLEX NUMBERS, Hans Schwerdtfeger. Illuminating, widely praised book on analytic geometry of circles, the Moebius transformation, and two-dimensional non-Euclidean geometries. 200pp. 5⅝ x 8¼. 63830-8

DIFFERENTIAL GEOMETRY, Heinrich W. Guggenheimer. Local differential geometry as an application of advanced calculus and linear algebra. Curvature, transformation groups, surfaces, more. Exercises. 62 figures. 378pp. 5⅜ x 8½. 63433-7

CURVATURE AND HOMOLOGY: Enlarged Edition, Samuel I. Goldberg. Revised edition examines topology of differentiable manifolds; curvature, homology of Riemannian manifolds; compact Lie groups; complex manifolds; curvature, homology of Kaehler manifolds. New Preface. Four new appendixes. 416pp. 5⅜ x 8½. 40207-X

History of Math

THE WORKS OF ARCHIMEDES, Archimedes (T. L. Heath, ed.). Topics include the famous problems of the ratio of the areas of a cylinder and an inscribed sphere; the measurement of a circle; the properties of conoids, spheroids, and spirals; and the quadrature of the parabola. Informative introduction. clxxxvi+326pp; supplement, 52pp. 5⅜ x 8½. 42084-1

A SHORT ACCOUNT OF THE HISTORY OF MATHEMATICS, W. W. Rouse Ball. One of clearest, most authoritative surveys from the Egyptians and Phoenicians through 19th-century figures such as Grassman, Galois, Riemann. Fourth edition. 522pp. 5⅜ x 8½. 20630-0

THE HISTORY OF THE CALCULUS AND ITS CONCEPTUAL DEVELOPMENT, Carl B. Boyer. Origins in antiquity, medieval contributions, work of Newton, Leibniz, rigorous formulation. Treatment is verbal. 346pp. 5⅜ x 8½. 60509-4

THE HISTORICAL ROOTS OF ELEMENTARY MATHEMATICS, Lucas N. H. Bunt, Phillip S. Jones, and Jack D. Bedient. Fundamental underpinnings of modern arithmetic, algebra, geometry, and number systems derived from ancient civilizations. 320pp. 5⅜ x 8½. 25563-8

A HISTORY OF MATHEMATICAL NOTATIONS, Florian Cajori. This classic study notes the first appearance of a mathematical symbol and its origin, the competition it encountered, its spread among writers in different countries, its rise to popularity, its eventual decline or ultimate survival. Original 1929 two-volume edition presented here in one volume. xxviii+820pp. 5⅜ x 8½. 67766-4

GAMES, GODS & GAMBLING: A History of Probability and Statistical Ideas, F. N. David. Episodes from the lives of Galileo, Fermat, Pascal, and others illustrate this fascinating account of the roots of mathematics. Features thought-provoking references to classics, archaeology, biography, poetry. 1962 edition. 304pp. 5⅜ x 8½. (Available in U.S. only.) 40023-9

OF MEN AND NUMBERS: The Story of the Great Mathematicians, Jane Muir. Fascinating accounts of the lives and accomplishments of history's greatest mathematical minds–Pythagoras, Descartes, Euler, Pascal, Cantor, many more. Anecdotal, illuminating. 30 diagrams. Bibliography. 256pp. 5⅜ x 8½. 28973-7

HISTORY OF MATHEMATICS, David E. Smith. Nontechnical survey from ancient Greece and Orient to late 19th century; evolution of arithmetic, geometry, trigonometry, calculating devices, algebra, the calculus. 362 illustrations. 1,355pp. 5⅜ x 8½. Two-vol. set. Vol. I: 20429-4 Vol. II: 20430-8

A CONCISE HISTORY OF MATHEMATICS, Dirk J. Struik. The best brief history of mathematics. Stresses origins and covers every major figure from ancient Near East to 19th century. 41 illustrations. 195pp. 5⅜ x 8½. 60255-9

Physics

OPTICAL RESONANCE AND TWO-LEVEL ATOMS, L. Allen and J. H. Eberly. Clear, comprehensive introduction to basic principles behind all quantum optical resonance phenomena. 53 illustrations. Preface. Index. 256pp. 5⅜ x 8½. 65533-4

QUANTUM THEORY, David Bohm. This advanced undergraduate-level text presents the quantum theory in terms of qualitative and imaginative concepts, followed by specific applications worked out in mathematical detail. Preface. Index. 655pp. 5⅜ x 8½. 65969-0

ATOMIC PHYSICS: 8th edition, Max Born. Nobel laureate's lucid treatment of kinetic theory of gases, elementary particles, nuclear atom, wave-corpuscles, atomic structure and spectral lines, much more. Over 40 appendices, bibliography. 495pp. 5⅜ x 8½. 65984-4

A SOPHISTICATE'S PRIMER OF RELATIVITY, P. W. Bridgman. Geared toward readers already acquainted with special relativity, this book transcends the view of theory as a working tool to answer natural questions: What is a frame of reference? What is a "law of nature"? What is the role of the "observer"? Extensive treatment, written in terms accessible to those without a scientific background. 1983 ed. xlviii+172pp. 5⅜ x 8½. 42549-5

AN INTRODUCTION TO HAMILTONIAN OPTICS, H. A. Buchdahl. Detailed account of the Hamiltonian treatment of aberration theory in geometrical optics. Many classes of optical systems defined in terms of the symmetries they possess. Problems with detailed solutions. 1970 edition. xv+360pp. 5⅜ x 8½. 67597-1

PRIMER OF QUANTUM MECHANICS, Marvin Chester. Introductory text examines the classical quantum bead on a track: its state and representations; operator eigenvalues; harmonic oscillator and bound bead in a symmetric force field; and bead in a spherical shell. Other topics include spin, matrices, and the structure of quantum mechanics; the simplest atom; indistinguishable particles; and stationary-state perturbation theory. 1992 ed. xiv+314pp. 6⅛ x 9¼. 42878-8

LECTURES ON QUANTUM MECHANICS, Paul A. M. Dirac. Four concise, brilliant lectures on mathematical methods in quantum mechanics from Nobel Prize–winning quantum pioneer build on idea of visualizing quantum theory through the use of classical mechanics. 96pp. 5⅜ x 8½. 41713-1

THIRTY YEARS THAT SHOOK PHYSICS: The Story of Quantum Theory, George Gamow. Lucid, accessible introduction to influential theory of energy and matter. Careful explanations of Dirac's anti-particles, Bohr's model of the atom, much more. 12 plates. Numerous drawings. 240pp. 5⅜ x 8½. 24895-X

ELECTRONIC STRUCTURE AND THE PROPERTIES OF SOLIDS: The Physics of the Chemical Bond, Walter A. Harrison. Innovative text offers basic understanding of the electronic structure of covalent and ionic solids, simple metals, transition metals and their compounds. Problems. 1980 edition. 582pp. 6⅛ x 9¼. 66021-4